MW00843856

CSR, Sustainability, Ethics & Governance

Series Editors

Samuel O. Idowu, London Metropolitan University, London, UK

René Schmidpeter, M3TRIX Institute of Sustainable Business, Cologne, Germany

In recent years the discussion concerning the relation between business and society has made immense strides. This has in turn led to a broad academic and practical discussion on innovative management concepts, such as Corporate Social Responsibility, Corporate Governance and Sustainability Management. This series offers a comprehensive overview of the latest theoretical and empirical research and provides sound concepts for sustainable business strategies. In order to do so, it combines the insights of leading researchers and thinkers in the fields of management theory and the social sciences – and from all over the world, thus contributing to the interdisciplinary and intercultural discussion on the role of business in society. The underlying intention of this series is to help solve the world's most challenging problems by developing new management concepts that create value for business and society alike. In order to support those managers, researchers and students who are pursuing sustainable business approaches for our common future, the series offers them access to cutting-edge management approaches.

CSR, Sustainability, Ethics & Governance is accepted by the Norwegian Register for Scientific Journals, Series and Publishers, maintained and operated by the Norwegian Social Science Data Services (NSD)

Samuel O. Idowu • Mary T. Idowu •
Abigail O. Idowu

Editors

Corporate Social Responsibility in the Health Sector

CSR and COVID-19 in Global Health Service Institutions

 Springer

Editors
Samuel O. Idowu
Guildhall School of Business and Law
London Metropolitan University
London, UK

Mary T. Idowu
Godstone, Surrey, UK

Abigail O. Idowu
London, UK

ISSN 2196-7075 ISSN 2196-7083 (electronic)
CSR, Sustainability, Ethics & Governance
ISBN 978-3-031-23260-2 ISBN 978-3-031-23261-9 (eBook)
https://doi.org/10.1007/978-3-031-23261-9

© The Editor(s) (if applicable) and The Author(s), under exclusive license to Springer Nature Switzerland AG 2023
This work is subject to copyright. All rights are solely and exclusively licensed by the Publisher, whether the whole or part of the material is concerned, specifically the rights of translation, reprinting, reuse of illustrations, recitation, broadcasting, reproduction on microfilms or in any other physical way, and transmission or information storage and retrieval, electronic adaptation, computer software, or by similar or dissimilar methodology now known or hereafter developed.
The use of general descriptive names, registered names, trademarks, service marks, etc. in this publication does not imply, even in the absence of a specific statement, that such names are exempt from the relevant protective laws and regulations and therefore free for general use.
The publisher, the authors, and the editors are safe to assume that the advice and information in this book are believed to be true and accurate at the date of publication. Neither the publisher nor the authors or the editors give a warranty, expressed or implied, with respect to the material contained herein or for any errors or omissions that may have been made. The publisher remains neutral with regard to jurisdictional claims in published maps and institutional affiliations.

This Springer imprint is published by the registered company Springer Nature Switzerland AG
The registered company address is: Gewerbestrasse 11, 6330 Cham, Switzerland

Dedicated to All Those Who Passed On Due to COVID-19 Pandemic Globally

Foreword

COVID-19 caused a myriad of changes—both good and bad. It caused 6.53 *million* deaths globally, caused the world's collective gross domestic product to fall by (at least) 3.4%, and has led to increased poverty among the most vulnerable populations and increased obesity and suicidal depression among the young. On the other hand, it also caused improved delivery of educational, health, legal, financial, and other services by digital means, and, by allowing some workers to work from home, has likely resulted in improved work–life balance and family cohesiveness in the most developed countries—though that may be offset by a decrease in work ethic as some first-world workers have become too accustomed to government-provided funds and have either withdrawn from work or are "quietly quitting".

In 2021, my international CSR institute suddenly had to find a new meeting place for its annual conference because of COVID. My institution, the Southern University Law Center, offered the use of its facilities, and I designed and directed the conference, historically one of the largest such global CSR conferences. Realising that COVID-19 had made travel difficult and travel funds unavailable for many, I ran the conference on a hybrid basis: both in person and via synchronous video means. Thanks to top-of-the-line digital equipment and support staff, a happy consequence was the willingness and ability of CSR experts from all over the globe to attend and share their research, even though such attendance was of necessity at an awkward time of day for some. Though smaller than pre-COVID, we still had an abundance of real-time, global idea-sharing from experts from around the world and from widely disparate fields, something that would have been unlikely prior to COVID.

CSR in the Health Sector provides a similar global experience in printed form. Written by experts from around the globe, it explores how COVID-19 affected relationships among health-care providers, government, and corporations in a wide variety of different countries and from a wide variety of perspectives. Contributing scholars include accountants, public health experts, information technology professionals and engineers, ergonomics, debt management, community wealth building, and housing services experts, union organisers, social and policy theorists, town planners, legal scholars, philosophers, ethicists, business management experts, medical doctors, and health administrators. Countries represented include four continents (Europe, Africa, Asia, and South America) and range from the most developed (the UK, Turkey) to one of the poorest in the Western Hemisphere (Bolivia).

The result is a wealth of information from a wonderful breadth of perspectives on the timeliest sustainability topic coming out of COVID-19: how to better handle pandemics to protect health as well as preserve economic vitality. *CSR in the Health Sector* provides researchers with several successful and unsuccessful approaches to dealing with such emergencies which can be used to develop best practices, allowing researchers to learn from what has been a very painful global experience. A common theme: COVID's most likely promising side effect was the innumerable and innovative alliances among governments, corporations, and health-care providers. Those improvements should be built upon to provide sustainable improvements around the globe.

Law Emerita, Southern U. Law Center,   Nadia E. Nedzel J.D., LL.M.
Baton Rouge, LA, USA

Preface

The pandemic that besieged our world in late 2019—the Corona virus 2019 commonly referred to as COVID-19—has seriously affected all countries of the world in a manner not seen for about 100 years. The pandemic has reshaped our world in an unimaginable way. More than two million global citizens have lost their lives because of the pandemic; hospitals and other health service institutions that cater for people were brought to a point of no return. Private hospitals in some parts of the world were turning prospective patients away because there were no beds for them or health personnel to look after these prospective patients. It was an unforgettable experience that will be remembered for centuries to come. How have health service institutions, governments, corporate entities, individual citizens, and society in general coped or are still coping with the pandemic? This book intends to increase our readers' understanding of how things have shaped up or still shaping up in this area.

The pandemic has transformed global lives in different areas. We are now able to do many things through the Internet. Meetings—local, nation, and international—are now held freely using different software, lectures are held remotely by institutions to students based in different parts of the world, medical practitioners meet their patients remotely and prescribe medication, conferences are held remotely with attendees taking part in these conferences without stepping out of their homes, and many other great things global communities are doing innovatively as a result of the unwanted pandemic. A few good things have come out of the evil COVID-19!

London, UK   Samuel O. Idowu, PhD
Godstone, Surrey, UK   Mary T. Idowu
London, UK   Abigail O. Idowu
October 2022

Acknowledgements

We wish to express our gratitude to all those fantastic contributors who have helped us in this great book. It is evident that without their hard work, it would have been impossible for us to have this end product. The lead editor expresses his gratitude to all these great individuals who have contributed to the success of this book, which is something he cherishes immensely as two of his daughters are on the editorial team of the book. The three editors are grateful to those of you who have stood by them with their impressive chapters in this book. Many of them despite their busy schedules felt obliged to help them in putting together this very fine informative addition to the literature on how global health services institutions have coped and are still coping with COVID-19. We are also grateful to Professor Nadia Nwdzel for her great Foreword to the book. Thank you all.

We would like to thank our publishing team at Springer headed by the Executive Editor, Christian Rauscher; Barbara Bethke; and other members of the publishing team who have supported this project and all my other projects.

We have dedicated the book to all those who sadly passed on as a result of the COVID-19 pandemic which besieged the world in November 2019.

Finally, we apologise for any errors or omissions that may appear anywhere in the book; please be assured that no harm was intended to anybody.

Contents

Corporate Social Responsibility and COVID-19 Pandemic
in Four Continents: An Introduction.......................... 1
Samuel O. Idowu, Mary T. Idowu, and Abigail O. Idowu

Part I CSR and COVID-19 Pandemic in Europe

COVID-19 Pandemic Management from a Sustainability Viewpoint:
An Analysis for Austria, the European Union, and the WHO......... 13
Ursula A. Vavrik

Corporate Social Responsibility Initiatives and Programs
in the Health System of Greece due to the Pandemic of COVID-19.... 93
Φ. Ioannis Panagiotopoulos

CSR Manifestations in Health Care Facilities in Poland During
the COVID-19 Pandemic.................................... 111
Anna Cierniak-Emerych, Ewa Mazur-Wierzbicka, Piotr Napora,
and Sylwia Szromba

Corporate Social Responsibility: A Solution for Resilience During
the COVID-19 Pandemic in Romania......................... 127
Silvia Puiu

Responsiveness, Strategy and Health as Diplomacy: The Unlikely
Case of Serbia... 145
Milan Todorovic

Corporate Social Responsibility and Profitability in Spanish
Private Health Care During the COVID-19 Period................ 173
María del Carmen Valls Martínez, Rafael Soriano Román,
Mayra Soledad Grasso, and Pedro Antonio Martín-Cervantes

Saving Lives and Minds: Understanding Social Value and the Role of Anchor Institutions in Supporting Community and Public Health before and after COVID-19 193
Julian Manley, Craig Garner, Emma Halliday, Julie Lee, Louise Mattinson, Mick Mckeown, Ioannis Prinos, Kate Smyth, and Jonathan Wood

Corporate Social Responsibility and Coping with COVID-19 Pandemic in the Global Health Service Institutions: The United Kingdom 219
Mohammed Ali and Courtney Grant

Responsible Innovation During the COVID-19 Pandemic: A Case Study from Türkiye 243
Gizem Aras Beger, Gönenç Dalgıç Turhan, and Gülen Rady

Part II   CSR and COVID-19 Pandemic in Africa

Grappling with COVID-19: The Implications for Ghana 263
Sam Sarpong

Corporate Social Responsibility and the Impact of COVID-19 on Healthcare Institutions in Nigeria 279
Gloria O. Okafor, Amaka E. Agbata, Innocent C. Nnubia, and Sunday C. Okaro

The Private Sector's Role in Strengthening Public Hospitals in Zambia During the Coronavirus (COVID-19) Pandemic: A Corporate Social Responsibility (CSR) Perspective 297
Isaac Kabelenga and Ndangwa Noyoo

Part III   CSR and COVID-19 Pandemic in Asia

Business Responses to COVID-19 Through CSR: A Study of Selected Companies in India 317
Sumona Ghosh

The Rippling Effect of COVID-19 in Malaysia: Now and Then 339
Sam Sarpong and Ali Saleh Alarussi

Part IV   CSR and COVID-19 Pandemic in South America

Corporate Social Responsibility in Bolivia: Hospital Responses to the COVID-19 Pandemic 355
Boris Christian Herbas-Torrico, Carlos Alejandro Arandia-Tavera, and Alessandra Villarroel-Vargas

Index 373

About the Editors

Samuel O. Idowu, PhD Samuel O Idowu is Senior Lecturer in Accounting and Corporate Social Responsibility at the Guildhall School of Business and Law, London Metropolitan University, where he is currently the course leader for the MSc Corporate Social Responsibility & Sustainability and Advanced Diploma in Professional Development (ADPD) Corporate Social Responsibility & Sustainability. Samuel is Professor of CSR and Sustainability at the Nanjing University of Finance and Economics, China. He is a fellow member of the Chartered Governance Institute, a fellow of the Royal Society of Arts, a Liveryman of the Worshipful Company of Chartered Governance Institute, and a named freeman of the City of London. He is the Deputy CEO and Vice President of the Global Corporate Governance Institute, an international network of CSR scholars. Samuel has published over 50 articles in both professional and academic journals and contributed chapters in several edited books and is the Editor-in-Chief of three major global reference books by Springer—the *Encyclopedia of Corporate Social Responsibility* (ECSR), the *Dictionary of Corporate Social Responsibility* (DCSR), and the *Encyclopaedia of Sustainable Management* (ESM), and he is Series Editor for Springer's *CSR, Sustainability, Ethics and Governance* books. Samuel is Editor-in-Chief of the *International Journal of Corporate Social Responsibility* and the *American Journal of Economics and Business Administration.* Samuel has been in academia for more than 30 years winning one of the Highly Commended Awards of Emerald Literati Network Awards for Excellence in 2008 and 2014. In 2010, one of his edited books was placed in 18th position out of forty top Sustainability books by Cambridge University Programme for Sustainability Leadership, and in 2016, one of his books won the outstanding Business Reference Book of the Year of the American Library Association. In 2018, he won a CSR Leadership Award in Cologne, Germany, and in 2019, he won the 101 Most Impactful CSR Leaders Award in Mumbai, India. Samuel is on the Editorial Advisory Boards of the *International Journal of Business Administration* and *Amfiteatru Economic Journal.* He has been researching in the field of CSR since 1983 and has attended and presented papers at several national and international conferences and workshops on CSR. Samuel has made a number of keynote speeches at international conferences and workshops and written the *fore-word* to a number of leading books in the field of CSR and sustainable development. And he has examined a few PhD theses in the UK, Australia, South Africa, the Netherlands, and New Zealand.

Mary T. Idowu Mary T Idowu holds a bachelor's degree in law with French. She is a highly experienced NHS senior leader having worked in the NHS for almost 15 years. This experience spans across national, regional, and local transformation, innovation, and quality improvement, and she was working at the heart of the national COVID response. She is skilled in working in strategy and operational delivery across commissioning, acute services, and systems (STPs, ICSs, and Cancer Systems). This includes across the national Medical Directorate, managing clinical services at a regional level (including cancer, audiology, MSK, gastroenterology, ophthalmology), and managing operational services (including cancer, pathology, radiology, cardiology, and endoscopy).

She has been warded the "Inspiring Women in Leadership" scholarship (2021) from Warwick Business School for "outstanding female candidates who are enthusiastic, engaging and inspiring role models" in organisational change; has sustained record of achievement in programme and change management through developing and delivering national and local policy, strategy, and priorities; and has highly developed leadership skills supporting organisational development change using evidenced-based tools and methodologies including behaviour change interventions, business process engineering, knowledge transfer application, and partnership working, which included the Disasters Emergency Committee, Action Aid, and the BBC. Mary is a recognised quality improvement expert, e.g. a review for the Foreign Commonwealth Development Office (FCDO) for the Better Health Programme, where she provided strategic advice and technical expertise for South Africa's leadership development plan for coordinating quality improvement initiatives. She is also Chair and Lay Panel Member for the Nursing and Midwifery Council Fitness to Practise Committees.

Abigail O. Idowu Abigail holds a Bachelor of Science in Psychology and Childhood & Society from Roehampton University. Prior to becoming an independent Emotional Health Consultant, she was the Director of Administration within Financial Services, but her professional focus became centred in specialising in the complex comprehension of people. Her specialist approach of using practical conversational methods to attain healing, clarity of thought, and mental and emotional balance, known as Core Rehabilitation, is well sought after.

Core rehabilitation is exactly what it says on the tin. It accesses the core of who a person is to rehabilitate a dysfunctional part of the life. It is a combi-tool of counselling and therapy; it combines the best attributes of both methods to bring about positive change. This is achieved by bringing into alignment the main governing members that control the flow of our life experience. By using the clients' goal objectives as the conversational anchor, the root-diagram technique helps connect conscious and subconscious thoughts, emotional patterns, and behaviour, which helps regain empowerment and widens the sphere of control in all areas of a person's life.

She is gifted in her ability to delve into one's objectives by presenting thought-provoking questions: What are the issues you are facing? What are emotional/mental barriers that are preventing you from moving forward? Abigail helps in the unfolding of knowing that the truth you stand under will govern your perspectives, and these perspectives will determine your mind experience. While we cannot control every circumstance, we can absolutely control our static perception and our motion perception of every experience we have, to ensure we do not develop negative roots that hinder our happiness. She has a gentle and effective way of creating an environment which is free of blame and judgement in order to uncover and uproot negative mental positions that work against the goal of freedom and happiness. Abigail is extremely passionate about giving people the simple tools necessary to help them reach the most progressive and uplifting mental state.

With the help of Abigail, people will walk away knowing that living life can be good, but feeling more convinced that living an abundant life is even better.

Corporate Social Responsibility and COVID-19 Pandemic in Four Continents: An Introduction

Samuel O. Idowu, Mary T. Idowu, and Abigail O. Idowu

The COVID-19 pandemic was a serious event that will never be forgotten on planet Earth even if another more serious event of this nature were to once again besiege planet Earth (we sincerely hope not). It was a war on us all with no conventional weapons used. It took so many lives. In fact, it was noted on the website of the *Worldometer* of Coronavirus that 6,531,169 people worldwide had died as a result of the virus. In the UK where these three editors live, every health service institution across the four countries—England, Northern Ireland, Scotland and Wales—faced unprecedented pressures. The UK government coronavirus website at the time of writing this piece notes that 177,977 people had died in the UK within 28 days of testing positive for the virus. A total of 204,015 people died with coronavirus cited on their death certificates as one of the causes of death as at the time of writing this introductory piece to the book. The havoc the virus caused in the UK as depicted above was similar to many countries globally, regardless of where those countries are based on planet Earth. In the USA, the disease took the lives of about 1,050,000 people. It was a never to be forgotten sad episode.

Needless to say, when people fall ill, the natural cause of action to take is to seek health assistance from health service providers, and this was what many people who were affected by the virus did. When people take this course of action, health service institutions would be impacted in several ways. One of these impacts during the COVID-19 pandemic led to the failure of many of these institutions in meeting their

S. O. Idowu (✉)
London Metropolitan University, London, UK
e-mail: s.idowu@londonmet.ac.uk

M. T. Idowu
NHS England, Godstone, Surrey, UK
e-mail: maryidowu@nhs.net

A. O. Idowu
Abigail Olamide Advisory, London, UK

© The Author(s), under exclusive license to Springer Nature Switzerland AG 2023
S. O. Idowu et al. (eds.), *Corporate Social Responsibility in the Health Sector*, CSR, Sustainability, Ethics & Governance, https://doi.org/10.1007/978-3-031-23261-9_1

"normal" day-to-day service provisions. In the UK, many urgent and non-urgent operations were cancelled, and about six million such operations were shelved. Commentators are noting that catching up with these backlogs could take several years. Many of those affected by these cancellations may not be lucky to be around for that long. Will those whose operations etc. had been cancelled to deal with those affected by COVID-19 survive to be helped if they remained untreated for several years? Would CSR and what it stands for not have failed in a serious way? Are operations not still been cancelled because of COVID-19? These and many other questions on the impact of COVID-19 pandemic are what this book hopes to provide answers to.

The book was able to generate interests from 23 countries in five continents around the globe, but it was only able to explore its theme in terms of 17 countries in four continents. A few countries were not able to successfully meet our requirements. This explains why the book has been divided into four parts and 15 chapters, two of these chapters from one country in Europe. Part I looks at the European perspectives of the theme of the book in nine chapters—these from eight countries. Part II looks at the African perspectives of the impact of COVID-19 on health service institutions from three countries. Part III looks at the theme in two Asian countries. The final part espouses the impact of the pandemic in one South American country. The remainder of this introductory chapter looks at what each of the participating countries says about how the pandemic has impacted or still impacting these 14 countries in terms of health service providers and corporate social responsibility.

The very first chapter of the book from Austria by a prolific writer and scholar of global repute Professor Ursula A Vivarik titles her chapter as "*COVID-19 Pandemic Management from Sustainability Viewpoint—Austria, the EU and the World Health Organisation*". In the chapter, she argues that COVID-19 pandemic challenged the world in many respects because it started as a global health crisis, and it was able to threaten the global economy and societies, leading to many unpleasant side effects, and even affecting democracies and human rights standards. Against this backdrop, Vivarik notes that the chapter was planned to examine global governance by the World Health Organization (WHO) and leadership at European and national levels. The chapter analyses the Austrian case in more detail, also in relation to Switzerland and the Nordic countries.

The second chapter by Panagiotopoulos Φ. Ioannis, a Greek engineer and CSR scholar, notes that the coronavirus consists of Scylla of health threat and the Charybdis of economic recession which is comparable to the Spanish Flu of 1918 and the Great Depression of 1929, respectively. Ioannis explains that the pandemic has caused more than 500 million cases and more than 6 million deaths globally up to April 2022. In Greece, a country which has experienced economic depression of more than a decade since 2009, which started in the wake of the global financial crisis of 2008–2009, the advent of coronavirus appears as its continuation with much more intense and complicated characteristics. A lot of firms have had to apply urgent corporate social responsibility (CSR) programmes to protect their employees and empower the national health system. This necessity marked the creation of a new universal urgent CSR type called critical CSR (chapter "Corporate Social

Responsibility Initiatives and Programs in the Health System of Greece due to the Pandemic of COVID-19"). Unequivocally, the biggest burden was placed on the health sector which consequently faced a novel coronavirus which called for new medicines, new vaccines and enormous capacity for hospitalised cases, especially in intensive care units (ICUs). The chapter was able to trace the impacts of the virus on Greek health service providers in a scholarly manner.

Professor Anna Cierniak-Emerych and three of her colleagues—Ewa Mazur-Wierzbicka, Piotr Napora and Sylwia Szromba—from Poland in the third chapter of the book entitled "*Corporate Social Responsibility Manifestations in Health Care Facilities in Poland During COVID-19 Pandemic*" provide an intriguing chapter on how the pandemic impacted the Polish health service institutions. The chapter describes some aspects of corporate social responsibility currently under discussion by both scientists and economic entities in Poland and elsewhere. Cierniak-Emerych *et al* argue that albeit CSR is a relatively new term in Poland, because it was only recently introduced into the public discourse, it has emerged into prominence during the COVID-19 pandemic, in both business enterprises and healthcare institutions. The third chapter also explains how CSR was implemented in the Polish healthcare institutions and how it influenced the socio-economic situation as a result of the Polish history and in terms of the actions taken by various social actors and, recently, by the COVID-19 pandemic. An essential reading for anyone wanting to fully understand how the pandemic has impacted on the Polish health service institutions.

In fourth chapter, Silvia Puiu, a scholar of repute from Romania in a chapter titled "*Corporate Social Responsibility—A solution for Resilience during the COVID-19 Pandemic in Romania*," highlights the way private and public service health institutions in Romania coped with the start of the COVID-19 pandemic and afterwards. Puiu notes that the way each country around the globe reacted to the pandemic crisis was different, simply because there was no recent precedent in the history to guide them. The chapter explores the impact of the pandemic on both the employees and the organisations during the crisis from the beginning to the point when humanity was provided the required vaccine to deal with the disease. The research methodology used in the chapter consists of applying PLS-SEM method based on the answers of 156 employees from different backgrounds, thus offering a wider perspective of the way corporate social responsibility helped in the fight against the pandemic. The results of the research by Puiu are helpful for a better understanding of how crises, including health issues, could be faced with resilience.

The fifth chapter is from a good friend and colleague of the lead editor—Milan Torodovic on "Responsiveness, Strategy and Health as Diplomacy—The Unlikely Case of Serbia" notes that the virus had caught the world unawares. Torodovic argues that it still continues to be contended whether the world will ever be truly comparable with the pre-2020 standards preceding the *Event*. Torodovic notes that no one is arguing that the world has not seen its fair share of plagues over the centuries—the *Spanish Flu* that plagues the world in 1918 is still fresh in our minds albeit many of us were not on this planet 104 years ago. Torodovic recounts the incidence of another brutal infection that swept the globe and killed more people than the Great War a century ago—the *Spanish Flu*. What has changed since is that, once gladly built then mainly disappeared, the notion of a welfare state and its ethics mobilised governments and many national and supranational stakeholders into action.

Serving as evidence that money is neither necessarily the prerequisite nor the answer is the case of Serbia, notes Torodovic. Arguably, a nation with very limited resources or influence on the world stage which is in between powerful trading nations and a customary portrayal in western press which seems less than flattering decades after the troubles, this small south European country responded rather better than many in the world's far wealthiest countries per capita, notes Torodovic. He went on to say that the teething problems with the pandemic response seem eerily commonplace: denial, then panic, blunt instruments and eventual realisation that a subtle strategy is the best tool against the invisible enemy. Not to underestimate the effects of COVID on all levels of supply chains, on labour disruption, an increased sense of international distrust and blame apportioning, it appears that often individual actors left to their own devices used such predicament as an opportunity. The chapter explained five facts that set Serbia apart in responding to the pandemic: (1) a brutally effective response in mid-March 2020 which created a hermetic curfew, especially on the over 65s, lasting for over 7 weeks; (2) the self-imposed national and international lockdown, then more limiting than the coinciding, widely reported one in Italy; (3) empirical data showing the causal impact of sudden loosening of restrictions over summer, leading to (4) a carefully coordinated strategy that included largely well-managed, less onerous partial restrictions; and (5) an unprecedented effort aimed at mass vaccination both of its own citizens and foreigners. The latter has a set of diplomatic characteristics: citizens of the region with even the loosest connection to Serbia felt invited to get vaccinated together with its own, by any number of jabs obtained by the country's leadership from any and all brand actors, manufacturers and states willing to oblige. This is a must-read chapter by anyone wanting to know more about Serbia's actions in dealing with the pandemic.

In the sixth chapter on "Corporate Social Responsibility and Profitability in Spanish Private Healthcare during COVID-19 Period" by four great scholars from Spain led by Maria del Carmen Valls Martínez argues that in recent decades, there has been a growing demand for companies to take responsibility for the adverse social and environmental effects caused by their activities. As a result, they are no longer only accountable for their economic performance but also their non-financial activities—*corporate social responsibility*. The chapter notes that an important pillar of corporate social responsibility is gender diversity in the company as a whole and on the board of directors in particular. The hospital sector, whose activity in itself requires social responsibility, was later than other sectors in its disclosure, the chapter notes. This research aims to analyse the influence of corporate social responsibility, together with gender diversity on boards of directors and the COVID-19 pandemic, on the profitability of Spanish private hospitals. In addition, the study notes that the dissemination of COVID-19 pandemic throughout the Spanish territory, relating to wealth and risk. The chapter used the data corresponding to the period 2017–2020 to analyse multiple linear regression analysis, cluster analysis and factor analysis. The results show that those socially responsible hospitals reported higher profitability, but no causal relationship has been established. Gender diversity negatively influences the profitability of the private hospital sector, although it can be considered non-significant. The COVID-19 pandemic significantly affected the profitability of hospitals, causing a sharp drop. The spread of the COVID-19 pandemic was mainly influenced not only by the population density of the territories but also by public health investment, showing a greater propensity to control the pandemic in those regions that allocate more funds to health care. Indeed, an interesting chapter on the impact of the pandemic on the profitability of health services providers during the pandemic.

Julian Manley and eight others in the seventh chapter on *"Saving lives and minds: Understanding social values and the role of anchor institutions in supporting community and public health before and after COVID-19"* focus their chapter on a community in the UK. Their chapter notes that there are great disparities in health between places in the UK. People living in poorer areas are dying on average 9 years earlier than in wealthy areas, largely due to regional economic differences, including high unemployment, low wages and social inequality, unrest and injustice that accompany economic disadvantage. The city of Preston is in the north-west of England and has been developing a community wealth building project known as the *Preston Model*, which shows signs of successfully increasing and retaining local wealth. The anchor institutions—large local organisations that are "anchored" in places, such as hospitals, universities, housing associations and local government—have developed social value policies and policies of cooperation with their communities that attend to a heightened awareness of corporate social responsibility and enhanced working relationships with local communities in order to turn around local fortunes in an allied economic and health initiative, they argue. *Corporate social responsibility* is the essence of cooperation and cooperatives and is a central feature of the *Preston Model* the chapter notes. Ultimately, CSR within the *Preston Model* is concerned with quality employment. The pandemic has highlighted the need for CSR and cooperation. This chapter brings together researchers from the University of Central Lancashire, Lancaster University and stakeholders from two of the anchor institutions—the Lancashire Teaching Hospitals NHS Foundation Trust and Community Gateway Association—to combine an academic framework, including local responses to interviews and participatory community groups in Preston, with two major anchor institutions as case studies, an undoubtedly interesting chapter we recommend to all our readers to go through.

The penultimate chapter in Part I is also from the UK by two health service experts in the UK—Mohammed Ali and Courtney Grant. In the chapter, the authors describe how the private sector, the third sector and philanthropists have carried out their corporate social responsibility (CSR) in supporting the UK's health service institutions during the COVID-19 pandemic. The chapter notes that the COVID-19 pandemic has created excess demand on the National Health Service's (NHS) resources, particularly on its essential equipment, medicines and workforce. The COVID-19 pandemic has also highlighted inequalities in health faced by poorer households from institutional services outside of the NHS. During the COVID-19 pandemic, Ali and Grant note that the aforementioned stakeholders have provided support to the NHS, as well as to disadvantaged communities, and wider society.

Health inequalities have become more visible due to the COVID-19 pandemic, as social determinants of health have disproportionately exacerbated COVID-19 mortality rates in a number of ethnic minority communities. Moreover, routine data on COVID-19 fatalities have found a correlation with age, with elderly people being more adversely affected. The two authors have used Carroll 1991 Pyramid of CSR to divide the chapter into four main parts that focus on the four major aspects of CSR: philanthropic, legal, ethical and economic responsibilities. Using CSR and health system policies, practices and cases, this approach is framed in the context of the NHS and health stakeholders during the COVID-19 pandemic. The aforementioned aspects of CSR in the UK are detailed using publicly available data. This overview is intended for researchers, health practitioners, students, policymakers, civic authorities, the private sector and the third sector and is intended to aid CSR planning for future waves of the COVID-19 pandemic and for different future pandemics.

The final chapter in Part I is from three great Turkish scholars based in the beautiful city of Izmir—Gizem Ara Berger, Gönenç Dalgic Turhan and Gülen Rady. The chapter argues that companies in almost all sectors and countries have faced the challenges of an urgent transition due to the rapid spread of COVID-19 pandemic. Alongside day-to-day operational adjustments, many companies have also had to make great efforts to mitigate the adverse impacts of COVID-19 on society by taking a socially responsible stance. Companies with a high commitment to society and the environment have successfully embraced their notion of corporate social responsibility (CSR) with innovation. During the pandemic, responsible innovation (RI), as one of the most important tools of CSR, has become an important way of generating societal benefits, argue the three young scholars. Above all, the health sector has experienced diverse versions of RI. Accordingly, this study that emanated to the chapter discusses how RI in the Turkish healthcare system has assisted in coping with the pandemic. The chapter uses the case of Abdi İbrahim, a pioneering Turkish pharmaceutical company to argue its case. The case provides useful insights into how health sector companies have handled the pandemic in responsible ways. The case also shows how responses could be made more rapid and effective in future pandemics and other global health crises.

The first chapter from the three participating countries in Africa is by Professor Sarpong, a great friend of the lead editor. Sarpong's chapter is titled *"Grappling with COVID-19: The Implications for Ghana"*. The chapter notes that the emergence of COVID-19 has had substantive economic, health and societal impacts across the world. It has created major disruptions in many economies, illuminated governments' failures and exposed major vulnerabilities in our social settings. Sarpong argues that in Ghana, many interventions have been made substantially within the health, economic and social standings of the citizens following these challenges. The chapter brings to the fore the complex challenges the country faced and or continues to face in the light of this. The chapter details the supporting measures that the government took to protect the poor and the vulnerable. Aside from that, it explores the myths, misconceptions and responses associated with the pandemic, which to a large extent impacted how the pandemic was perceived. Another key issue that the chapter looks at includes the role played by business leaders in the fight against the pandemic.

Nigeria, the country with the largest economy and most populous country in the continent, occupies the 11th chapter of the book. The chapter was authored by four professors from a university named after one of the greatest leaders of the country after independence. Okafor, Agbata, Nubia and Okaro in the chapter entitled "CSR and the Impact of COVID-19 in Health Service Institutions in Nigeria" examine the corporate social responsibility (CSR) activities in the healthcare institutions in Nigeria, before and during the COVID-19 era. The extent to which the healthcare institutions are seen to be socially responsible and the CSR from profit-oriented companies to hospitals were examined. The chapter reviewed annual reports of 20 listed companies from 2017 to 2020 and websites of 46 healthcare institutions. The results show that before the COVID-19 era, there was poor CSR from profit-oriented companies to healthcare institutions, but there was a huge change during the COVID-19 era, and most of the CSR activities reported by these business organisations were committed to the healthcare institutions. Majority of the healthcare institutions reported on employee-related issues in the workplace. Other reported CSR activities found in the websites of healthcare institutions but not by majority are the level of their ethical behaviour and their relationships with the community. Reports on the management of toxic wastes and relationships with patients were scarcely found. The study concludes that CSR has not penetrated the healthcare institutions in Nigeria and proposes increased resource allocations to the healthcare system from both government and private companies. The study also encourages healthcare institutions to willingly report on their socially responsible activities.

The final chapter from Africa emanates from Zambia by two great scholars born in this great African country formerly known as North Rhodesia was ruled over by one of the great fathers of Africa Kenneth Kaunda from independence until he retired from politics due to old age in June 1998. Professors Kabelenga and Noyoo in the 12th chapter note that their chapter is based on a desktop research study which focused on the coronavirus (COVID-19) pandemic in Zambia that had and continues to negatively impact the country's public hospitals. Indeed, after the outbreak of COVID-19 in the country many people were hospitalised for treatment and palliative care. Thus, a sharp rise in COVID-19 cases resulted in an unprecedented high demand for testing kits, personal protective equipment (PPEs) for both medical staff and patients; hospital beds, oxygen for COVID-19 patients and medicine; among other things. During the pandemic, public hospitals were under tremendous strain and pressure. Up till the present moment, Kabelenga and Noyoo argue that these institutions are still struggling to meet the increased demand for hospital care. Despite the foregoing challenges, in the same period, something which is referred to as COVID-19 Emergency Corporate Social Responsibility (ECSR) emerged in Zambia, Kabelenga and Noyoo, note. This has served as one way to strengthen public hospitals to cope with the increased number of patients. To this end, through COVID-19 ECSR, the private sector supported public hospitals by donating *inter alia*, money, PPEs, oxygen concentrators, medicine and food. The study carried out by Kabelenga and Noyoo revealed that the private sector's contributions during the pandemic had helped to improve and maintain the health of Zambians, after fortifying public health systems, so that they coped with the increased demand for health services and other shocks. The chapter explores how ECSR had assisted public hospitals in Zambia to deal with the ramifications of the COVID-19 pandemic.

Moving swiftly on to Part III where the chapters of the two countries in Asia that participated in the project are housed, let us see how the first country India coped and is still coping with the impact of COVID-19 pandemic. Professor Sumona Ghosh in her chapter entitled *"Business Responses to COVID-19 Pandemic through CSR: A study of selected companies in India"* argues that on 30 January 2020, the World Health Organization (WHO) declared COVID-19 a public health emergency of international concern. Ghosh argues that such a global health crisis has resulted in restructuring of resources in terms of both speed and scale of mobilisation. Corporate social responsibility (CSR) is playing a crucial role in the age of this pandemic COVID-19, where business is trying their best to cope with this tremendous challenging time. On 23 March 2020, the Indian government declared that all expenditures incurred on activities related to COVID-19 would be regarded to be CSR expenditure. Since the announcement of the PM CARES Fund and its inclusion in Schedule VII of the Companies Act, 2013, through a subsequent amendment, a huge amount of funding has also been directed from corporates to the PM CARES Fund. The chapter gave an opportunity to study the business responses to COVID-19 through the lens of CSR of the top 50 companies ranked on the basis of market capitalisation for the years 2019–2020 to 2020–2021 by constructing a Corporate Health Disclosure Index (CHDI). Ghosh's study showed that business response towards health during COVID-19 was average. These businesses have mostly concentrated on short-run plans, mostly supporting healthcare infrastructure, assisting in vaccination programmes and contributing to PM CARES Fund. Certainly, an interesting piece of research from the world's second most populous nation.

The second chapter in Part II came from Professor Sarpong from Ghana who is based in Malaysia and Professor Alarussi. In their chapter on Malaysia, they argue that the last few years have seen COVID-19 taking centre stage in the lives of Malaysians. Although the cases have gone down considerably, the severe impact it left in its trail has led to various changes in the lives of the people and the operations of various institutions in the country. The chapter takes a holistic view of the COVID-19 situation in Malaysia. It documents the issues the country faced during the prolonged COVID-19 crisis. Particularly, it highlights the health, socio-economic and humanitarian crises that bedevilled Malaysia in the course of the pandemic. It also provides an insight into the interventions the government made in its response to the pandemic.

Part IV from South America with only one chapter from Bolivia by three scholars led by Professor Herbas-Torrico argues that the coronavirus (COVID-19) pandemic has impacted the health, the economy and the social fabric of Bolivia. The world is grappling with the consequences of the COVID-19 pandemic on firms, workers, consumers, communities and each other. Due to these consequences, people worldwide are committed to working together and supporting one another in all possible ways. Using stakeholder theory and corporate social responsibility (CSR), the chapter explores hospital responses to the COVID-19 pandemic in Bolivia from the perspective of hospital managers, hospital staff and patients. The study used quantitative and qualitative analyses to understand how CSR initiatives, operational difficulties and healthcare quality services of Bolivian hospitals were used for the COVID-19 pandemic.

The above introduction to the 15 chapters of this book in our view has laid the foundation for our readers to build on in order to fully understand what was envisaged when the idea of the book was conceived and how the end product in the form of this book culminated. It is hoped that the chapters add to your knowledge of how COVID-19 impacted corporate and civil societies in the countries featured in the book.

Samuel O. Idowu is a senior lecturer in Accounting and Corporate Social Responsibility at the Guildhall School of Business and Law, London Metropolitan University, where he is currently the course leader for the MSc Corporate Social Responsibility & Sustainability and Advanced Diploma in Professional Development (ADPD) Corporate Social Responsibility & Sustainability. Samuel is a professor of CSR and Sustainability at Nanjing University of Finance & Economics, China. He is a fellow member of the Chartered Governance Institute, a fellow of the Royal Society of Arts, a liveryman of the Worshipful Company of Chartered Governance Institute and a named freeman of the City of London. He is the deputy CEO and vice president of the Global Corporate Governance Institute an international network of CSR scholars. Samuel has published over 50 articles in both professional and academic journals and contributed chapters in several edited books and is the editor-in-chief of three major global reference books by Springer—the *Encyclopedia of Corporate Social Responsibility* (ECSR), the *Dictionary of Corporate Social Responsibility* (DCSR) and the *Encyclopaedia of Sustainable Management* (ESM)—and he is a series editor for Springer's CSR, Sustainability, Ethics and Governance books. Samuel is an editor-in-chief of the International Journal of Corporate Social Responsibility and the American Journal of Economics and Business Administration. Samuel has been in academia for more than 30 years winning one of the Highly Commended Awards of Emerald Literati Network Awards for Excellence in 2008 and 2014. In 2010, one of his edited books was placed in 18th position out of forty top Sustainability books by Cambridge University Programme for Sustainability Leadership, and in 2016, one of his books won the outstanding Business Reference Book of the year of the American Library Association. In 2018, he won a CSR Leadership Award in Cologne, Germany, and in 2019, he won the 101 Most Impactful CSR Leaders Award in Mumbai, India. Samuel is on the Editorial Advisory Boards of the International Journal of Business Administration and Amfiteatru Economic Journal. He has been researching in the field of CSR since 1983 and has attended and presented papers at several national and international conferences and workshops on CSR. Samuel has made a number of keynote speeches at international conferences and workshops and written the *foreword* to a number of leading books in the field of CSR and Sustainable Development. He has examined a few PhD theses in the UK, Australia, South Africa, the Netherlands and New Zealand.

Mary T. Idowu holds a bachelor's degree in Law with French. She is a highly experienced NHS senior leader having worked in the NHS for almost 15 years. This experience spans across national, regional and local transformation, innovation and quality improvement, and she was working at the heart of the national COVID response. She is skilled in working in strategy and operational delivery across commissioning, acute services and systems (STPs, ICSs and cancer systems). This includes across the national Medical Directorate, managing clinical services at a regional level (including cancer, audiology, MSK, gastroenterology, and ophthalmology) and managing operational services (including cancer, pathology, radiology, cardiology and endoscopy).

She was awarded the "Inspiring Women in Leadership" scholarship (2021) from Warwick Business School for "outstanding female candidates who are enthusiastic, engaging and inspiring role models" in organisational change. She sustained a record of achievement in programme and change management through developing and delivering national and local policy, strategy and priorities. She has highly developed leadership skills supporting organisational development change using evidenced-based tools and methodologies including behaviour change interventions, business process engineering, knowledge transfer application and partnership working, which has included the Disasters Emergency Committee, Action Aid and the BBC. Mary is a recognised quality improvement expert, e.g. a review for the Foreign Commonwealth Development Office (FCDO) for the Better Health Programme, where she provided strategic advice and technical expertise for South Africa's leadership development plan for coordinating quality improvement initiatives. She is also a Chair and Lay Panel Member for the Nursing and Midwifery Council Fitness to Practise Committees.

Abigail O. Idowu holds a Bachelor of Science in Psychology and Childhood & Society from Roehampton University. Prior to becoming an independent Emotional Health Consultant, she was the Director of Administration within Financial Services, but her professional focus became centred on specialising in the complex comprehension of people. Her specialist approach of using practical conversational methods to attain healing, clarity of thought and mental and emotional balance, known as core rehabilitation, is well sought after.

Core rehabilitation is exactly what it says on the tin. It is accessing the core of who a person is to rehabilitate a dysfunctional part of life. It is a combi tool of counselling and therapy, and it combines the best attributes of both methods to bring about positive change. This is achieved by bringing into alignment the main governing members that control the flow of our life experience. By using the clients' goal objectives as the conversational anchor, the root diagram technique helps connect conscious and subconscious thoughts, and emotional patterns and behaviour, which helps regain empowerment and widens the sphere of control in all areas of a persons' life.

She is gifted in her ability to delve into ones objectives by presenting thought-provoking questions such as what are the issues you are facing? What are emotional/mental barriers that are preventing you from moving forward? Abigail helps in the unfolding of knowing that the truth you stand under will govern your perspectives, and these perspectives will determine your mind experience. While we cannot control every circumstance, we can absolutely control our static perception and in motion perception on every experience we have, to ensure we do not develop negative roots that hinder our happiness. Her gentle and effective way of creating an environment which is free of blame and judgement in order to uncover and uproot negative mental positions that work against the goal of freedom and happiness. Abigail is extremely passionate about giving people the simple tools necessary to help them reach the most progressive and uplifting mental state.

With the help of Abigail, people will walk away knowing that living life can be good, but feeling more convicted that living an abundant life is even better.

Part I
CSR and COVID-19 Pandemic in Europe

Austria
 Greece
 Poland
 Romania
 Serbia
 Spain
 United Kingdom 1
 United Kingdom 2
 Turkey

COVID-19 Pandemic Management from a Sustainability Viewpoint: An Analysis for Austria, the European Union, and the WHO

Ursula A. Vavrik

1 Introduction

More than 2 years after the outbreak of the COVID-19 pandemic, an interim analysis of international and national pandemic management appears to be opportune. In a first step, we investigate WHO guidance which seemed to have revealed certain weaknesses. Several questions remain about timing, adequacy, effectiveness and the proportionality of measures and guidance, in particular when focusing on an early warning and prevention strategy, and from a sustainable development perspective.

Major uncertainties persist about eventual unpleasant consequences of the worldwide vaccination program which has been based on mere emergency admissions so far, with drastically shortened scientific and medical protocols in comparison to the current state of the art in medicine. Research on long-term implications or side effects of vaccines as demanded in current international medical protocols just does not exist. The WHO recommended that COVID-19 vaccination program might thus be considered as the best available policy option, but probably not as an evidence-based policy strategy including sustainability considerations.

In a second step, the role of the European Union in pandemic management is scrutinized, followed by a case study about Austria. Policies are screened from a sustainability perspective, in particular lockdowns and the vaccination policy. Alternative (and probably underexploited) medical treatments such as ozone therapy will be debated, and conclusions and recommendations are drawn for the future, even though they may be premature to do so.

In this respect, we would like to address the following questions:

The chapter was finalized in June 2022.

U. A. Vavrik (✉)
NEW WAYS Center for Sustainable Development, Vienna, Austria

© The Author(s), under exclusive license to Springer Nature Switzerland AG 2023
S. O. Idowu et al. (eds.), *Corporate Social Responsibility in the Health Sector*, CSR, Sustainability, Ethics & Governance, https://doi.org/10.1007/978-3-031-23261-9_2

- How well and efficient was the guidance States received from the WHO before and throughout the pandemic, and how were sustainability considerations taken into account?
- Similarly, how well did leaders at European-level perform with regards to pandemic management, and were sustainability considerations raised at all?
- What differences or similarities could be observed at national pandemic management? Which countries performed better than others and what were the main approaches of frontrunners? Did their decisions foster sustainability more, and if so, in what respect?
- What did Austrian politicians do well, where was room for improvement, and how sustainable were policy decisions?
- How did the pandemic alter our society, in particular from a human rights perspective—which is a key part of a sustainable development policy? What about the notions of democracy, democratic debate, freedom, freedom of expression, freedom of thought, freedom of science, transparency, and the like?
- How well were policy decisions balanced from a sustainability angle, i.e., balancing measures for controlling epidemic development with economic and social development of the society, and applying the precautionary principle?
- Can it be argued that a worldwide vaccination policy has helped the survival of mankind rather than touching upon a new planetary boundary with unknown consequences for the human race?

2   I. The Role of the WHO in the Global Health Crisis: Was the Necessary Guidance Delivered in Time?

From the very beginning of the COVID-19 pandemic, the world would have probably needed a better **pandemic prevention strategy.** A general lack of pandemic prevention and preparedness was also found in the most recent OECD analysis (OECD, 2022), along with a certain lack of transparency and inclusive decision-making processes. The study also stated that *"issues relating to policies' proportionality are still under-explored."* A study in the Lancet likewise comes to the insight that global health governance *"highlighted profound weaknesses, inadequate preparation, coordinations and accountability hampered the collective response of nations at each stage. Changes to the global health architecture are necessary to mitigate the health and socioeconomic damage of the ongoing pandemic, and to prepare for the next major global threat to health."* One of the measures proposed by the World Health Assembly was the establishment of a Pandemic Treaty, which was discussed for the first time by this Assembly in December 2021 (Williamson et al., 2022). We already called for better international health guidance from the WHO and a more effective early warning system in April 2020 (Vavrik, 2020b).

Major shortcomings relate to the **knowledge about the origin of the SARS-CoV-2 virus, the timely, effective information and communication about potential** *threats and efficient countermeasures*, the *definition of a COVID death*, and last but not least the *lack of sustainability considerations* at most times.

An assessment of the European Institute for Security Studies describing the outbreak of the COVID-19 virus stated that the WHO announced a global health emergency on January 30, 2020 (Gaub & Boswinkel, 2020). This was about 3 months after the first outbreaks of the virus were reported by Chinese authorities at the beginning of November 2019 (Himmel & Frey, 2022), even though official warnings were expressed by Taiwan already 1 month earlier (Mazumdaru, 2020). On several parameters, WHO declared COVID-19 much later a pandemic than the swine flu (Gaub & Boswinkel, 2020). In July 2021, China rejected a WHO plan to further investigate the origins of the pandemic (NPR, 2021). The United States and Australia (Karp & Davidson, 2020), scientists from several countries and also the European Union argued for further in-depth investigation on the origin of the virus, but were so far unheard (EEAS, 2021; Pezenik, 2022; Raynolde, 2021). Moreover, WHO Director General Tedros Adhanom Ghebreyesus was criticized for aligning too closely with China, particularly in this pandemic (Mazumdaru, 2020).

2.1 A Governance Issue?

The international community is urgently requested to think of measures to avoid such a situation from happening again. Suspected countries must open their doors to independent international investigations for the well-being of all nations, i.e., enforceable by the WHO. Even though the origin of the SARS-Cov-2 virus is still not clear, it appears to be common knowledge that the virus spread from China. In case of transmission from certain wild species at a marketplace or a laboratory accident, representing two of the prevailing theories about the coronavirus outbreak (Ni & Burger, 2021; Regan, 2022; Tech2, 2021), how can it be justified that millions of people had to die without almost any discussion about sanctioning the responsible persons, organizations or States? How adequate, responsible, and sustainable is such a governance behavior? And why does the WHO not do more to remediate this situation or at least try to urgently find solutions to avoid recurrence of similar catastrophic pandemics?

Criticism related to the fact that investigations about the origin of the virus did not follow international standards was not independent, and no sanctions were taken into consideration. A laboratory accident *"is most likely but least probed COVID origin,"* a memo of the US State Department said (Kopp, 2022). Hence, for the safety of the world population and especially future generations, it may be imperative to foresee *an international sanction system* for provoked global health threats, be them intentional or unintentional, as it exists for instance for hazardous environmental accidents.

As already mentioned above, the *WHO early warning system* did not function well enough. Would Europe and the world have been informed about potential health threats 1 or even 3 months earlier, how many lives could have been saved?

Combined with that early warning appeal, a precautionary healthcare protocol for a potential pandemic threat could have been installed worldwide as early as November 2019. But the WHO was very late in issuing prevention recommendations to combat the SARS-CoV-2 virus such as to wear masks and to keep the right distance (2 m instead of one), the two most helpful measures together with enforced hygiene such as repeated long hand washing. Similarly, it could have been WHO or international guidance from the beginning to generate standardized data sheets about every COVID-19 patient, aimed at facilitating applied research on pandemic containment. Furthermore, WHO gave some diet advice for self-quarantine and during the pandemic, explicitly stating that "*While no food or dietary supplements can prevent or cure COVID-19 infection, healthy diets are important for supporting immune systems. Good nutrition can also reduce the likelihood of developing other health problems, including obesity, heart disease, diabetes and some types of cancer.*" Unfortunately, the issue date of this advice was not traceable on the WHO website (WHO, 2020b, 2022a). Presumably, at no stage had the WHO published guidance related to precautionary diet preferences such as alkaline food stopping viruses from developing, or on avoiding obesity as precautionary measure, given the strong correlation between obesity and severe COVID-19 cases. On the other hand, already almost 100 years ago, Dr. Otto Warburg, Nobel Prize Winner for his work on respiratory enzymes and cancer, stated that "No disease can exist in an alkaline environment" (Anderson, 2017) and has discovered that "An alkaline body can absorb up to 20 times more oxygen than an acidic body and . . . . .that diseased bodies are acidic bodies" (Prairy Naturals, 2017). Similarly, Prof. Vormann, founder and leader of the Institute of Prevention and Nutrition near Munich highlighted in an interview "*that it is fascinating to observe how viruses can be irreversibly inactivated by changing the cell milieu from acid to alkaline,*" to a pH value above 7 (IPEV, 2020). The same news was published by the Austrian Press Agency (APA, 2020). In a similar vein, Indian doctors refer to the importance of keeping the body in an alkaline stage (pH of 7.4). "*A well-oxygenated balanced alkaline body has the adequate immunity to successfully fight diseases.*" Dr. Eapen Koshy emphasized in his book "*Survive Covid: By Staying Alkaline*" (Times of India, 2021). Another group of Indian scientists equally state that "No one with an alkaline body balance would succumb to the COVID-19 virus" (Khandar et al., 2021). Schwallenberg researched that an alkaline diet could reduce morbidity and mortality in chronic diseases (Schwallenberg, 2012). The Kenyan government issued dietary recommendations to prevent and overcome COVID-19, relying also on "*supplementation with Vitamin C, zinc, Vitamins A, B6, D, E, iron, Folate and fiber*" (Republic of Kenya, 2020). Focusing on prevention, Italian, Swiss, and Rumanian clinics were applying ozone therapy as prevention and treatment already in the early phase of the pandemic (see chapter IV below).

In that respect, non-existent WHO directions on alternative prevention measures such as alkaline diet or oxygenated ozone therapy, etc., represent a missed opportunity to apply the precautionary principle and act in accordance with sustainability guidelines for the world as a whole.

Moreover, promoting a ***worldwide vaccination program without sufficient scientific evidence on long-term consequences may raise scientific concerns for the sustainable development at planetary level*** (see Velot, 2021). As Dr. Velot, scientific director of the scientific committee of independent information research of the genetic genius (CRIIGEN) pointed out, it was a strategic scientific and medical error to vaccinate the whole population, since vaccinating during an pandemic facilitates the development of the virus and its mutations (Velot, 2021). Similar thoughts were expressed by various experts and scientists from around the world (see Kupferschmidt, 2021; Salomon, 2021). Even cohabitations between injected genetic materials like the mRNA vaccines and other viruses are possible. One should have concentrated on the persons at risk and not vaccinate younger people, since it was not necessary from a scientific perspective, even counterproductive, Velot deemed. In the UK and Israel, the two countries with highest vaccination figures, the new Delta variant has spread more rapidly. Dr. Velot and others also missed the absence of a scientific debate with differing opinions (Velot, 2021). As Seneff/Nigh are pointing out, to date merely the British Medical Journal offers a platform for a more diverse debate on measures to combat the SARS-CoV-2 pandemic (Seneff & Nigh, 2021). Dr. Velot asked why medical instances do not reflect on why we have so many deaths of cancer or other comorbidities. Those diseases show much higher death tolls than SARS-CoV-2. He admitted at the same time that environmental causes including high pesticides use are part of the answer. In this context, he also questioned the proportionality of political measures and stated that COVID-19 is just hiding another important disease, the one of our sick environment. Likewise, he emphasized that if too many people are vaccinated, the human immune system may not be able to withstand other upcoming, maybe even more disastrous variants or mutations or even other viral or bacterial infections, which may lead to worrying consequences. Since long-term side effects or consequences are not known, and the appearance of new mutations and variants is not predictable, it appears to be scientifically unwise to vaccinate the whole population worldwide because this does not leave a fallback option in case some unexpected developments would appear (Velot, 2021).

Seneff/Neigh argue alongside similar lines and developing their findings further could lead to the assumption that mass vaccination could even entail a threat for the survival of mankind. Due to the absence of long-term trials and manifold unprecedented aspects of COVID-19 and subsequent vaccine development for use in the general population, the COVID-19 vaccination program with mRNA vaccines can therefore be considered a worldwide experiment with novel technologies and possible unintended consequences (see Seneff & Nigh, 2021). Given the potential of predictable and unpredictable negative consequences, Seneff/Neigh *"conclude with a plea to governments and the pharmaceutical industry to consider exercising greater caution in the current undertaking to vaccinate as many people as possible."* One predictable negative consequence of mass vaccination could be the more rapid appearance of novel SARS-CoV-2 variants, in which science may not be quick enough to respond with effective vaccines. It was already shown that the Pfizer vaccine *"was only 2/3 as effective against the South African strain as against the original strain"* (Liu et al., 2021, in Seneff & Nigh, 2021). Another potential consequence could lead to genetic transformation or genetic engineering processes in which the new gene could be transferred through sperm cells to future generations. Thus, *"the mRNA vaccines are an experimental gene therapy with the potential to incorporate the code for the SARS-CoV-2 spike protein into human DNA"* (Seneff & Nigh, 2021). Was the general public sufficiently informed about that specific consequence? Probably not, or at least not in a transparent way. Seneff and Neigh therefore urge governments to be more cautious in the future in applying biotech-nologies. At the same time, governments could be encouraged to advise the population on how to boost their immune systems, which was partly done by WHO and FAO. Nevertheless, focus should also be given to raising vitamin D levels (e.g., through sunlight), eating mainly organic whole food and food containing vitamins A, C, and K2 (see Seneff & Nigh, 2021). In our search, we did not find this particular health governance on pandemic prevention by WHO or FAO.

The **WHO definition of a COVID death** and the derived excess mortality figures also raise some issues of scientific uncertainties. Besides that, not all countries apply the same definition, and definitions changed over time. Already at the beginning of the pandemic, the WHO referred to a COVID death as every person who died with or probably with COVID-19, *"unless there was a clear alternative cause of death that cannot be related to COVID-19"* (WHO, 2020a). The Public Health England defines a COVID-19 death as a person that died within 28 days after a laboratory confirmed positive COVID-19 result, without recovery in between (Heneghan & One, 2020; Newton, 2020). As of May 2022, countries reported 5.4 million excess deaths due to COVID-19 to the WHO, while the latter considered this figure may be three times higher, amounting to about 14.9 million (Jerving, 2022; WHO, 2021). Other research, however, came to different conclusions about excess mortality. Kuhbander presented a study of Michael Höhle from the University of Stockholm investigating COVID-19 excess death figures in Bavaria/Germany in 2020, which showed even a slight negative excess death toll when considering population growth and age distribution together with the number of deaths (Kuhbandner, 2021). With respect to world data, available figures do not show significant changes in the death rate between 2019 and 2022 either, although the same website states that deaths *"2020 and later are UN projections and do not include any impacts of the COVID-19 virus"* (Macrotrends, 2022). Maybe the most recent figure of the WHO of about 14.9 million COVID-19 deaths has therefore to be looked at with caution. The question though remains: Why has the COVID-19 pandemic received so much attention while for other even more deadly diseases the WHO did not recommend stricter measures—to avoid ten million cancer deaths or 25 million deaths of cardiovascular diseases in 2020? And how big was the additional international effort to confront those diseases?

Mortality rates of SARS-CoV-2 vary among countries. According to the John Hopkins Corona Virus Resource Center, best performers were Australia/South Korea/Iceland with 0.1%, followed by Denmark/Cyprus/Maldives with 0.2%, the Netherlands/Serbia/Malaysia/Israel with 0.3%, Finland/Luxembourg/Estonia/

Switzerland/Vietnam/Kuwait/Mauritius with 0.4%, and Austria/Germany/Liechten-stein with 0.5% (John Hopkins, 2022).

Mortality rates also vacillate enormously with age and sex. Data available from New York City from 2020 showed 14% mortality for 80+ age and 0.2–0.4% for ages 10–50 years; a normal influenza in comparison has a rate of about 0.1% (Worldometer, 2020).

In this perspective, the *vaccine policy* proposed and led by the WHO also needs to be examined more carefully. Was it necessary to vaccinate almost all age groups or would not it have been enough to vaccinate persons with particular health anteced-ents and particularly vulnerable or high-risk cases? Whereas the WHO COVID-19 advice for the public page on vaccines refers to hygiene standards, physical distanc-ing, as well as to getting vaccinated as the best possible way to prevent COVID-19 (WHO, 2022b), there is nowhere any COVID-19 diet recommendation, for espe-cially alkaline nutrition to prevent viral infections, and also as alternative to vacci-nation with similar or maybe even better results. The WHO advice that vaccination is the best possible way to avoid a COVID-19 infection maybe particularly true for persons with 80+ years of age, since the risk for a lethal COVID-19 infection raises with age and was most prevalent at the age groups 80+ (about 70% in Switzerland) and 70+ (about 90% in Switzerland), but for the younger generations, scientific evidence is less obvious. In Austria, median age for people dying from COVID-19 was 77 years for men and 82 years for women (see also Statista, 2022a, 2022b; Statistik Austria, 2022a).

In 2021, the US Center of Disease Control changed the general *definition for vaccines* for messenger RNA vaccines such as Pfizer or Moderna to be included as vaccines, which would not have been possible with the old definition (Loe, 2022). While before 2021 a vaccine was defined as "*a product that stimulates a person's immune system to produce immunity to a specific disease, protecting that person from that disease,*" the new definition reads as "*a preparation that is used to stimulate the body's immune response against diseases.*" Also the Webster Dictio-nary had to change the definition, adding that such a preparation could be either "*a) an antigenic preparation . . . ..or b) a preparation of a genetic material (such as a strand of a synthesized messenger RNA). . .*" (Loe, 2022). These new vaccines thus contain genetically modified material, which was never the case with any other vaccine before. Both aspects seem at least to a certain extent irritating, especially since they were not communicated openly and transparently enough, neither by the WHO nor governments, nor the media. Even though it is more understandable now than 2 years ago that this whole vaccination exercise was more of an experiment in which even vaccine developers were "learning on the job," it would have seemed fairer to inform the public better and earlier about definitions, novelties, conse-quences, and possible harm to individuals and the society as a whole. But of course, this was not possible since the relevant research was not available. There were many uncertainties, policy changes, and rectifications throughout the pandemic. In the beginning, people were told that with a vaccination there will be free access to economic, cultural, and social life. Meanwhile, several revisions of these statements took place, and in some parts of the world, governments talk or even demand a fourth or even a fifth booster. As an Austrian pharmacologist stated, these vaccines are "strong," never before had he seen those negative effects from a vaccine like it was the case with the COVID-19 vaccines (Lang, 2022).

Beyond that, scientific uncertainty persists on the question whether COVID-19 vaccines contain the toxic component graphene or not. Pablo Campra presented evidence in his research in which the toxic material graphene was detected in COVID-19 vaccines (Campra, 2021). A parliamentary question to the Commission on the same issue, however, resulted in an answer denying any presence of graphene in COVID-19 vaccines (European Parliament, 2022). Moreover, the WHO remained quiet on this issue.

A recent study published in the British Medical Journal pointed out that **"Mandatory COVID-19 vaccine policies** ... have provoked considerable social and political resistance, suggesting that the have unintended harmful consequences and may not be ethical, scientifically justified, and effective. <They> prove to be both counterproductive and damaging to public health. <The> framework synthesizes insight from behavioral psychology (reactance, cognitive dissonance, stigma, and distrust), politics and law (effects on civil liberties, polarization, and global governance), socioeconomics (effects on equality, health system capacity, and social well-being), and the integrity of science and public health (the erosion of public health ethics and regulatory public oversight). <The authors> question the effectiveness and consequence of coercive vaccination policy in pandemic response and urge the research community and policy makers to return to non-discriminatory, trust-based public health approaches"(Bardosh et al., 2022).

When analyzing WHO policy guidance on COVID-19 through *sustainability* lenses, the WHO Director General's mission aspires to achieve the UN SDGs. Yet, looking very closely into the policy proposed, we miss the forward-looking application of the precautionary principle, one of the core principles of sustainable development policy, as already explained in more detail earlier. Health problems should be avoided at first; treatment is only the second best option. This leads us again to the already reported fact of major failures with respect to an effective early warning, prevention, and preparedness strategy for such types of pandemics.

Recommendation 1: Review and improvement of the WHO early warning system regarding pandemics or other global health threats to impede rapid spreading, such as systematic, coordinated, and connected alert systems with immediate action recommendations, for governments and all medical institutions, including hospitals and medical staff worldwide.

Recommendation 2: Establish a mechanism or agreement that under certain conditions each WHO Member or at least a group of members could ask for an independent investigation at the assumed source of the pandemic, at any moment, to be executed not later than within 1 week. The concerned country is obliged to accept this request and does not have the right to delay, hamper, dilute, mask, or circumvent such investigations.

Recommendation 3: An international sanction system or mechanism needs to be put in place, providing for proportionate damage payments by the country in which the pandemic took its origin, to affected countries, in case of intended or unintended consequences caused by this pandemic. International and European law on hazard-ous environmental accidents hereby could deliver guidance.

Recommendation 4: The precautionary principle should be put more promi-nently into the center of WHO activities and pandemic prevention strategies as one of the core principles of sustainable development.

Recommendation 5: The WHO is urged to make a sustained effort for the improvement of public health worldwide through relevant campaigns including for alkaline, whole, organic, and other healthy food and taking more advantage of local and indigenous knowledge to overcome most prevalent diseases like cancer and cardiovascular diseases.

3 II. European Governance and its Contribution to Global Health Crisis Management

In particular from a European Member State's perspective, the involvement of the European Union in pandemic crisis management was perceived as somewhat helpful and pertinent, at least to a certain extent. However, given failures at WHO and at Member State's level, more decisive European governance could have made the difference to an efficient preventive containment strategy, including sustainability aspects and protecting European values, which was unfortunately not really the case.

Even though health policy so far has mainly been a competence of Member States only, the European Commission in its coordinating role took several initiatives such as buying intensive care medical equipment, masks, or vaccines for all Member States that were willing to join into this common undertaking (European Commis-sion, 2020). This made a huge positive difference especially for smaller countries like Austria that with regards to enormously competitive market conditions, espe-cially for the mask and vaccine markets, would not have been able to negotiate those price levels and the quantities needed within the best possible time frame. Member States could opt for the vaccines of their choice and the quantities needed and thereby benefit from the common order and purchase through the European Commission.

In July 2020, the European Council agreed on a huge pandemic recovery fund dotted with EUR 750 billion for 2021–2027 to help curtail economic shocks (Alvarez, 2020; European Council, 2022). Within the EU Multi-annual Financial Framework 2021–2027, EUR 1.8 trillion including EUR 800 billion for the EU Next Generation Fund was dedicated to COVID crisis response and sustainable and resilient recovery. Other European activities encompass the launch of the first EU social bond EU SURE of EU 17 billion—listed at the Luxembourg Stock Exchange—the participation at the COVAX initiative with EUR 1 billion, disburse-ment of EUR 1 billion to Team Europe Initiative on Africa, EU humanitarian air bridges to Sudan, Venezuela, Peru, Latin America and the Caribbean, and the portfolio for 10 most promising treatments for COVID-19 (European Commission, 2022a).

Additionally, entrance rules into European countries were up to a certain extent discussed and decided at European level. Sometimes, especially in the case of working commuters for agricultural and healthcare jobs, these European agreements posed problems, also to Austria. While Austria depended on those workers in the various sectors, a lot of cross-border commuters were excluded from their jobs so badly needed in Austria.

European-wide coordination and exchange of experience, knowledge, and best practice during the pandemic were a positive aspect with regards to European-level activities. Also, the focus on solidarity and a more resilient Europe post-coronavirus was part of a successful strategy (European Commission, 2022a).

However, we missed guidance fostering the focus on preventive policy decisions, the debate on various possible policy approaches in regard to combatting the virus, or emphasis on more sustainable solutions for all, or best public health policy for all, and even more importantly the protection of European values across Europe.

The European Center of Disease Control has a major role to play in this context and, at least with respect to the forth shot, issued a valuable recommendation only for the elder generation.

With respect to the aforementioned lack of WHO governance, we would have expected a more demanding EU position, challenging WHO guidance and showing the way toward more sustainable policy decision making by requesting an urgent update of the WHO pandemic prevention strategy, a new mechanism to detect and inform about pandemic outbreaks, as well as a pandemic sanction mechanism that would come into force if a country would undermine its obligation to rapidly inform the world about a new epidemic threat or variant. Similarly, the European Commission could have been stricter with Member States asking them for more epidemical data transparency and availability, fostering focus on protecting European values, democracies and human rights, and demanding proportionate policy measures with the least negative impact for economy and society, leading to a more resilient Europe.

Recommendation 6: The European Union and where appropriate the European Commission are invited to strengthen their role and influence in global crisis and pandemic management based on the precautionary principle and on European values. In particular, the EU should requisition the WHO to develop a proper global pandemic prevention and preparedness strategy including a sanction mechanism, as well as insist on another independent international investigation on the origin of the SARS-CoV-2 virus.

4 III. Austria's Performance in COVID-19 Crisis Management

The recent report of the Austrian Court of Auditors gave the government quite an alarming attestation with regards to the performance in COVID-19 pandemic management. Several points of mismanagement and failures were detected and illustrated, and finally, solutions for urgent rectification of those policy failures are provided. The title of the report expressed that the challenges of pandemic crisis management, although known sometimes for several years still remain unsolved. At the same time, the government did not take advantage of lessons learned since the emergence of the pandemic in order to improve or develop its crisis management capacity. During the last years, the various Ministers of Health, highest authorities for the protection of public health, did not execute their responsibilities accordingly in ensuring the prevention and the management of the pandemic in the best possible way. In particular, they neglected to adapt the national pandemic prevention plan before the pandemic along WHO guidelines and to adjust the outdated pandemic law to modern requirements. Further, coordination failures were depicted, especially within the government at national level, but also in cooperation with the regional level. One example was the parallel development of pandemic key data, also of the interior ministry in cooperation with the regional governments, even though this was not their competence, which led to differing official key data and statements, and thus to confusion, impeding evidence-based policy making and ultimately undermining the credibility of the government (ÖRH, 2022a). Only recently, the government had to make a late notification of more than 3000 additional COVID deaths (Schmidt-Vierthaler, 2022).

Moreover, criticism was expressed in relation to the unfilled key positions before and even during the pandemic, as e.g. the position of the Directorate General for Public Health which was vacant already in 2019, and appointed only at the end of 2020. Similarly, the new nomination of the members of the Highest Sanitary Council, which should have taken place in early 2020, was realized with delay in March 2021.

Finally, the Court of Auditors urged the government to rapidly develop a pandemic plan and a new crisis management mechanism ensuring the efficient distribution of responsibilities and the effective cooperation of all stakeholders involved (ÖRH, 2022a).

Compared to other European countries, Austria seemed to have managed the COVID19—crisis relatively better than others, although several Nordic countries or some Asian countries still showed better results. It will only be possible at a later stage to evaluate whether the measures to fight the pandemic were at all times adequate, effective, and proportional—especially with respect to the damage caused to economy and society—and complying with common ethics and human rights standards. Probably, much more studies are necessary, and data transparency and availability need to be improved.

In this respect, we would like to investigate about the adequacy, effectiveness, and proportionality of the policy looking into more detail into certain parameters and issues like a) death toll, b) law framework and finances, c) political governance to steer economy and society, d) vaccine policy and tests, e) public health status and scientific debate, f) data availability and access to information, and g) proportionality, protection, and respect of human rights like freedom, freedom of choice, freedom of expression, freedom of choice of medical treatment, freedom of science, inclusiveness and diversity in decision-making processes, as well as fostering competitiveness and scientific debate.

4.1   Death Toll

One of the overall goals of governments certainly was to save as many lives as possible, but at what cost? "*At all cost*" was the answer of our Federal Chancellor right in the beginning of the pandemic. Austria's death toll with almost 19,000 COVID-19 deaths up to May 2022 at the time of writing of this analysis, was in relative terms, compared to worldwide figures, not too high. For the year 2020, Statistik Austria reported 6491 COVID deaths and for 2021 7857 (Statistik Austria, 2022b). It may be interesting to note that the number of total deaths in Austria grew from more than 83,000 in 2019 to 91,599 in 2020 and fell again to 90,434 in 2021, whereas cardiovascular diseases accounted for 31,302 cases, cancer for 20,659 cases, and other diseases for 26,274 cases in 2021 (Statistik Austria, 2022b).

Considering that in an earlier study, experts calculated that in a year of a heavy influenza, there could be up to 6000 deaths (ORF, 2010), and the COVID-19 cases are not too far away from that figure. At least for the currently prevailing Omicron variant, people compare it with a heavier flu. This is probably why most European or Western governments ceased all or most of the COVID-19 measures, hoping that the population would get immunized.

With respect to the age of COVID deaths in Austria, the highest percentage was for ages between 75 and 85 years; average age for men was 77 years and for women 82 years (BVZ, 2021). This leads us to the question why governments have chosen to vaccinate the whole population and not only the most vulnerable ones.

Compared to other countries, e.g., Switzerland, Sweden, or other Nordic countries, Austria performed worse than all those countries with respect to the parameter death toll. In terms of deaths per million people, Austria showed 2049 cases, whereas Switzerland had 1598, Denmark 1090, Sweden 1854, Germany 1654, Ireland 1442, the Netherlands 1297, Norway 571, and Finland 814 cases (Worldometer, 2022a).

With respect to the deaths related to COVID-19, or excess mortality, Statistic Austria reported a rise of about 9% in 2021 (BVZ, 2021). While writing this chapter, Statistik Austria changed the information officially available to the public and the more detailed statistical raw data suddenly was no longer available with a direct link from Google, but only through the Statistik Austria webpage, with some additional clicks (see Annex 1). This highlights again the huge data transparency and availability problem scientists face, since official data often can neither be accessed nor is it transparent or robust enough.

4.2 Law Framework and Finances

In March 2020, the Austrian Parliament voted for a comprehensive legislative package designed to confront the COVID-19 pandemic, also transferring legislative power to the president, with the support of at least two opposition parties. Even if the Parliament transferred part of its competences, it still remained with the control function over the government, just to a lesser extent, which very soon already created disputes with the government. In the same vein, the Minister of Finance was empowered in a first step to spend around 40 billion of Euros to manage the pandemic, almost without any parliamentary control. This sum was then upgraded to more than 73 billion of Euros, whereas until June 2021 about half of these funds were spent as follows (see ÖRH, 2022b):

- Short-time work 8.5 bio Euros (Labor market).
- COFAG outage bonus 2.4 bio Euros (Economy).
- COFAG Curfew turnover compensation 2.2 bio Euros (Economy).
- Hardship Fund 1.8 bio Euros (Economy).
- COFAG Curfew turnover compensation II 1 bio Euros (Economy).
- COFAG fixed cost grant 2 bio Euros (Economy).

Whereas the government had the initial support of two opposition parties, this situation entailed several parliamentary disputes and debates about whether the money was spent accordingly. Some issues are still under investigation such as the COVID-19 support for the People's Party (ÖVP) senior association of EUR two million.

4.3 Political Governance to Steer Economic and Social Development

After four lockdowns of several weeks, the economy was both heavily challenged and weakened, and government support arrived rather late, for the cultural sector even later. The fact that, in Austria, the third Minister of Health is currently operating since the pandemic started underlines the difficulties in pandemic management encountered by the government. Even though economic financial support was rather generous, it took long and was bureaucratic and complicated, and thus heavily criticized. Main measures for economic support included a credit system and direct support for enterprises on the basis of lost gains, as well as support for short-time work, later on also for associations and artists. The full list of measures can be found on the official websites (BMSGPK, 2022).

In our neighboring country Switzerland, a new system of government support for the economy was developed within a very short time and very effectively (Swissinfo, 2022a), earlier and more effective than in Austria. Unfortunately, our government did not take advantage of that knowledge although it would have been possible.

Generally speaking, the lockdown and social restriction measures appeared too strict to the wider public; e.g., people were not allowed to sit on park banks or public parks, restaurants were even closed, and outdoor sports were not allowed, which was especially hard and cumbersome for families with children. Several thousand fines were issued, amounting up to 3000 Euros: A mother had to pay 500 Euros, because two children played football in a park. At a later stage, at least a part of these fines were paid back due to inconsistence with the law. Yet, to some extent, Austria was more liberal than France or Italy, where citizens even had more restrictions for outdoor mobility. Sweden, in comparison, seemed like heaven (Riedel, 2020). Almost no regulations prevailed there; instead, the government relied on the wisdom and individual responsibility of its citizens and gave merely recommendations and advice. There were no lockdowns, restaurants, parks and sports facilities were open, there were no restrictions for the economy or schools, public and outdoor life was almost normal. With some thought, the Austrian set of measures to combat the pandemic could have been better tailored: Physical distancing was limited to 1 m instead of 2 m, masks were not promoted during the first months, and outdoor sport was unnecessarily restricted. We published some suggestions for the government already in April 2020 which became reality only much later thereafter (see Vavrik, 2020a, 2020b).

In particular in the beginning of the pandemic, the government overestimated the curve of COVID deaths, probably influenced by developments in Italy and by reading the mathematical models not carefully enough, while at the same time not providing sufficient or transparent data to experts and not seeking or involving a balanced expert advice. Sustainability concerns did not seem to be of importance (see also Vavrik, 2020b). The COVID-19 government advisory body was mainly composed of virologists and mathematicians, for some time had only one public health expert, and never had a sustainability expert.

At its worst stage, the economic downturn amounted to 7.4% in the first half of 2020, the deepest recession since the Second World War (ECFIN, 2021), and lockdowns and closed schools lead to elevated psychological and health problems. So-called reduced intensive care capacities in hospitals on the other hand delayed a huge number of operations. Actually, Austria providing one of the most generous healthcare systems worldwide, including a very high number of intensive care beds, reduced the number of beds during the COVID-19 pandemic, due to lack of trained personnel (John, 2021). In fact, the highest degree of capacity utilization for intensive care beds was about 60% in November 2020 and about 70% for normal hospital beds in April 2021, with an average of about 30–40% throughout the pandemic (Statista, 2022c). Unfortunately, during the last 2 years, no efforts were made to train more intensive care personnel, which would have proved to be useful to be able to have three shifts during intense times of the pandemic and to avoid burnouts of hospital staff. As experts suggest, it cannot be excluded that either new variants or new mutations or even new pandemics would arise in the future.

In comparison to Austria, Switzerland applied less stricter measures and had only one lockdown of 8 weeks based on recommendations to the public without a formal obligation to stay at home. Even though Switzerland disposes of less intensive care bed capacity, its death toll was much less. As already mentioned, policy support for economic recovery was much more effective and better organized (Sager & Mavrot, 2020; Swissinfo, 2022b).

4.4 Vaccine Policy and Tests

Austria was the first country worldwide aspiring to impose mandatory vaccination as of March 1, 2022. Finally, some days later on March 9, 2022, this decision was revised and suspended (Agence France Press, 2022), and finally, on June 24, 2022, the government announced the end of mandatory vaccination (ORF, 2022c). This highly unproportional policy measure was heavily criticized by the public, the Parliament received numerous statements from informed and concerned citizens (see also Sprenger, 2021a), and this law led to the biggest wave of demonstrations since about 20 years. It probably was a desperate attempt to raise the number of fully vaccinated people, which in May 2022 amounted to about 74%. Taking the example of Switzerland again, the proportion of the population vaccinated there is even lower, amounting to 69%, and *"the FOPH as well as the Federal Commission for Vaccination do not recommend a second booster jab"* (Swissinfo, 2022a), showing a much lower death toll of about 13,000 cases in May 2022 compared to almost 19,000 cases in Austria (Worldometer, 2022a).

Pandemic management in general, but especially the vaccination policy, was responsible for collapsing rates of public support for the government in office. During the last 2 years, two health ministers had to resign, and especially, the right-wing party (FPÖ) which represents about 20% of the population now could gain terrain, by promoting an anti-vaccination policy. In addition, a new political party, MFG Austria—People Freedom Fundamental Rights, the so-called anti-vaccination party, emerged in February 2021, with unprecedented amazing results of up to 20% at communal elections and about 10% on average at some regional elections (ORF, 2022a; Stacher, 2021). In the capital of Styria, Graz, citizens were so desperate about the government course during the pandemic that for the first time in history, the communist party surprisingly won almost 30% and their female candidate got the post of the mayor, a double landslide victory (Müller & Tomaselli, 2021).

On the positive side, intensive testing was part of a successful COVID-19 containment strategy. Up to May 2022, in Austria there were 188.5 million of PCR or antigen tests processed (Statista, 2022d). With more than 20 million tests per million inhabitants, Austria is worldwide number two after the world's test champion Denmark with about 21 million tests per million inhabitants (Worldometer, 2022b). The city of Vienna became famous for its Gurgel tests, probably a unique system around the world: one could make these tests at home in front of a running internet camera, deliver them to the next supermarket, and the results were received some 6–12 h later, all free of charge. Even today, the newest regulation in this respect foresees five free PCR or antigen tests per month for each citizen without symptoms. With symptoms, they are free of charge anyway. Gurgel tests by the way are much less invasive than nasopharyngeal tests (The Local, 2021). Some studies question the efficiency of such PCR tests for all patients, stating that they provide little added diagnostic value and prescribe a strategy based on clinical presentation only (The AP-HP/Universities/Inserm COVID-19 Research Collaboration, 2021). The interpretation via the CT value is debatable as well. Whereas it seems common to regard a CT value above 30 as "positive," at the Chinese Olympics a value of 35 was requested and in Vienna even a value of 39 (ORF, 2022b).

4.5 Public Health Policy and Scientific Debate

The longer the pandemic lasted, the more failures in Austria's public health strategy were visible. Unfortunately, the public health perspective only seemed to play a minor role in Austria's pandemic management. This was for instance obvious through the composition of the pandemic advisory board to the Austrian Government/Ministry of Health and Social Affairs, in which virologists and mathematicians formed the majority as already mentioned. After 2 years of pandemic management, it became obvious that Nordic countries like Sweden or Denmark strongly focusing on a public health approach showed much better results in pandemic management than Austria.

A public health approach is described as "...*the collective action for sustained population—wide health improvement*" (Beaglehole R. et al., Lancet 2004, 363, 2084–86, in Sprenger, 2021b) or "*the science and art of preventing disease, prolonging life and promoting health through he organized efforts and informed choices of society, organizations, public and private, communities and individuals*" (Charles Edwards A. Winslow, 1920, in Sprenger, 2021b). Evidence-based public health hereby relies on funded scientific knowledge to improve a nation's health situation in which illnesses and measures are approached from a systemic and transparent perspective leading to explicit, justified, and well-grounded measures and decisions. The availability of data for science and data management as well as a good early warning system embedded in the epidemiological detection scheme with systemic data gathering from infected patients would have been essential, was, however, neglected and ignored in Austria (Sprenger, 2021b), and is still not satisfactory even more than 2 years after the outbreak of the pandemic. Sprenger also highlighted the uneven distribution patterns throughout the pandemic, especially from a social perspective, where people in poorer social conditions or in nursing homes were proportionally stronger affected by the pandemic. In Austria, about 50% of all COVID deaths happened in nursing homes, whereby people in nursing homes constitute only 2% of the total population. Further, about 40% of the COVID-19 deaths were people with more than 84 years of age, the meridian age of death averaged about 82 years. Similarly, people with underlying health conditions and a higher body mass index (BMI) were more concerned, the latter by a factor of 4 (Sprenger, 2021b).

Other studies led to similar results and showed that population health played a determinant role in COVID-19 pandemic death tolls. In UK, 70% of COVID patients were overweight and obese (Gaub & Boswinkel, 2020). Likewise, in the USA, where about 40% of the whole population are obese, a study found that within 17,000 people hospitalized, 77% of these COVID patients were overweight. With obesity and SARS-CoV-2 virus, chances to enter an intensive care unit were 74% and chances to die 48%. *"The biology of obesity includes impaired immunity, chronic inflammation, and blood that's prone to clot, all of it can worsen COVID-19"* (Wadman, 2020). The USA accounting for more than 3000 COVID-19 deaths per millions of people also shows much higher figures than other wealthy nations. There are only very few countries worldwide that have even higher figures here, such as Peru, Ukraine, and Georgia (Worldometer, 2022a). High COVID death data in the USA mirror a poor public health situation, presumably also fueled by bad nutrition habits, and high levels of pesticides and GMOs in agriculture and food.

Before that background information, it appears astonishing that the risk factor of nutrition was not debated so much. It would have been interesting to receive warnings from WHO or government level that fighting obesity, representing a particular risk factor for COVID patients, could have saved lots of lives worldwide. To our knowledge, this did not happen. Neither WHO nor government websites seem to have stressed this correlation nor the fact that healthy, vegetable, and alkaline diets and physical exercise (see also Palmai, 2021) may be main determinants to overcoming the pandemic, since healthy diets and physical exercise stimulate the immune system and therefore play an important role. Instead, people were confined in lockdowns which definitely worsened the public health status of populations. On the contrary, Nordic countries had more liberal measures, particularly Sweden had no lockdowns, Switzerland only 1 of 8 weeks, based on recommendations, and in both Sweden and Denmark kindergartens and schools were open most of the time.

Successful pandemic management requires a multidimensional approach in which multidimensional damage needs to be minimized and measures are in total more useful than harmful, as well as proportional (Sprenger, 2021b).

Communication about COVID-19 measures has deteriorated throughout the last 2 years, especially with respect to the information about the usefulness of vaccines and the treatment of unvaccinated people. Government's spokespersons politicized, polarized, orchestrated, and disunited what entailed a massive loss of confidence in the government, Sprenger stresses (Sprenger, 2021b). Although Austria disposed of a national tool to assess health consequences, this has been completely forgotten during the pandemic. Similarly, the health objectives as stipulated in 2012 or the specific strategy for the health of children and youth were not applied, even though health figures are very highly ranked in most government agreements and represent the highest intrinsic value for Austrians, according to Springier (Sprenger, 2021b). Instead, every fourth child showed stress symptoms, and at a certain moment, psychiatric youth clinics were overcrowded, since schools and kindergartens were kept closed for long periods and lockdowns hindered them to get the necessary and healthy free moving space. People also suffered from a perceived loss of freedom (Sprenger, 2021b). Finally, Sprenger highlighted the importance of social cohesion as a lesson from the pandemic, and that *"social security is the most reliable basis for democracy"* (in Sprenger, 2021b).

Austrian statistics reveal that death tolls in cardiovascular diseases or cancer were a multiple factor compared to COVID-19 deaths. Nevertheless, operations were postponed from the very beginning to free beds for COVID-19 patients. Thus, this fact also raises certain concerns about the proportionality of measures. In this respect, it would be interesting to investigate how much money governments spend to avoid these types of deaths and why there are not more impeding measures to avoid lung cancer or to avoid obesity, diabetes, and cardiovascular diseases? Several European countries such as Belgium, Finland, France, Hungary, Ireland, Latvia, Monaco, Norway, Portugal, and the UK have introduced a tax on soft drinks for that reason.

With regards to the possibility of ***scientific debate*** on how to best overcome the pandemic, this debate seemed often very one-sided, not broad enough and certainly not inspired or enforced. In the beginning of the pandemic, Sweden appeared to be a complete outsider but turned out to be one of the frontrunners and winners in crisis management at a later stage. Apparently, public health experts in Austria were not heard enough, and even worse, those who had different opinions were either excluded or sidelined or even could lose their job as Professor Sönnichsen from the MedUni Vienna (RT, 2021). And most probably, these were no isolated cases. In other countries, similar developments took place. In another vein, sustainability concerns did rarely enter the scientific debate, and the Austrian government was not really keen on taking up advice in this respect (see Vavrik, 2020a)(Vavrik, 2020b).

4.6   Data Availability and Access to Information

Even today, in May 2022, more than 2 years after the outbreak of the pandemic, scientists declare remaining difficulties regarding availability and comparability of key statistical data in view of the COVID-19 pandemic. We highlighted this situation as well very early (see Vavrik, 2020a, 2020b). Due to two different main statistical systems in Austria with differing definitions of a COVID death, Austria suddenly had to correct its death toll figure upwards by more than 3400 cases (Schmidt-Vierthaler, 2022).

4.7 Proportionality of Measures and Respect of Human Rights: Democracies at Stake?

The outbreak of the COVID-19 pandemic has led to numerous societal changes and developments in governance and policy patterns. A first major impact relates to the **empowerment of governments**, mostly through the proclamation of the state of emergency, being concomitant with diminished parliamentary power or influence and therefore impeding democratic and inclusive decision-making processes (see also Brown et al., 2020). The Austrian government for instance replaced the existing epidemic law, which would have been a financial ruin for the government, with a comprehensive legislative package to lead the economy and society through the pandemic. As soon as vaccines were available on the market, citizens were sooner or later obliged to get vaccinated since otherwise access to work, restaurants, cultural events, or other leisure activities was almost impossible. No other policy since the existence of the European Union has had such a totalitarian touch as this pandemic management policy, or more particular the mass vaccination objective, even though the European Commission made it clear on its website that "EU law does not require compulsory vaccination," vaccination being a Member States' competence (European Commission, 2022b). However, the *freedom of choice* with respect to vaccine intake did only exist in a very limited way. Whereas in the beginning there existed the so-called 3G rule "Geimpft, Getestet, Genesen" (vaccinated, tested or recovered), later on the 2G rule was introduced: Testing was no longer possible for some activities such as going to a restaurant, or to a cultural or sports event.

During several lockdowns, citizens were deprived by their **freedom of movement**. **The right to demonstrate** was often not granted, and *freedom of expression* was often not possible, since conventional media were relatively uncritical of the government course and would not allow discussions or articles from people thinking outside the box. These phenomena were not restricted to Austria, and they happened everywhere around the globe, with maybe the exception of some Nordic countries.

In this respect, *core principles like equality were disrespected,* since suddenly the unvaccinated were treated as different and sometimes—erroneously—even as the responsible ones for fueling the pandemic. Austria had one lockdown only for the unvaccinated which was probably unique worldwide, even though some studies revealed that virus transmission was very similar in vaccinated and unvaccinated people (Paredes, 2022).

Human rights were also violated insofar as personal data should have been protected. With the introduction of the green national or European passport, it was evident whether someone was vaccinated or not while medical data so far had been considered as personal and confidential data.

All in all, the pandemic jeopardized well-founded European values, a rather disturbing fact. It is therefore legitimate to examine *the proportionality of all those measures* or whether governments did exaggerate to some extent, for whatever reason.

Given the lack of a proper pandemic prevention strategy both at WHO level as well as at national level, going along with a relatively massive lack of information and availability of data, proportionality of measures can be queried. The population had to suffer because of the neglect of politicians. Similarly, the usefulness of lockdowns with all its negative consequences for economy and society including the loss of freedom and mobility as well as entailed problems for personal and public health can equally be contested. Since lockdowns were inexistent in Sweden and only very short in Switzerland, they can be considered as too drastic measures and could or should have been avoided at most. Other measures relate to restrictions with regards to the Omicron variant, which was already known by the scientific community as much more benign. Moreover, the issue of closing schools or testing children massively has proven to be unnecessarily over-precautious, and thus, unproportional since young people were much less affected by the SARS-CoV-2 virus, as evidence showed (see above). More studies on this issue are awaited. Furthermore, massive vaccination appeared to be unproportional, since this hampered human rights, especially freedom of choice of treatment which corresponded to public health ethics so far. Secondly, newer studies reveal a high degree of uncertainty concerning the usefulness or even harmfulness of this mass vaccination and unintended negative consequences (Bardosh et al., 2022).

Policy measures for the elder generation were not proportional, on the other hand, since they were not protected well enough, which happened almost everywhere in Europe. Both in Sweden, France, or Austria, the death toll among people in retirement homes was unproportionally very high. Policy measures failed there. Another unproportional and not evidence-based policy measure was probably the recommended physical distance to be respected. For a long time, the Austrian government promoted the 1-m small elephant inter-personal distance recommendation, although some studies already described the virus transmission of two and more meters in early 2020 (Setti et al., 2020). A 2-m distance recommendation would have been more appropriate as applied in Sweden and other countries and probably could have saved more lives (see also Vavrik, 2020b).

Finally, we can conclude that democracies have been heavily shaken by the pandemic and countermeasures lacked proportionality and scientific evidence to a high degree in many respects.

Recommendation 7: The Austrian Government is urged to keep its pandemic prevention policy in line with best available practice at international level. This includes the rapid elaboration of a proper, adequate pandemic prevention plan, as requested by the WHO, and the revision and update of the epidemic law. It is also prompted to use all its influence at WHO level to ensure global public health guidance in the framework of the UN SDGs.

Recommendation 8: Before and during pandemics, the government is called upon applying the precautionary principle and to strongly focus on the proportionality of measures, in particular with a view to avoid backsliding to lower levels of already achieved policy standards, i.e., regarding democracy, transparency, inclusiveness, or human rights standards. This also implies balancing epidemiological goals with economic, environmental, and societal requirements, aspiring to achieve high standards of public health.

Recommendation 9: In this respect, advisory bodies to the government are to be composed of all the necessary scientific and diverse knowledge leading to adequate, efficient, and balanced policy decisions.

Recommendation 10: Furthermore, all alternative treatments to a specific health threat should be included in the debate and explicitly offered to the public. The government should also foster campaigns for healthy and sustainable food and nutrition, also as part of a public health prevention strategy.

Recommendation 11: In general, policy should be evidence-based, based on the precautionary principle, prevention-oriented, and at the same time respecting all personal freedoms as outlined in the Charter of Fundamental Rights of the European Union. In this context, our recommendation to abolish the established but suspended law obliging citizens to vaccinate was considered on June 24, 2022.

Recommendation 12: The government and media are requested to foster diversity also in public scientific debates as well as to ensure data transparency and data availability, especially during new epidemiological threats, thus facilitating research and evidence-based policy decisions for rapid countermeasures and solutions.

Recommendation 13: Finally, the government is invited to make the national economy less dependent on external shocks, i.e., by diversifying energy supply or developing more reserves and/or national capacity of critical resources as gas or equipment such as face masks or clothing for biochemical threats.

5   IV. The Ozone Therapy to Combat COVID-19: A Promising Approach although Not Recommended So Far by the WHO

While literature offers more than 280 papers with most promising results for the effectiveness of ozone therapy, WHO does not list this particular therapy as a possible anti-COVID treatment. Similarly, also in Austria, public information on ozone therapy is scarce. Ozone therapy was also never reported on in the main conventional media as a successful new therapy, at least not to my knowledge, even though scientific evidence of the antiviral activity and immunological role of ozone in COVID-19 patients already existed in early 2021. At least on the website of the Austrian society of ozone therapy, a statement and an auspicious report of the International Committee of Ozone Therapy can be found (International Scientific Committee ISCO3, 2020). Two Austrian websites, however, while claiming to do fact checks surprisingly completely deny the existence of positive effects of ozone therapy for COVID-19 patients (Cochrane, 2021; VKI, 2021). This appears to be even more astonishing, since several notable doctors have been already applying ozone therapy in Austria such as Dr. Lahodny, its founder.

At the same time, Swiss, Italian, or Romanian clinics and hospitals have already integrated ozone therapy into their daily routine protocols (Aeskulap, 2022; Cattel et al., 2021; Marini et al., 2020; WFOT, 2020). Meanwhile, hundreds of studies have demonstrated a whole series of treatment advantages. Current findings read as follows: *"Some data suggests the possible role of ozone therapy in SARS, either as a mono therapy or, more realistically, as an adjunct to standard treatment regimes…..Of 280 articles found on ozone therapy, 13 were selected and narratively reviewed. Ozone exerts antiviral activity through the inhibition of viral replication and direct inactivation of the viruses. Ozone is an antiviral drug enhancer and not an alternative to antiviral drugs. Combined treatment with involving ozone and antivirals demonstrated a reduction in inflammation and lung damage…Systemic ozone therapy seems useful in controlling inflammation, stimulating immunity and antiviral activity and providing protection from acute coronary syndromes and ischeamia repercussion damage"* (Cattel et al., 2021). Other studies refer to ozone therapy for prevention, treatment, and recovery with encouraging results. In particular, ozone therapy *"reduces inflammation indices, decreases the time of assisted respiration, decreases C-reactive protein, improves oxygen saturation, could decrease mortality, and makes polymerase chain reaction tests negative in shorter periods"* (Martinez-Sanchez, 2022). Again another study concludes that *"the antioxidants, anti-inflammatory and anti-thrombotic properties of oxygen-ozone therapy might be crucial against COVID-19-induced hyper-inflammation, immunodeficiency, hyper coagulability and poor response to therapies. Based on the first published studies, we, thus, propose O2/O3 treatment as a promising adjuvant therapy in mild to severe cases of SARS-CoV2 infection, and we call for its consideration in the clinical practice"* (Varesi et al., 2022). Cenzi et al. describe therapeutic advantages of ozone therapy with reference to SARS-CoV-2 in much detail and even report *"that air pollution strongly correlates to increased SARS-CoV-2 morbidity and mortality,"* also due to ground-level ozone (Cenci et al., 2022).

The question remains why the WHO does not mention any of these developments on its website, neither as mere information nor as a recommendation. Evidence exists also from China. From a sustainability perspective, one would expect the WHO to inform the public about every promising treatment or even recommend and foster it.

Recommendation 14: Develop scientific evidence on the usefulness of treatment with ozone therapy against COVID-19 further with the aim to include it in official anti-COVID health protocols.

Annex 1: Gestorbene 2020 nach Todesursachen, Alter und Geschlecht

Lfd. Nr.	Todesursache (Pos. Nr. ICD10)	Gestorbene insgesamt	Gestorbene im Alter von ... bis unter ... Jahren					Gestorbene im Alter von ... bis unter ... Jahren		
			unter	1–2	2–3	3–4	4–5	1–5	5–10	10–15
		Grundzahlen -								
	Alle Todesursachen (A00–Y89)	**91.599**	**262.**	**17.**	**15.**	**8.**	**10.**	**50.**	**27.**	**38.**
1	Infektiöse und parasitäre Krankheiten (A00–B99, U07–U10)	7.405	4.	–	–	–	–	–	–	–
2	Tuberkulose (A15–A19, B90)	40.	–	–	–	–	–	–	–	–
3	AIDS (HIV-Krankheit) (B20–B24)	36.	–	–	–	–	–	–	–	–
4	Virushepatitis (B15–B19, B94.2)	134.	–	–	–	–	–	–	–	–
5	COVID-19 (U07–U10)	6.491	–	–	–	–	–	–	–	–
6	Neubildungen (C00–D48)	21.803	1.	1.	1.	1.	4.	7.	10.	12.
7	**Bösartige Neubildungen (C00–C97)**	20.969	1.	1.	1.	1.	4.	7.	10.	11.
8	Bösart.Neubild. der Lippe, der Mundhöhle und des Rachens (C00–C14)	529.	–	–	–	–	–	–	–	–
9	Bösart. Neubild. der Speiseröhre (C15)	424.	–	–	–	–	–	–	–	–
10	Bösart. Neubild. des Magens (C16)	753.	–	–	–	–	–	–	–	–
11	Bösart. Neubild. des Dünndarms (C17)	72.	–	–	–	–	–	–	–	–
12	Bösart. Neubild. des Colon, des Rektums und des Anus (C18–C21)	2.132	–	–	–	–	–	–	–	–
13	Bösart. Neubild. der Leber (C22)	847.	–	–	–	–	1.	1.	2.	–
14	Bösart. Neubild. der Gallenblase und-wege (C23,C24)	296.	–	–	–	–	–	–	–	–
15	Bösart. Neubild. der Bauchspeicheldrüse (C25)	1.863	–	–	–	–	–	–	–	–
16	Bösart. Neubild. des Kehlkopfes (C32)	146.	–	–	–	–	–	–	–	–
17	Bösart. Neubild. der Luftröhre, Bronchien und Lunge (C33–C34)	4.047	–	–	–	–	–	–	–	–
18	Bösartiges Melanom der Haut (C43)	394.	–	–	–	–	–	–	–	–
19	Bösart. Neubild. der Brustdrüse (C50)	1.663	–	–	–	–	–	–	–	–
20	Bösart. Neubild. der Zervix uteri (C53)	143.	–	–	–	–	–	–	–	–

(continued)

(continued)

Lfd. Nr.	Todesursache (Pos. Nr. ICD10)	Gestorbeneinsgesamt	Gestorbene im Alter von ... bis unter ... Jahren					Gestorbene im Alter von ... bis unter ... Jahren		
			unter	1–2	2–3	3–4	4–5	1–5	5–10	10–15
21	Bösart. Neubild. der anderen Teile der Gebärmutter (C54–C55)	292.	–	–	–	–	–	–	–	–
22	Bösart. Neubild. des Ovariums (C56)	492.	–	–	–	–	–	–	–	–
23	Bösart. Neubild. der Prostata (C61)	1.398	–	–	–	–	–	–	–	–
24	Bösart. Neubild. der Niere (C64)	375.	–	–	–	–	–	–	1.	–
25	Bösart. Neubild. der Harnblase (C67)	555.	–	–	–	–	–	–	–	–
26	Bösart. Neubild. des Gehirns und zentralen Nervensystems (C70–C72)	619.	–	–	–	–	1.	1.	4.	5.
27	Bösart. Neubild. der Schilddrüse (C73)	85.	–	–	–	–	–	–	–	–
28	Morbus Hodgkin und Lymphome (C81–C85)	653.	–	–	–	–	–	–	–	–
29	Leukämie (C91–C95)	851.	–	–	1.	1.	1.	3.	1.	3.
30	Sonstige bösartige Neubildungen des lymphatischen und blutbildenen Gewebes (C88, C90, C96)	371.	–	–	–	–	–	–	–	–
31	Sonst. bösartige Neubildungen (Rest von C00–C97)	1.969	–	1.	–	–	1.	2.	2.	3.
32	Neubildungen, ausgenommen bösartige (D00–D48)	834.	–	–	–	–	–	–	–	1.
33	Krankheiten des Blutes, der blutbildenden Organe sowie best. Störungen mit Beteiligung des Immunsystems (D50–D89)	204.	2.	–	–	–	–	–	–	2.
34	Endokrine, Ernährungs- und Stoffwechselkrankheiten (E00–E90)	3.849	4.	1.	2.	–	–	3.	–	1.
35	Diabetes mellitus (E10–E14)	2.855	–	–	–	–	–	–	–	–
36	Psychische Krankheiten (F01–F99)	3.198	–	–	–	–	–	–	–	1.
37	Demenz (F01,F03)	2.452	–	–	–	–	–	–	–	–
38	Störungen durch Alkohol (F10)	564.	–	–	–	–	–	–	–	–

(continued)

(continued)

Lfd. Nr.	Todesursache (Pos. Nr. ICD10)	Gestorbene- eins- gesamt	Gestorbene im Alter von ... bis unter ... Jahren					Gestorbene im Alter von ... bis unter ... Jahren		
			unter	1–2	2–3	3–4	4–5	1–5	5–10	10–15
39	Drogenabhängigkeit, Toxikomanie (F11–F16, F18–F19)	131.	–	–	–	–	–	–	–	1.
40	Krankheiten des Nervensystems und der Sinnesorgane (G00–H95)	3.434	4.	4.	1.	2.	–	7.	5.	1.
41	Morbus Parkinson (G20)	1.093	–	–	–	–	–	–	–	–
42	Alzheimer Krankheit (G30)	1.184	–	–	–	–	–	–	–	–
43	**Krankheiten des Herz- Kreislaufsystems (I00– I99)**	32.678	3.	–	–	–	1.	1.	–	1.
44	Ischämische Herzkrankheit (I20–I25)	13.445	1.	–	–	–	–	–	–	–
45	Akuter Myokardinfarkt (I21–I22)	4.583	1.	–	–	–	–	–	–	–
46	Sonst. ischämische Herzkrankheiten (I20, I23– I25)	8.862	–	–	–	–	–	–	–	–
47	Andere Herzkrankheiten (I30–I33, I39–I51)	6.728	1.	–	–	–	–	–	–	–
48	Zerebrovaskuläre Krankheiten (I60–I69)	4.737	–	–	–	–	–	–	–	1.
49	**Krankheiten der Atmungsorgane (J00– J99)**	4.850	–	2.	2.	–	–	4.	2.	2.
50	Influenza (J09, J10–J11)	299.	–	2.	–	–	–	2.	2.	2.
51	Pneumonie (J12–J18)	902.	–	–	2.	–	–	2.	–	–
52	Chronische Krankheiten der unteren Atemwege (J40–J47)	3.192	–	–	–	–	–	–	–	–
53	Asthma (J45–J46)	74.	–	–	–	–	–	–	–	–
54	Sonstige chronische Krankheiten der unteren Atemwege (J40–J44, J47)	3.118	–	–	–	–	–	–	–	–
55	**Krankheiten der Verdauungsorgane (K00–K92)**	3.205	1.	–	–	–	–	–	–	–
56	Magen-, Duodenal- und Gastrojejunalgeschwür (K25–K28)	161.	–	–	–	–	–	–	–	–
57	Chronische Leberkrankheit und - zirrhose (K70, K73–K74)	1.370	–	–	–	–	–	–	–	–
58	Krankheiten der Haut und der Unterhaut (L00–L99)	63.	–	–	–	–	–	–	–	–

(continued)

(continued)

Lfd. Nr.	Todesursache (Pos. Nr. ICD10)	Gestorbeneinsgesamt	Gestorbene im Alter von ... bis unter ... Jahren					Gestorbene im Alter von ... bis unter ... Jahren		
			unter	1–2	2–3	3–4	4–5	1–5	5–10	10–15
59	Krankheiten des Muskel-Skelett-Systems und des Bindegewebes (M00–M99)	344.	–	–	–	–	–	–	1.	–
60	Chronische Polyarthritis und Arthrose (M05–M06, M15–M19)	56.	–	–	–	–	–	–	–	–
61	Krankheiten des Urogenitalsystems (N00–N99)	2.295	–	–	–	–	–	–	–	–
62	Krankheiten der Niere und des Ureters (N00–N29)	1.642	–	–	–	–	–	–	–	–
63	Komplikationen in der Schwangerschaft, bei der Geburt und im Wochenbett (O00–O99)	2.	–	–	–	–	–	–	–	–
64	Perinatale Affektionen (P00–P96)	153.	152.	1.	–	–	–	1.	–	–
65	Angeborene Fehlbildungen, Deformitäten und Chromosomenanomalien (Q00–Q99)	261.	72.	5.	4.	4.	1.	14.	7.	8.
66	Symptome und schlecht bezeichnete Affektionen (R00–R99)	2.950	14.	1.	1.	–	–	2.	–	–
67	Plötzlicher Kindstod (R95)	9.	9.	–	–	–	–	–	–	–
68	Ungenau bezeichnete und unbekannte Todesursachen (R96–R99)	1.799	4.	1.	1.	–	–	2.	–	–
69	**Verletzungen und Vergiftungen (V01–Y89)**	4.905	5.	2.	4.	1.	4.	11.	2.	10.
70	Unfälle (V01–X59, Y85–Y86)	3.093	2.	1.	2.	–	4.	7.	–	5.
71	Transportmittelunfälle (V01–V99, Y85)	369.	1.	–	2.	–	1.	3.	–	2.
72	Unfälle durch Sturz (W00–W19)	986.	1.	–	–	–	1.	1.	–	–
73	Unfall durch Ertrinken und Untergehen (W65–W74)	28.	–	–	–	–	–	–	–	–
74	Unfälle durch Vergiftungen (X40–X49)	31.	–	–	–	–	–	–	–	–
75	Selbsttötung und Selbstbeschädigung (X60–X84, Y87.0)	1.072	–	–	–	–	–	–	–	3.

(continued)

(continued)

Lfd. Nr.	Todesursache (Pos. Nr. ICD10)	Gestor-beneins-gesamt	Gestorbene im Alter von … bis unter … Jahren					Gestorbene im Alter von … bis unter … Jahren		
			unter	1–2	2–3	3–4	4–5	1–5	5–10	10–15
76	Mord, tätlicher Angriff (X85–Y09, Y87.1)	38.	3.	–	1.	1.	–	2.	1.	–
77	Ereignisse, dessen nähere Umstände unbestimmt sind (Y10–Y34, Y87.2)	437.	–	1.	1.	–	–	2.	1.	2.
78	**Sonstige Krankheiten (A00–B99, D00–H95, L00–R99, U07–U10)**	24.992	252.	12.	8.	6.	1.	27.	13.	14.
	Grundzahlen -									
	Alle Todesursachen (A00–Y89)	**45.372**	**138.**	**9.**	**11.**	**4.**	**7.**	**31.**	**17.**	**26.**
1	Infektiöse und parasitäre Krankheiten (A00–B99, U07–U10)	3.839	1.	–	–	–	–	–	–	–
2	Tuberkulose (A15–A19, B90)	26.	–	–	–	–	–	–	–	–
3	AIDS (HIV-Krankheit) (B20–B24)	27.	–	–	–	–	–	–	–	–
4	Virushepatitis (B15–B19, B94.2)	76.	–	–	–	–	–	–	–	–
5	COVID-19 (U07–U10)	3.403	–	–	–	–	–	–	–	–
6	Neubildungen (C00–D48)	11.769	–	1.	1.	1.	3.	6.	6.	9.
7	**Bösartige Neubildungen (C00–C97)**	11.383	–	1.	1.	1.	3.	6.	6.	8.
8	Bösart.Neubild. der Lippe, der Mundhöhle und des Rachens (C00–C14)	389.	–	–	–	–	–	–	–	–
9	Bösart. Neubild. der Speiseröhre (C15)	332.	–	–	–	–	–	–	–	–
10	Bösart. Neubild. des Magens (C16)	438.	–	–	–	–	–	–	–	–
11	Bösart. Neubild. des Dünndarms (C17)	43.	–	–	–	–	–	–	–	–
12	Bösart. Neubild. des Colon, des Rektums und des Anus (C18–C21)	1.236	–	–	–	–	–	–	–	–
13	Bösart. Neubild. der Leber (C22)	625.	–	–	–	–	1.	1.	1.	–
14	Bösart. Neubild. der Gallenblase und-wege (C23,C24)	154.	–	–	–	–	–	–	–	–
15	Bösart. Neubild. der Bauchspeicheldrüse (C25)	934.	–	–	–	–	–	–	–	–
16	Bösart. Neubild. des Kehlkopfes (C32)	126.	–	–	–	–	–	–	–	–

(continued)

U. A. Vavrik

(continued)

Lfd. Nr.	Todesursache (Pos. Nr. ICD10)	Gestorbeneinsgesamt	Gestorbene im Alter von … bis unter … Jahren					Gestorbene im Alter von … bis unter … Jahren		
			unter	1–2	2–3	3–4	4–5	1–5	5–10	10–15
17	Bösart. Neubild. der Luftröhre, Bronchien und Lunge (C33–C34)	2.412	–	–	–	–	–	–	–	–
18	Bösartiges Melanom der Haut (C43)	234.	–	–	–	–	–	–	–	–
19	Bösart. Neubild. der Brustdrüse (C50)	17.	–	–	–	–	–	–	–	–
20	Bösart. Neubild. der Zervix uteri (C53)	–	–	–	–	–	–	–	–	–
21	Bösart. Neubild. der anderen Teile der Gebärmutter (C54–C55)	–	–	–	–	–	–	–	–	–
22	Bösart. Neubild. des Ovariums (C56)	–	–	–	–	–	–	–	–	–
23	Bösart. Neubild. der Prostata (C61)	1.398	–	–	–	–	–	–	–	–
24	Bösart. Neubild. der Niere (C64)	219.	–	–	–	–	–	–	–	–
25	Bösart. Neubild. der Harnblase (C67)	383.	–	–	–	–	–	–	–	–
26	Bösart. Neubild. des Gehirns und zentralen Nervensystems (C70–C72)	338.	–	–	–	–	1.	1.	3.	4.
27	Bösart. Neubild. der Schilddrüse (C73)	36.	–	–	–	–	–	–	–	–
28	Morbus Hodgkin und Lymphome (C81–C85)	361.	–	–	–	–	–	–	–	–
29	Leukämie (C91–C95)	457.	–	–	1.	1.	–	2.	1.	2.
30	Sonstige bösartige Neubildungen des lymphatischen und blutbildenen Gewebes (C88, C90, C96)	205.	–	–	–	–	–	–	–	–
31	Sonst. bösartige Neubildungen (Rest von C00–C97)	1.046	–	1.	–	–	1.	2.	1.	2.
32	Neubildungen, ausgenommen bösartige (D00–D48)	386.	–	–	–	–	–	–	–	1.
33	Krankheiten des Blutes, der blutbildenden Organe sowie best. Störungen mit Beteiligung des Immunsystems (D50–D89)	81.	2.	–	–	–	–	–	–	1.

(continued)

(continued)

Lfd. Nr.	Todesursache (Pos. Nr. ICD10)	Gestor-beneins-gesamt	Gestorbene im Alter von . . . bis unter . . . Jahren					Gestorbene im Alter von . . . bis unter . . . Jahren		
			unter	1–2	2–3	3–4	4–5	1–5	5–10	10–15
34	Endokrine, Ernährungs- und Stoffwechselkrankheiten (E00–E90)	1.904	–	1.	1.	–	–	2.	–	1.
35	Diabetes mellitus (E10–E14)	1.405	–	–	–	–	–	–	–	–
36	Psychische Krankheiten (F01–F99)	1.357	–	–	–	–	–	–	–	–
37	Demenz (F01,F03)	792.	–	–	–	–	–	–	–	–
38	Störungen durch Alkohol (F10)	457.	–	–	–	–	–	–	–	–
39	Drogenabhängigkeit, Toxikomanie (F11–F16, F18–F19)	84.	–	–	–	–	–	–	–	–
40	Krankheiten des Nervensystems und der Sinnesorgane (G00–H95)	1.554	3.	2.	1.	2.	–	5.	4.	1.
41	Morbus Parkinson (G20)	626.	–	–	–	–	–	–	–	–
42	Alzheimer Krankheit (G30)	371.	–	–	–	–	–	–	–	–
43	**Krankheiten des Herz-Kreislaufsystems (I00–I99)**	14.766	2.	–	–	–	1.	1.	–	1.
44	Ischämische Herzkrankheit (I20–I25)	7.233	1.	–	–	–	–	–	–	–
45	Akuter Myokardinfarkt (I21–I22)	2.773	1.	–	–	–	–	–	–	–
46	Sonst. ischämische Herzkrankheiten (I20, I23–I25)	4.460	–	–	–	–	–	–	–	–
47	Andere Herzkrankheiten (I30–I33, I39–I51)	2.689	1.	–	–	–	–	–	–	–
48	Zerebrovaskuläre Krankheiten (I60–I69)	2.062	–	–	–	–	–	–	–	1.
49	**Krankheiten der Atmungsorgane (J00–J99)**	2.583	–	1.	–	–	–	1.	1.	–
50	Influenza (J09, J10–J11)	156.	–	1.	–	–	–	1.	1.	–
51	Pneumonie (J12–J18)	425.	–	–	–	–	–	–	–	–
52	Chronische Krankheiten der unteren Atemwege (J40–J47)	1.758	–	–	–	–	–	–	–	–
53	Asthma (J45–J46)	30.	–	–	–	–	–	–	–	–
54	Sonstige chronische Krankheiten der unteren Atemwege (J40–J44, J47)	1.728	–	–	–	–	–	–	–	–

(continued)

(continued)

Lfd. Nr.	Todesursache (Pos. Nr. ICD10)	Gestor-beneins-gesamt	Gestorbene im Alter von ... bis unter ... Jahren					Gestorbene im Alter von ... bis unter ... Jahren		
			unter	1–2	2–3	3–4	4–5	1–5	5–10	10–15
55	**Krankheiten der Verdauungsorgane (K00–K92)**	1.741	1.	–	–	–	–	–	–	–
56	Magen-, Duodenal- und Gastrojejunalgeschwür (K25–K28)	73.	–	–	–	–	–	–	–	–
57	Chronische Leberkrankheit und -zirrhose (K70, K73–K74)	975.	–	–	–	–	–	–	–	–
58	Krankheiten der Haut und der Unterhaut (L00–L99)	24.	–	–	–	–	–	–	–	–
59	Krankheiten des Muskel-Skelett-Systems und des Bindegewebes (M00–M99)	120.	–	–	–	–	–	–	–	–
60	Chronische Polyarthritis und Arthrose (M05–M06, M15–M19)	22.	–	–	–	–	–	–	–	–
61	Krankheiten des Urogenitalsystems (N00–N99)	903.	–	–	–	–	–	–	–	–
62	Krankheiten der Niere und des Ureters (N00–N29)	647.	–	–	–	–	–	–	–	–
63	Komplikationen in der Schwangerschaft, bei der Geburt und im Wochenbett (O00–O99)	–	–	–	–	–	–	–	–	–
64	Perinatale Affektionen (P00–P96)	84.	84.	–	–	–	–	–	–	–
65	Angeborene Fehlbildungen, Deformitäten und Chromosomenanomalien (Q00–Q99)	126.	33.	2.	4.	1.	1.	8.	6.	5.
66	Symptome und schlecht bezeichnete Affektionen (R00–R99)	1.532	9.	–	1.	–	–	1.	–	–
67	Plötzlicher Kindstod (R95)	7.	7.	–	–	–	–	–	–	–
68	Ungenau bezeichnete und unbekannte Todesursachen (R96–R99)	1.180	1.	–	1.	–	–	1.	–	–
69	**Verletzungen und Vergiftungen (V01–Y89)**	2.989	3.	2.	3.	–	2.	7.	–	8.
70	Unfälle (V01–X59, Y85–Y86)	1.684	1.	1.	2.	–	2.	5.	–	4.
71	Transportmittelunfälle (V01–V99, Y85)	283.	–	–	2.	–	1.	3.	–	2.

(continued)

(continued)

Lfd. Nr.	Todesursache (Pos. Nr. ICD10)	Gestor-beneins-gesamt	Gestorbene im Alter von . . . bis unter . . . Jahren					Gestorbene im Alter von . . . bis unter . . . Jahren		
			unter	1–2	2–3	3–4	4–5	1–5	5–10	10–15
72	Unfälle durch Sturz (W00–W19)	532.	1.	–	–	–	–	–	–	–
73	Unfall durch Ertrinken und Untergehen (W65–W74)	22.	–	–	–	–	–	–	–	–
74	Unfälle durch Vergiftungen (X40–X49)	18.	–	–	–	–	–	–	–	–
75	Selbsttötung und Selbstbeschädigung (X60–X84, Y87.0)	838.	–	–	–	–	–	–	–	2.
76	Mord, tätlicher Angriff (X85–Y09, Y87.1)	15.	2.	–	–	–	–	–	–	–
77	Ereignisse, dessen nähere Umstände unbestimmt sind (Y10–Y34, Y87.2)	302.	–	1.	1.	–	–	2.	–	2.
78	**Sonstige Krankheiten (A00–B99, D00–H95, L00–R99, U07–U10)**	11.910	132.	5.	7.	3.	1.	16.	10.	9.
		Grundzahlen -								
	Alle Todesursachen (A00–Y89)	**46.227**	**124.**	**8.**	**4.**	**4.**	**3.**	**19.**	**10.**	**12.**
1	Infektiöse und parasitäre Krankheiten (A00–B99, U07–U10)	3.566	3.	–	–	–	–	–	–	–
2	Tuberkulose (A15–A19, B90)	14.	–	–	–	–	–	–	–	–
3	AIDS (HIV-Krankheit) (B20–B24)	9.	–	–	–	–	–	–	–	–
4	Virushepatitis (B15–B19, B94.2)	58.	–	–	–	–	–	–	–	–
5	COVID-19 (U07–U10)	3.088	–	–	–	–	–	–	–	–
6	Neubildungen (C00–D48)	10.034	1.	–	–	–	1.	1.	4.	3.
7	**Bösartige Neubildungen (C00–C97)**	9.586	1.	–	–	–	1.	1.	4.	3.
8	Bösart.Neubild. der Lippe, der Mundhöhle und des Rachens (C00–C14)	140.	–	–	–	–	–	–	–	–
9	Bösart. Neubild. der Speiseröhre (C15)	92.	–	–	–	–	–	–	–	–
10	Bösart. Neubild. des Magens (C16)	315.	–	–	–	–	–	–	–	–
11	Bösart. Neubild. des Dünndarms (C17)	29.	–	–	–	–	–	–	–	–
12	Bösart. Neubild. des Colon, des Rektums und des Anus (C18–C21)	896.	–	–	–	–	–	–	–	–

(continued)

(continued)

Lfd. Nr.	Todesursache (Pos. Nr. ICD10)	Gestor-beneins-gesamt	Gestorbene im Alter von … bis unter … Jahren					Gestorbene im Alter von … bis unter … Jahren		
			unter	1–2	2–3	3–4	4–5	1–5	5–10	10–15
13	Bösart. Neubild. der Leber (C22)	222.	–	–	–	–	–	–	1.	–
14	Bösart. Neubild. der Gallenblase und-wege (C23,C24)	142.	–	–	–	–	–	–	–	–
15	Bösart. Neubild. der Bauchspeicheldrüse (C25)	929.	–	–	–	–	–	–	–	–
16	Bösart. Neubild. des Kehlkopfes (C32)	20.	–	–	–	–	–	–	–	–
17	Bösart. Neubild. der Luftröhre, Bronchien und Lunge (C33–C34)	1.635	–	–	–	–	–	–	–	–
18	Bösartiges Melanom der Haut (C43)	160.	–	–	–	–	–	–	–	–
19	Bösart. Neubild. der Brustdrüse (C50)	1.646	–	–	–	–	–	–	–	–
20	Bösart. Neubild. der Zervix uteri (C53)	143.	–	–	–	–	–	–	–	–
21	Bösart. Neubild. der anderen Teile der Gebärmutter (C54–C55)	292.	–	–	–	–	–	–	–	–
22	Bösart. Neubild. des Ovariums (C56)	492.	–	–	–	–	–	–	–	–
23	Bösart. Neubild. der Prostata (C61)	–	–	–	–	–	–	–	–	–
24	Bösart. Neubild. der Niere (C64)	156.	–	–	–	–	–	–	1.	–
25	Bösart. Neubild. der Harnblase (C67)	172.	–	–	–	–	–	–	–	–
26	Bösart. Neubild. des Gehirns und zentralen Nervensystems (C70–C72)	281.	1.	–	–	–	–	–	1.	1.
27	Bösart. Neubild. der Schilddrüse (C73)	49.	–	–	–	–	–	–	–	–
28	Morbus Hodgkin und Lymphome (C81–C85)	292.	–	–	–	–	–	–	–	–
29	Leukämie (C91–C95)	394.	–	–	–	–	1.	1.	–	1.
30	Sonstige bösartige Neubildungen des lymphatischen und blutbildenen Gewebes (C88, C90, C96)	166.	–	–	–	–	–	–	–	–
31	Sonst. bösartige Neubildungen (Rest von C00–C97)	923.	–	–	–	–	–	–	1.	1.

(continued)

(continued)

Lfd. Nr.	Todesursache (Pos. Nr. ICD10)	Gestorbeneinsgesamt	Gestorbene im Alter von ... bis unter ... Jahren					Gestorbene im Alter von ... bis unter ... Jahren		
			unter	1–2	2–3	3–4	4–5	1–5	5–10	10–15
32	Neubildungen, ausgenommen bösartige (D00–D48)	448.	–	–	–	–	–	–	–	–
33	Krankheiten des Blutes, der blutbildenden Organe sowie best. Störungen mit Beteiligung des Immunsystems (D50–D89)	123.	–	–	–	–	–	–	–	1.
34	Endokrine, Ernährungs- und Stoffwechselkrankheiten (E00–E90)	1.945	4.	–	1.	–	–	1.	–	–
35	Diabetes mellitus (E10–E14)	1.450	–	–	–	–	–	–	–	–
36	Psychische Krankheiten (F01–F99)	1.841	–	–	–	–	–	–	–	1.
37	Demenz (F01,F03)	1.660	–	–	–	–	–	–	–	–
38	Störungen durch Alkohol (F10)	107.	–	–	–	–	–	–	–	–
39	Drogenabhängigkeit, Toxikomanie (F11–F16, F18–F19)	47.	–	–	–	–	–	–	–	1.
40	Krankheiten des Nervensystems und der Sinnesorgane (G00–H95)	1.880	1.	2.	–	–	–	2.	1.	–
41	Morbus Parkinson (G20)	467.	–	–	–	–	–	–	–	–
42	Alzheimer Krankheit (G30)	813.	–	–	–	–	–	–	–	–
43	**Krankheiten des Herz-Kreislaufsystems (I00–I99)**	17.912	1.	–	–	–	–	–	–	–
44	Ischämische Herzkrankheit (I20–I25)	6.212	–	–	–	–	–	–	–	–
45	Akuter Myokardinfarkt (I21–I22)	1.810	–	–	–	–	–	–	–	–
46	Sonst. ischämische Herzkrankheiten (I20, I23–I25)	4.402	–	–	–	–	–	–	–	–
47	Andere Herzkrankheiten (I30–I33, I39–I51)	4.039	–	–	–	–	–	–	–	–
48	Zerebrovaskuläre Krankheiten (I60–I69)	2.675	–	–	–	–	–	–	–	–
49	**Krankheiten der Atmungsorgane (J00–J99)**	2.267	–	1.	2.	–	–	3.	1.	2.

(continued)

(continued)

Lfd. Nr.	Todesursache (Pos. Nr. ICD10)	Gestor- beneins- gesamt	Gestorbene im Alter von … bis unter … Jahren					Gestorbene im Alter von … bis unter … Jahren		
			unter	1–2	2–3	3–4	4–5	1–5	5–10	10–15
50	Influenza (J09, J10–J11)	143.	–	1.	–	–	–	1.	1.	2.
51	Pneumonie (J12–J18)	477.	–	–	2.	–	–	2.	–	–
52	Chronische Krankheiten der unteren Atemwege (J40–J47)	1.434	–	–	–	–	–	–	–	–
53	Asthma (J45–J46)	44.	–	–	–	–	–	–	–	–
54	Sonstige chronische Krankheiten der unteren Atemwege (J40–J44, J47)	1.390	–	–	–	–	–	–	–	–
55	**Krankheiten der Verdauungsorgane (K00–K92)**	1.464	–	–	–	–	–	–	–	–
56	Magen-, Duodenal- und Gastrojejunalgeschwür (K25–K28)	88.	–	–	–	–	–	–	–	–
57	Chronische Leberkrankheit und - zirrhose (K70, K73–K74)	395.	–	–	–	–	–	–	–	–
58	Krankheiten der Haut und der Unterhaut (L00–L99)	39.	–	–	–	–	–	–	–	–
59	Krankheiten des Muskel-Skelett-Systems und des Bindegewebes (M00–M99)	224.	–	–	–	–	–	–	1.	–
60	Chronische Polyarthritis und Arthrose (M05–M06, M15–M19)	34.	–	–	–	–	–	–	–	–
61	Krankheiten des Urogenitalsystems (N00–N99)	1.392	–	–	–	–	–	–	–	–
62	Krankheiten der Niere und des Ureters (N00–N29)	995.	–	–	–	–	–	–	–	–
63	Komplikationen in der Schwangerschaft, bei der Geburt und im Wochenbett (O00–O99)	2.	–	–	–	–	–	–	–	–
64	Perinatale Affektionen (P00–P96)	69.	68.	1.	–	–	–	1.	–	–
65	Angeborene Fehlbildungen, Deformitäten und Chromosomenanomalien (Q00–Q99)	135.	39.	3.	–	3.	–	6.	1.	3.
66	Symptome und schlecht bezeichnete Affektionen (R00–R99)	1.418	5.	1.	–	–	–	1.	–	–
67	Plötzlicher Kindstod (R95)	2.	2.	–	–	–	–	–	–	–

(continued)

(continued)

Lfd. Nr.	Todesursache (Pos. Nr. ICD10)	Gestor-beneins-gesamt	Gestorbene im Alter von ... bis unter ... Jahren					Gestorbene im Alter von ... bis unter ... Jahren		
			unter	1–2	2–3	3–4	4–5	1–5	5–10	10–15
68	Ungenau bezeichnete und unbekannte Todesursachen (R96–R99)	619.	3.	1.	–	–	–	1.	–	–
69	**Verletzungen und Vergiftungen (V01–Y89)**	1.916	2.	–	1.	1.	2.	4.	2.	2.
70	Unfälle (V01–X59, Y85–Y86)	1.409	1.	–	–	–	2.	2.	–	1.
71	Transportmittelunfälle (V01–V99, Y85)	86.	1.	–	–	–	–	–	–	–
72	Unfälle durch Sturz (W00–W19)	454.	–	–	–	–	1.	1.	–	–
73	Unfall durch Ertrinken und Untergehen (W65–W74)	6.	–	–	–	–	–	–	–	–
74	Unfälle durch Vergiftungen (X40–X49)	13.	–	–	–	–	–	–	–	–
75	Selbsttötung und Selbstbeschädigung (X60–X84, Y87.0)	234.	–	–	–	–	–	–	–	1.
76	Mord, tätlicher Angriff (X85–Y09, Y87.1)	23.	1.	–	1.	1.	–	2.	1.	–
77	Ereignisse, dessen nähere Umstände unbestimmt sind (Y10–Y34, Y87.2)	135.	–	–	–	–	–	–	1.	–
78	**Sonstige Krankheiten (A00–B99, D00–H95, L00–R99, U07–U10)**	13.082	120.	7.	1.	3.	–	11.	3.	5.
		Auf 100.000 der Bevölkerung gleichen Alters und Geschlechts -								
	Alle Todesursachen (A00–Y89)	1.027	313.4	19.8	17.	8.9	11.3	14.2	6.3	9.
1	Infektiöse und parasitäre Krankheiten (A00–B99, U07–U10)	83.	4.8	–	–	–	–	–	–	–
2	Tuberkulose (A15–A19, B90)	.4	–	–	–	–	–	–	–	–
3	AIDS (HIV-Krankheit) (B20–B24)	.4	–	–	–	–	–	–	–	–
4	Virushepatitis (B15–B19, B94.2)	1.5	–	–	–	–	–	–	–	–
5	COVID-19 (U07–U10)	72.8	–	–	–	–	–	–	–	–
6	Neubildungen (C00–D48)	244.5	1.2	1.2	1.1	1.1	4.5	2.	2.3	2.8
7	**Bösartige Neubildungen (C00–C97)**	235.1	1.2	1.2	1.1	1.1	4.5	2.	2.3	2.6
8	Bösart.Neubild. der Lippe, der Mundhöhle und des Rachens (C00–C14)	5.9	–	–	–	–	–	–	–	–

(continued)

(continued)

Lfd. Nr.	Todesursache (Pos. Nr. ICD10)	Gestor-beneins-gesamt	Gestorbene im Alter von ... bis unter ... Jahren					Gestorbene im Alter von ... bis unter ... Jahren		
			unter	1–2	2–3	3–4	4–5	1–5	5–10	10–15
9	Bösart. Neubild. der Speiseröhre (C15)	4.8	–	–	–	–	–	–	–	–
10	Bösart. Neubild. des Magens (C16)	8.4	–	–	–	–	–	–	–	–
11	Bösart. Neubild. des Dünndarms (C17)	.8	–	–	–	–	–	–	–	–
12	Bösart. Neubild. des Colon, des Rektums und des Anus (C18–C21)	23.9	–	–	–	–	–	–	–	–
13	Bösart. Neubild. der Leber (C22)	9.5	–	–	–	–	1.1	.3	.5	–
14	Bösart. Neubild. der Gallenblase und-wege (C23,C24)	3.3	–	–	–	–	–	–	–	–
15	Bösart. Neubild. der Bauchspeicheldrüse (C25)	20.9	–	–	–	–	–	–	–	–
16	Bösart. Neubild. des Kehlkopfes (C32)	1.6	–	–	–	–	–	–	–	–
17	Bösart. Neubild. der Luftröhre, Bronchien und Lunge (C33–C34)	45.4	–	–	–	–	–	–	–	–
18	Bösartiges Melanom der Haut (C43)	4.4	–	–	–	–	–	–	–	–
19	Bösart. Neubild. der Brustdrüse (C50)	18.6	–	–	–	–	–	–	–	–
20	Bösart. Neubild. der Zervix uteri (C53)	1.6	–	–	–	–	–	–	–	–
21	Bösart. Neubild. der anderen Teile der Gebärmutter (C54–C55)	3.3	–	–	–	–	–	–	–	–
22	Bösart. Neubild. des Ovariums (C56)	5.5	–	–	–	–	–	–	–	–
23	Bösart. Neubild. der Prostata (C61)	15.7	–	–	–	–	–	–	–	–
24	Bösart. Neubild. der Niere (C64)	4.2	–	–	–	–	–	–	.2	–
25	Bösart. Neubild. der Harnblase (C67)	6.2	–	–	–	–	–	–	–	–
26	Bösart. Neubild. des Gehirns und zentralen Nervensystems (C70–C72)	6.9	1.2	–	–	–	1.1	.3	.9	1.2
27	Bösart. Neubild. der Schilddrüse (C73)	1.	–	–	–	–	–	–	–	–
28	Morbus Hodgkin und Lymphome (C81–C85)	7.3	–	–	–	–	–	–	–	–
29	Leukämie (C91–C95)	9.5	–	–	1.1	1.1	1.1	.9	.2	.7

(continued)

(continued)

Lfd. Nr.	Todesursache (Pos. Nr. ICD10)	Gestor- beneins- gesamt	Gestorbene im Alter von ... bis unter ... Jahren					Gestorbene im Alter von ... bis unter ... Jahren		
			unter	1–2	2–3	3–4	4–5	1–5	5–10	10–15
30	Sonstige bösartige Neubildungen des lymphatischen und blutbildenen Gewebes (C88, C90, C96)	4.2	–	–	–	–	–	–	–	–
31	Sonst. bösartige Neubildungen (Rest von C00–C97)	22.1	–	1.2	–	–	1.1	.6	.5	.7
32	Neubildungen, ausgenommen bösartige (D00–D48)	9.4	–	–	–	–	–	–	–	.2
33	Krankheiten des Blutes, der blutbildenden Organe sowie best. Störungen mit Beteiligung des Immunsystems (D50–D89)	2.3	2.4	–	–	–	–	–	–	.5
34	Endokrine, Ernährungs- und Stoffwechselkrankheiten (E00–E90)	43.2	4.8	1.2	2.3	–	–	.9	–	.2
35	Diabetes mellitus (E10–E14)	32.	–	–	–	–	–	–	–	–
36	Psychische Krankheiten (F01–F99)	35.9	–	–	–	–	–	–	–	.2
37	Demenz (F01,F03)	27.5	–	–	–	–	–	–	–	–
38	Störungen durch Alkohol (F10)	6.3	–	–	–	–	–	–	–	–
39	Drogenabhängigkeit, Toxikomanie (F11–F16, F18–F19)	1.5	–	–	–	–	–	–	–	.2
40	Krankheiten des Nervensystems und der Sinnesorgane (G00–H95)	38.5	4.8	4.7	1.1	2.2	–	2.	1.2	.2
41	Morbus Parkinson (G20)	12.3	–	–	–	–	–	–	–	–
42	Alzheimer Krankheit (G30)	13.3	–	–	–	–	–	–	–	–
43	**Krankheiten des Herz-Kreislaufsystems (I00–I99)**	366.4	3.6	–	–	–	1.1	.3	–	.2
44	Ischämische Herzkrankheit (I20–I25)	150.8	1.2	–	–	–	–	–	–	–
45	Akuter Myokardinfarkt (I21–I22)	51.4	1.2	–	–	–	–	–	–	–
46	Sonst. ischämische Herzkrankheiten (I20, I23–I25)	99.4	–	–	–	–	–	–	–	–

(continued)

(continued)

Lfd. Nr.	Todesursache (Pos. Nr. ICD10)	Gestor- beneins- gesamt	Gestorbene im Alter von ... bis unter ... Jahren					Gestorbene im Alter von ... bis unter ... Jahren		
			unter	1–2	2–3	3–4	4–5	1–5	5–10	10–15
47	Andere Herzkrankheiten (I30–I33, I39–I51)	75.4	1.2	–	–	–	–	–	–	–
48	Zerebrovaskuläre Krankheiten (I60–I69)	53.1	–	–	–	–	–	–	–	.2
49	**Krankheiten der Atmungsorgane (J00–J99)**	54.4	–	2.3	2.3	–	–	1.1	.5	.5
50	Influenza (J09, J10–J11)	3.4	–	2.3	–	–	–	.6	.5	.5
51	Pneumonie (J12–J18)	10.1	–	–	2.3	–	–	.6	–	–
52	Chronische Krankheiten der unteren Atemwege (J40–J47)	35.8	–	–	–	–	–	–	–	–
53	Asthma (J45–J46)	.8	–	–	–	–	–	–	–	–
54	Sonstige chronische Krankheiten der unteren Atemwege (J40–J44, J47)	35.	–	–	–	–	–	–	–	–
55	**Krankheiten der Verdauungsorgane (K00–K92)**	35.9	1.2	–	–	–	–	–	–	–
56	Magen-, Duodenal- und Gastrojejunalgeschwür (K25–K28)	1.8	–	–	–	–	–	–	–	–
57	Chronische Leberkrankheit und -zirrhose (K70, K73–K74)	15.4	–	–	–	–	–	–	–	–
58	Krankheiten der Haut und der Unterhaut (L00–L99)	.7	–	–	–	–	–	–	–	–
59	Krankheiten des Muskel-Skelett-Systems und des Bindegewebes (M00–M99)	3.9	–	–	–	–	–	–	.2	–
60	Chronische Polyarthritis und Arthrose (M05–M06, M15–M19)	.6	–	–	–	–	–	–	–	–
61	Krankheiten des Urogenitalsystems (N00–N99)	25.7	–	–	–	–	–	–	–	–
62	Krankheiten der Niere und des Ureters (N00–N29)	18.4	–	–	–	–	–	–	–	–
63	Komplikationen in der Schwangerschaft, bei der Geburt und im Wochenbett (O00–O99)	–	–	–	–	–	–	–	–	–
64	Perinatale Affektionen (P00–P96)	1.7	181.8	1.2	–	–	–	.3	–	–

(continued)

(continued)

Lfd. Nr.	Todesursache (Pos. Nr. ICD10)	Gestor-beneins-gesamt	Gestorbene im Alter von ... bis unter ... Jahren					Gestorbene im Alter von ... bis unter ... Jahren		
			unter	1–2	2–3	3–4	4–5	1–5	5–10	10–15
65	Angeborene Fehlbildungen, Deformitäten und Chromosomenanomalien (Q00–Q99)	2.9	86.1	5.8	4.5	4.5	1.1	4.	1.6	1.9
66	Symptome und schlecht bezeichnete Affektionen (R00–R99)	33.1	16.7	1.2	1.1	–	–	.6	–	–
67	Plötzlicher Kindstod (R95)	.1	10.8	–	–	–	–	–	–	–
68	Ungenau bezeichnete und unbekannte Todesursachen (R96–R99)	20.2	4.8	1.2	1.1	–	–	.6	–	–
69	**Verletzungen und Vergiftungen (V01–Y89)**	55.	6.	2.3	4.5	1.1	4.5	3.1	.5	2.4
70	Unfälle (V01–X59, Y85–Y86)	34.7	2.4	1.2	2.3	–	4.5	2.	–	1.2
71	Transportmittelunfälle (V01–V99, Y85)	4.1	1.2	–	2.3	–	1.1	.9	–	.5
72	Unfälle durch Sturz (W00–W19)	11.1	1.2	–	–	–	1.1	.3	–	–
73	Unfall durch Ertrinken und Untergehen (W65–W74)	.3	–	–	–	–	–	–	–	–
74	Unfälle durch Vergiftungen (X40–X49)	.3	–	–	–	–	–	–	–	–
75	Selbsttötung und Selbstbeschädigung (X60–X84, Y87.0)	12.	–	–	–	–	–	–	–	.7
76	Mord, tätlicher Angriff (X85–Y09, Y87.1)	.4	3.6	–	1.1	1.1	–	.6	.2	–
77	Ereignisse, dessen nähere Umstände unbestimmt sind (Y10–Y34, Y87.2)	4.9	–	1.2	1.1	–	–	.6	.2	.5
78	**Sonstige Krankheiten (A00–B99, D00–H95, L00–R99, U07–U10)**	280.3	301.4	14.	9.1	6.7	1.1	7.7	3.	3.3
		Auf 100.000 der Bevölkerung gleichen Alters und Geschlechts -								
	Alle Todesursachen (A00–Y89)	1,033.9	321.4	20.5	24.3	8.6	15.4	17.1	7.7	12.
1	Infektiöse und parasitäre Krankheiten (A00–B99, U07–U10)	87.5	2.3	–	–	–	–	–	–	–
2	Tuberkulose (A15–A19, B90)	0.6	–	–	–	–	–	–	–	–
3	AIDS (HIV-Krankheit) (B20–B24)	0.6	–	–	–	–	–	–	–	–

(continued)

(continued)

Lfd. Nr.	Todesursache (Pos. Nr. ICD10)	Gestor-beneins-gesamt	Gestorbene im Alter von . . . bis unter . . . Jahren					Gestorbene im Alter von . . . bis unter . . . Jahren		
			unter	1–2	2–3	3–4	4–5	1–5	5–10	10–15
4	Virushepatitis (B15–B19, B94.2)	1.7	–	–	–	–	–	–	–	–
5	COVID-19 (U07–U10)	77.5	–	–	–	–	–	–	–	–
6	Neubildungen (C00–D48)	268.2	–	2.3	2.2	2.2	6.6	3.3	2.7	4.2
7	**Bösartige Neubildungen (C00–C97)**	259.4	–	2.3	2.2	2.2	6.6	3.3	2.7	3.7
8	Bösart.Neubild. der Lippe, der Mundhöhle und des Rachens (C00–C14)	8.9	–	–	–	–	–	–	–	–
9	Bösart. Neubild. der Speiseröhre (C15)	7.6	–	–	–	–	–	–	–	–
10	Bösart. Neubild. des Magens (C16)	10.0	–	–	–	–	–	–	–	–
11	Bösart. Neubild. des Dünndarms (C17)	1.0	–	–	–	–	–	–	–	–
12	Bösart. Neubild. des Colon, des Rektums und des Anus (C18–C21)	28.2	–	–	–	–	–	–	–	–
13	Bösart. Neubild. der Leber (C22)	14.2	–	–	–	–	2.2	.6	.5	–
14	Bösart. Neubild. der Gallenblase und-wege (C23,C24)	3.5	–	–	–	–	–	–	–	–
15	Bösart. Neubild. der Bauchspeicheldrüse (C25)	21.3	–	–	–	–	–	–	–	–
16	Bösart. Neubild. des Kehlkopfes (C32)	2.9	–	–	–	–	–	–	–	–
17	Bösart. Neubild. der Luftröhre, Bronchien und Lunge (C33–C34)	55.0	–	–	–	–	–	–	–	–
18	Bösartiges Melanom der Haut (C43)	5.3	–	–	–	–	–	–	–	–
19	Bösart. Neubild. der Brustdrüse (C50)	0.4	–	–	–	–	–	–	–	–
20	Bösart. Neubild. der Zervix uteri (C53)	–	–	–	–	–	–	–	–	–
21	Bösart. Neubild. der anderen Teile der Gebärmutter (C54–C55)	–	–	–	–	–	–	–	–	–
22	Bösart. Neubild. des Ovariums (C56)	–	–	–	–	–	–	–	–	–
23	Bösart. Neubild. der Prostata (C61)	31.9	–	–	–	–	–	–	–	–
24	Bösart. Neubild. der Niere (C64)	5.0	–	–	–	–	–	–	–	–

(continued)

(continued)

Lfd. Nr.	Todesursache (Pos. Nr. ICD10)	Gestorbene insgesamt	Gestorbene im Alter von ... bis unter ... Jahren					Gestorbene im Alter von ... bis unter ... Jahren		
			unter	1–2	2–3	3–4	4–5	1–5	5–10	10–15
25	Bösart. Neubild. der Harnblase (C67)	8.7	–	–	–	–	–	–	–	–
26	Bösart. Neubild. des Gehirns und zentralen Nervensystems (C70–C72)	7.7	–	–	–	–	2.2	.6	1.4	1.8
27	Bösart. Neubild. der Schilddrüse (C73)	0.8	–	–	–	–	–	–	–	–
28	Morbus Hodgkin und Lymphome (C81–C85)	8.2	–	–	–	–	–	–	–	–
29	Leukämie (C91–C95)	10.4	–	–	2.2	2.2	–	1.1	.5	.9
30	Sonstige bösartige Neubildungen des lymphatischen und blutbildenen Gewebes (C88, C90, C96)	4.7	–	–	–	–	–	–	–	–
31	Sonst. bösartige Neubildungen (Rest von C00–C97)	23.8	–	2.3	–	–	2.2	1.1	.5	.9
32	Neubildungen, ausgenommen bösartige (D00–D48)	8.8	–	–	–	–	–	–	–	.5
33	Krankheiten des Blutes, der blutbildenden Organe sowie best. Störungen mit Beteiligung des Immunsystems (D50–D89)	1.8	4.7	–	–	–	–	–	–	.5
34	Endokrine, Ernährungs- und Stoffwechselkrankheiten (E00–E90)	43.4	–	2.3	2.2	–	–	1.1	–	.5
35	Diabetes mellitus (E10–E14)	32.0	–	–	–	–	–	–	–	–
36	Psychische Krankheiten (F01–F99)	30.9	–	–	–	–	–	–	–	–
37	Demenz (F01,F03)	18.0	–	–	–	–	–	–	–	–
38	Störungen durch Alkohol (F10)	10.4	–	–	–	–	–	–	–	–
39	Drogenabhängigkeit, Toxikomanie (F11–F16, F18–F19)	1.9	–	–	–	–	–	–	–	–
40	Krankheiten des Nervensystems und der Sinnesorgane (G00–H95)	35.4	7.	4.6	2.2	4.3	–	2.8	1.8	.5
41	Morbus Parkinson (G20)	14.3	–	–	–	–	–	–	–	–
42	Alzheimer Krankheit (G30)	8.5	–	–	–	–	–	–	–	–

(continued)

(continued)

Lfd. Nr.	Todesursache (Pos. Nr. ICD10)	Gestor-beneins-gesamt	Gestorbene im Alter von ... bis unter ... Jahren					Gestorbene im Alter von ... bis unter ... Jahren		
			unter	1–2	2–3	3–4	4–5	1–5	5–10	10–15
43	**Krankheiten des Herz-Kreislaufsystems (I00–I99)**	336.5	4.7	–	–	–	2.2	.6	–	.5
44	Ischämische Herzkrankheit (I20–I25)	164.8	2.3	–	–	–	–	–	–	–
45	Akuter Myokardinfarkt (I21–I22)	63.2	2.3	–	–	–	–	–	–	–
46	Sonst. ischämische Herzkrankheiten (I20, I23–I25)	101.6	–	–	–	–	–	–	–	–
47	Andere Herzkrankheiten (I30–I33, I39–I51)	61.3	2.3	–	–	–	–	–	–	–
48	Zerebrovaskuläre Krankheiten (I60–I69)	47.0	–	–	–	–	–	–	–	.5
49	**Krankheiten der Atmungsorgane (J00–J99)**	58.9	–	2.3	–	–	–	.6	.5	–
50	Influenza (J09, J10–J11)	3.6	–	2.3	–	–	–	.6	.5	–
51	Pneumonie (J12–J18)	9.7	–	–	–	–	–	–	–	–
52	Chronische Krankheiten der unteren Atemwege (J40–J47)	40.1	–	–	–	–	–	–	–	–
53	Asthma (J45–J46)	0.7	–	–	–	–	–	–	–	–
54	Sonstige chronische Krankheiten der unteren Atemwege (J40–J44, J47)	39.4	–	–	–	–	–	–	–	–
55	**Krankheiten der Verdauungsorgane (K00–K92)**	39.7	2.3	–	–	–	–	–	–	–
56	Magen-, Duodenal- und Gastrojejunalgeschwür (K25–K28)	1.7	–	–	–	–	–	–	–	–
57	Chronische Leberkrankheit und -zirrhose (K70, K73–K74)	22.2	–	–	–	–	–	–	–	–
58	Krankheiten der Haut und der Unterhaut (L00–L99)	0.5	–	–	–	–	–	–	–	–
59	Krankheiten des Muskel-Skelett-Systems und des Bindegewebes (M00–M99)	2.7	–	–	–	–	–	–	–	–
60	Chronische Polyarthritis und Arthrose (M05–M06, M15–M19)	0.5	–	–	–	–	–	–	–	–
61	Krankheiten des Urogenitalsystems (N00–N99)	20.6	–	–	–	–	–	–	–	–

(continued)

(continued)

Lfd. Nr.	Todesursache (Pos. Nr. ICD10)	Gestorbeneinsgesamt	Gestorbene im Alter von ... bis unter ... Jahren					Gestorbene im Alter von ... bis unter ... Jahren		
			unter	1–2	2–3	3–4	4–5	1–5	5–10	10–15
62	Krankheiten der Niere und des Ureters (N00–N29)	14.7	–	–	–	–	–	–	–	–
63	Komplikationen in der Schwangerschaft, bei der Geburt und im Wochenbett (O00–O99)	–	–	–	–	–	–	–	–	–
64	Perinatale Affektionen (P00–P96)	1.9	195.6	–	–	–	–	–	–	–
65	Angeborene Fehlbildungen, Deformitäten und Chromosomenanomalien (Q00–Q99)	2.9	76.9	4.6	8.8	2.2	2.2	4.4	2.7	2.3
66	Symptome und schlecht bezeichnete Affektionen (R00–R99)	34.9	21.	–	2.2	–	–	.6	–	–
67	Plötzlicher Kindstod (R95)	0.2	16.3	–	–	–	–	–	–	–
68	Ungenau bezeichnete und unbekannte Todesursachen (R96–R99)	26.9	2.3	–	2.2	–	–	.6	–	–
69	**Verletzungen und Vergiftungen (V01–Y89)**	68.1	7.	4.6	6.6	–	4.4	3.9	–	3.7
70	Unfälle (V01–X59, Y85–Y86)	38.4	2.3	2.3	4.4	–	4.4	2.8	–	1.8
71	Transportmittelunfälle (V01–V99, Y85)	6.4	–	–	4.4	–	2.2	1.7	–	.9
72	Unfälle durch Sturz (W00–W19)	12.1	2.3	–	–	–	–	–	–	–
73	Unfall durch Ertrinken und Untergehen (W65–W74)	0.5	–	–	–	–	–	–	–	–
74	Unfälle durch Vergiftungen (X40–X49)	0.4	–	–	–	–	–	–	–	–
75	Selbsttötung und Selbstbeschädigung (X60–X84, Y87.0)	19.1	–	–	–	–	–	–	–	.9
76	Mord, tätlicher Angriff (X85–Y09, Y87.1)	0.3	4.7	–	–	–	–	–	–	–
77	Ereignisse, dessen nähere Umstände unbestimmt sind (Y10–Y34, Y87.2)	6.9	–	2.3	2.2	–	–	1.1	–	.9
78	**Sonstige Krankheiten (A00–B99, D00–H95, L00–R99, U07–U10)**	271.4	307.4	11.4	15.4	6.5	2.2	8.8	4.5	4.2

(continued)

(continued)

Lfd. Nr.	Todesursache (Pos. Nr. ICD10)	Gestor- beneins- gesamt	Gestorbene im Alter von ... bis unter ... Jahren					Gestorbene im Alter von ... bis unter ... Jahren		
			unter	1–2	2–3	3– 4	4–5	1–5	5– 10	10– 15
			Auf 100.000 der Bevölkerung gleichen Alters und Geschlechts -							
	Alle Todesursachen (A00–Y89)	1,020.7	304.9	19.1	9.3	9.2	7.	11.1	4.8	5.8
1	Infektiöse und parasitäre Krankheiten (A00–B99, U07–U10)	78.7	7.4	–	–	–	–	–	–	–
2	Tuberkulose (A15–A19, B90)	0.3	–	–	–	–	–	–	–	–
3	AIDS (HIV-Krankheit) (B20–B24)	0.2	–	–	–	–	–	–	–	–
4	Virushepatitis (B15–B19, B94.2)	1.3	–	–	–	–	–	–	–	–
5	COVID-19 (U07–U10)	68.2	–	–	–	–	–	–	–	–
6	Neubildungen (C00–D48)	221.5	2.5	–	–	–	2.3	.6	1.9	1.5
7	**Bösartige Neubildungen (C00–C97)**	211.7	2.5	–	–	–	2.3	.6	1.9	1.5
8	Bösart.Neubild. der Lippe, der Mundhöhle und des Rachens (C00–C14)	3.1	–	–	–	–	–	–	–	–
9	Bösart. Neubild. der Speiseröhre (C15)	2.0	–	–	–	–	–	–	–	–
10	Bösart. Neubild. des Magens (C16)	7.0	–	–	–	–	–	–	–	–
11	Bösart. Neubild. des Dünndarms (C17)	0.6	–	–	–	–	–	–	–	–
12	Bösart. Neubild. des Colon, des Rektums und des Anus (C18–C21)	19.8	–	–	–	–	–	–	–	–
13	Bösart. Neubild. der Leber (C22)	4.9	–	–	–	–	–	–	.5	–
14	Bösart. Neubild. der Gallenblase und-wege (C23,C24)	3.1	–	–	–	–	–	–	–	–
15	Bösart. Neubild. der Bauchspeicheldrüse (C25)	20.5	–	–	–	–	–	–	–	–
16	Bösart. Neubild. des Kehlkopfes (C32)	0.4	–	–	–	–	–	–	–	–
17	Bösart. Neubild. der Luftröhre, Bronchien und Lunge (C33–C34)	36.1	–	–	–	–	–	–	–	–
18	Bösartiges Melanom der Haut (C43)	3.5	–	–	–	–	–	–	–	–
19	Bösart. Neubild. der Brustdrüse (C50)	36.3	–	–	–	–	–	–	–	–

(continued)

(continued)

Lfd. Nr.	Todesursache (Pos. Nr. ICD10)	Gestor-beneins-gesamt	Gestorbene im Alter von . . . bis unter . . . Jahren					Gestorbene im Alter von . . . bis unter . . . Jahren		
			unter	1–2	2–3	3–4	4–5	1–5	5–10	10–15
20	Bösart. Neubild. der Zervix uteri (C53)	3.2	–	–	–	–	–	–	–	–
21	Bösart. Neubild. der anderen Teile der Gebärmutter (C54–C55)	6.4	–	–	–	–	–	–	–	–
22	Bösart. Neubild. des Ovariums (C56)	10.9	–	–	–	–	–	–	–	–
23	Bösart. Neubild. der Prostata (C61)	–	–	–	–	–	–	–	–	–
24	Bösart. Neubild. der Niere (C64)	3.4	–	–	–	–	–	–	.5	–
25	Bösart. Neubild. der Harnblase (C67)	3.8	–	–	–	–	–	–	–	–
26	Bösart. Neubild. des Gehirns und zentralen Nervensystems (C70–C72)	6.2	2.5	–	–	–	–	–	.5	.5
27	Bösart. Neubild. der Schilddrüse (C73)	1.1	–	–	–	–	–	–	–	–
28	Morbus Hodgkin und Lymphome (C81–C85)	6.4	–	–	–	–	–	–	–	–
29	Leukämie (C91–C95)	8.7	–	–	–	–	2.3	.6	–	.5
30	Sonstige bösartige Neubildungen des lymphatischen und blutbildenen Gewebes (C88, C90, C96)	3.7	–	–	–	–	–	–	–	–
31	Sonst. bösartige Neubildungen (Rest von C00–C97)	20.4	–	–	–	–	–	–	.5	.5
32	Neubildungen, ausgenommen bösartige (D00–D48)	9.9	–	–	–	–	–	–	–	–
33	Krankheiten des Blutes, der blutbildenden Organe sowie best. Störungen mit Beteiligung des Immunsystems (D50–D89)	2.7	–	–	–	–	–	–	–	.5
34	Endokrine, Ernährungs- und Stoffwechselkrankheiten (E00–E90)	42.9	9.8	–	2.3	–	–	.6	–	–
35	Diabetes mellitus (E10–E14)	32.0	–	–	–	–	–	–	–	–
36	Psychische Krankheiten (F01–F99)	40.6	–	–	–	–	–	–	–	.5
37	Demenz (F01,F03)	36.7	–	–	–	–	–	–	–	–

(continued)

(continued)

Lfd. Nr.	Todesursache (Pos. Nr. ICD10)	Gestor-beneins-gesamt	Gestorbene im Alter von ... bis unter ... Jahren					Gestorbene im Alter von ... bis unter ... Jahren		
			unter	1–2	2–3	3–4	4–5	1–5	5–10	10–15
38	Störungen durch Alkohol (F10)	2.4	–	–	–	–	–	–	–	–
39	Drogenabhängigkeit, Toxikomanie (F11–F16, F18–F19)	1.0	–	–	–	–	–	–	–	.5
40	Krankheiten des Nervensystems und der Sinnesorgane (G00–H95)	41.5	2.5	4.8	–	–	–	1.2	.5	–
41	Morbus Parkinson (G20)	10.3	–	–	–	–	–	–	–	–
42	Alzheimer Krankheit (G30)	18.0	–	–	–	–	–	–	–	–
43	**Krankheiten des Herz-Kreislaufsystems (I00–I99)**	395.5	2.5	–	–	–	–	–	–	–
44	Ischämische Herzkrankheit (I20–I25)	137.2	–	–	–	–	–	–	–	–
45	Akuter Myokardinfarkt (I21–I22)	40.0	–	–	–	–	–	–	–	–
46	Sonst. ischämische Herzkrankheiten (I20, I23–I25)	97.2	–	–	–	–	–	–	–	–
47	Andere Herzkrankheiten (I30–I33, I39–I51)	89.2	–	–	–	–	–	–	–	–
48	Zerebrovaskuläre Krankheiten (I60–I69)	59.1	–	–	–	–	–	–	–	–
49	**Krankheiten der Atmungsorgane (J00–J99)**	50.1	–	2.4	4.7	–	–	1.8	.5	1.
50	Influenza (J09, J10–J11)	3.2	–	2.4	–	–	–	.6	.5	1.
51	Pneumonie (J12–J18)	10.5	–	–	4.7	–	–	1.2	–	–
52	Chronische Krankheiten der unteren Atemwege (J40–J47)	31.7	–	–	–	–	–	–	–	–
53	Asthma (J45–J46)	1.0	–	–	–	–	–	–	–	–
54	Sonstige chronische Krankheiten der unteren Atemwege (J40–J44, J47)	30.7	–	–	–	–	–	–	–	–
55	**Krankheiten der Verdauungsorgane (K00–K92)**	32.3	–	–	–	–	–	–	–	–
56	Magen-, Duodenal- und Gastrojejunalgeschwür (K25–K28)	1.9	–	–	–	–	–	–	–	–
57	Chronische Leberkrankheit und -zirrhose (K70, K73–K74)	8.7	–	–	–	–	–	–	–	–

(continued)

(continued)

Lfd. Nr.	Todesursache (Pos. Nr. ICD10)	Gestorbeneins-gesamt	Gestorbene im Alter von ... bis unter ... Jahren					Gestorbene im Alter von ... bis unter ... Jahren		
			unter	1–2	2–3	3–4	4–5	1–5	5–10	10–15
58	Krankheiten der Haut und der Unterhaut (L00–L99)	0.9	–	–	–	–	–	–	–	–
59	Krankheiten des Muskel-Skelett-Systems und des Bindegewebes (M00–M99)	4.9	–	–	–	–	–	–	.5	–
60	Chronische Polyarthritis und Arthrose (M05–M06, M15–M19)	0.8	–	–	–	–	–	–	–	–
61	Krankheiten des Urogenitalsystems (N00–N99)	30.7	–	–	–	–	–	–	–	–
62	Krankheiten der Niere und des Ureters (N00–N29)	22.0	–	–	–	–	–	–	–	–
63	Komplikationen in der Schwangerschaft, bei der Geburt und im Wochenbett (O00–O99)	–	–	–	–	–	–	–	–	–
64	Perinatale Affektionen (P00–P96)	1.5	167.2	2.4	–	–	–	.6	–	–
65	Angeborene Fehlbildungen, Deformitäten und Chromosomenanomalien (Q00–Q99)	3.0	95.9	7.2	–	6.9	–	3.5	.5	1.5
66	Symptome und schlecht bezeichnete Affektionen (R00–R99)	31.3	12.3	2.4	–	–	–	.6	–	–
67	Plötzlicher Kindstod (R95)	–	4.9	–	–	–	–	–	–	–
68	Ungenau bezeichnete und unbekannte Todesursachen (R96–R99)	13.7	7.4	2.4	–	–	–	.6	–	–
69	**Verletzungen und Vergiftungen (V01–Y89)**	42.3	4.9	–	2.3	2.3	4.7	2.3	1.	1.
70	Unfälle (V01–X59, Y85–Y86)	31.1	2.5	–	–	–	4.7	1.2	–	.5
71	Transportmittelunfälle (V01–V99, Y85)	1.9	2.5	–	–	–	–	–	–	–
72	Unfälle durch Sturz (W00–W19)	10.0	–	–	–	–	2.3	.6	–	–
73	Unfall durch Ertrinken und Untergehen (W65–W74)	0.1	–	–	–	–	–	–	–	–
74	Unfälle durch Vergiftungen (X40–X49)	0.3	–	–	–	–	–	–	–	–

(continued)

(continued)

Lfd. Nr.	Todesursache (Pos. Nr. ICD10)	Gestor-beneins-gesamt	Gestorbene im Alter von … bis unter … Jahren					Gestorbene im Alter von … bis unter … Jahren		
			unter	1–2	2–3	3–4	4–5	1–5	5–10	10–15
75	Selbsttötung und Selbstbeschädigung (X60–X84, Y87.0)	5.2	–	–	–	–	–	–	–	.5
76	Mord, tätlicher Angriff (X85–Y09, Y87.1)	0.5	2.5	–	2.3	2.3	–	1.2	.5	–
77	Ereignisse, dessen nähere Umstände unbestimmt sind (Y10–Y34, Y87.2)	3.0	–	–	–	–	–	–	.5	–
78	**Sonstige Krankheiten (A00–B99, D00–H95, L00–R99, U07–U10)**	288.8	295.1	16.7	2.3	6.9	–	6.4	1.5	2.4

Gestorbene im Alter von ... bis unter ... Jahren

Lfd. Nr.	Todesursache (Pos. Nr. ICD10)	insgesamt	15–20	20–25	25–30	30–35	35–40	40–45	45–50	50–55	55–60	60–65	65–70	70–75	75–80	80–85	85–90	90–95	95 und älter	Lfd. Nr.
	Alle Todesursachen (A00–Y89)		122.	167.	196.	319.	406.	615.	963.	1.915	3.225	4.492	5.689	8.284	12.235	14.881	16.652	14.403	6.658	
1	Infektiöse und parasitäre Krankheiten (A00–B99, U07-U10)		2.	2.	3.	8.	17.	17.	37.	81.	138.	218.	349.	614.	999.	1.472	1.567	1.314	563.	1
2	Tuberkulose (A15–A19, B90)		–	1.	1.	–	–	1.	2.	1.	4.	–	3.	7.	2.	6.	4.	4.	4.	2
3	AIDS (HIV-Krankheit) (B20–B24)		–	–	–	–	3.	1.	7.	2.	5.	2.	2.	4.	6.	1.	1.	1.	1.	3
4	Virushepatitis (B15–B19,B94.2)		1.	–	–	1.	5.	2.	3.	6.	10.	14.	20.	9.	7.	9.	23.	18.	7.	4
5	COVID-19 (U07-U10)		1.	1.	2.	4.	6.	9.	19.	60.	105.	187.	290.	543.	880.	1.321	1.413	1.165	485.	5
6	Neubildungen (C00–D48)		22.	17.	26.	55.	106.	157.	311.	694.	1.376	1.830	2.283	3.022	3.791	3.492	2.598	1.466	527.	6
7	**Bösartige Neubildungen (C00–C97)**		22.	17.	25.	52.	103.	156.	308.	688.	1.354	1.803	2.243	2.928	3.680	3.320	2.418	1.341	482.	7
8	Bösart.Neubild. der Lippe, der Mundhöhle und des Rachens (C00-C14)		–	–	–	–	2.	2.	10.	34.	60.	70.	81.	88.	73.	56.	27.	17.	9.	8
9	Bösart. Neubild. der Speiseröhre (C15)		–	–	–	–	–	6.	11.	21.	41.	65.	70.	64.	66.	46.	21.	12.	1.	9
10	Bösart. Neubild. des Magens (C16)		–	1.	–	5.	6.	5.	16.	27.	50.	58.	68.	96.	128.	118.	105.	55.	15.	10
11	Bösart. Neubild. des Dünndarms (C17)		–	–	–	–	1.	–	2.	8.	2.	5.	4.	6.	17.	12.	11.	4.	–	11
12	Bösart. Neubild. des Colon, des Rektums und des Anus (C18-C21)		–	–	2.	3.	6.	17.	32.	67.	97.	159.	195.	252.	394.	363.	302.	176.	67.	12
13	Bösart. Neubild. der Leber (C22)		–	1.	–	1.	3.	2.	8.	21.	59.	86.	113.	149.	176.	127.	65.	24.	9.	13
14	Bösart. Neubild. der Gallenblase und-wege (C23,C24)		–	1.	–	–	1.	1.	6.	14.	12.	18.	32.	41.	51.	48.	44.	22.	5.	14
15	Bösart. Neubild. der Bauchspeicheldrüse (C25)		–	–	–	1.	6.	8.	15.	56.	136.	169.	220.	293.	354.	324.	183.	74.	24.	15
16	Bösart. Neubild. des Kehlkopfes (C32)		–	–	–	–	–	–	3.	5.	17.	19.	18.	23.	33.	14.	9.	4.	1.	16
17	Bösart. Neubild. der Luftröhre, Bronchien und Lunge (C33–C34)		–	–	4.	6.	20.	43.	152.	354.	526.	609.	740.	748.	496.	234.	99.	16.	17	
18	Bösartiges Melanom der Haut (C43)		1.	1.	2.	2.	3.	6.	10.	14.	26.	32.	29.	42.	64.	63.	54.	32.	14.	18

(continued)

(continued)

Lfd. Nr.	Todesursache (Pos. Nr. ICD10)	15–20	20–25	25–30	30–35	35–40	40–45	45–50	50–55	55–60	60–65	65–70	70–75	75–80	80–85	85–90	90–95	95 und älter	Lfd. Nr.
		Gestorbene im Alter von … bis unter … Jahren																	
19	Bösart. Neubild. der Brustdrüse (C50)	–	–	2.	9.	21.	31.	54.	63.	105.	102.	157.	184.	240.	249.	209.	145.	92.	19
20	Bösart. Neubild. der Zervix uteri (C53)	–	–	–	4.	6.	3.	9.	8.	16.	20.	17.	13.	6.	20.	9.	9.	3.	20
21	Bösart. Neubild. der anderen Teile der Gebärmutter (C54–C55)	–	–	1.	–	–	–	2.	6.	23.	21.	30.	38.	58.	49.	34.	23.	7.	21
22	Bösart. Neubild. des Ovariums (C56)	–	–	1.	–	7.	5.	9.	27.	36.	45.	47.	58.	85.	85.	56.	26.	5.	22
23	Bösart. Neubild. der Prostata (C61)	–	–	–	–	–	–	–	11.	23.	43.	95.	150.	267.	280.	304.	163.	62.	23
24	Bösart. Neubild. der Niere (C64)	1.	–	–	–	2.	1.	2.	6.	17.	20.	30.	62.	71.	67.	58.	28.	9.	24
25	Bösart. Neubild. der Harnblase (C67)	–	–	–	–	–	2.	2.	5.	21.	30.	37.	71.	86.	94.	112.	71.	24.	25
26	Bösart. Neubild. des Gehirns und zentralen Nervensystems (C70–C72)	9.	5.	3.	6.	21.	22.	26.	56.	82.	66.	63.	71.	83.	59.	28.	8.	–	26
27	Bösart. Neubild. der Schilddrüse (C73)	–	–	–	–	–	–	–	2.	4.	4.	5.	9.	19.	20.	11.	11.	–	27
28	Morbus Hodgkin und Lymphome (C81–C85)	2.	–	5.	4.	2.	5.	13.	16.	26.	42.	60.	71.	117.	149.	82.	48.	11.	28
29	Leukämie (C91–C95)	2.	–	2.	2.	2.	3.	9.	10.	25.	43.	54.	110.	159.	185.	128.	78.	32.	29
30	Sonstige bösartige Neubildungen des lymphatischen und blutbildenen Gewebes (C88, C90, C96)	–	–	–	1.	–	–	2.	10.	12.	22.	29.	55.	82.	88.	53.	13.	4.	30
31	Sonst. bösartige Neubildungen (Rest von C00–C97)	8.	8.	7.	10.	8.	17.	24.	49.	110.	138.	180.	242.	303.	308.	279.	199.	72.	31
32	Neubildungen, ausgenommen bösartige (D00–D48)	–	–	1.	3.	3.	1.	3.	6.	22.	27.	40.	94.	111.	172.	180.	125.	45.	32
33	Krankheiten des Blutes, der blutbildenden Organe sowie best. Störungen mit Beteiligung des Immunsystems (D50–D89)	–	–	–	–	1.	2.	3.	1.	4.	5.	5.	11.	31.	40.	42.	35.	20.	33
34	Endokrine, Ernährungs- und Stoffwechselkrankheiten (E00–E90)	2.	4.	9.	10.	14.	21.	31.	80.	130.	223.	286.	376.	517.	713.	730.	495.	200.	34
35	Diabetes mellitus (E10–E14)	–	2.	2.	1.	3.	6.	3.	18.	52.	114.	151.	264.	391.	599.	642.	437.	170.	35

#																			#
36	Psychische Krankheiten (F01–F99)	4.	5.	9.	14.	22.	34.	33.	64.	94.	99.	111.	148.	291.	464.	732.	715.	358.	36
37	Demenz (F01,F03)	–	–	–	3.	7.	16.	17.	–	2.	7.	15.	55.	205.	418.	703.	696.	351.	37
38	Störungen durch Alkohol (F10)	3.	4.	8.	11.	14.	18.	15.	47.	82.	78.	86.	87.	80.	36.	19.	6.	–	38
39	Drogenabhängigkeit, Toxikomanie (F11–F16, F18–F19)	6.	7.	11.	10.	10.	18.	23.	15.	10.	12.	9.	3.	2.	1.	2.	3.	–	39
40	Krankheiten des Nervensystems und der Sinnesorgane (G00–H95)	–	–	–	–	–	–	–	70.	83.	126.	145.	254.	537.	672.	710.	538.	197.	40
41	Morbus Parkinson (G20)	–	–	–	–	–	–	–	–	4.	13.	29.	85.	208.	277.	251.	171.	55.	41
42	Alzheimer Krankheit (G30)	–	–	–	–	–	–	–	–	5.	6.	10.	47.	147.	229.	326.	298.	116.	42
43	**Krankheiten des Herz-Kreislaufsystems (I00–I99)**	6.	10.	10.	22.	49.	115.	186.	347.	635.	940.	1.239	2.082	3.709	5.409	7.296	7.141	3.477	43
44	Ischämische Herzkrankheit (I20–I25)	–	1.	1.	6.	19.	53.	103.	194.	369.	529.	691.	1.070	1.678	2.287	2.799	2.497	1.147	44
45	Akuter Myokardinfarkt (I21–I22)	–	1.	1.	4.	14.	39.	72.	142.	257.	333.	394.	489.	625.	767.	697.	555.	193.	45
46	Sonst. ischämische Herzkrankheiten (I20, I23–I25)	–	1.	–	2.	5.	14.	31.	52.	112.	196.	297.	581.	1.053	1.520	2.102	1.942	954.	46
47	Andere Herzkrankheiten (I30–I33, I39–I51)	1.	4.	3.	9.	10.	23.	24.	43.	67.	112.	161.	275.	612.	1.014	1.628	1.795	946.	47
48	Zerebrovaskuläre Krankheiten (I60–I69)	2.	4.	2.	5.	9.	21.	24.	58.	88.	136.	192.	358.	638.	908.	1.048	909.	334.	48
49	**Krankheiten der Atmungsorgane (J00–J99)**	–	–	3.	4.	7.	5.	15.	42.	109.	236.	334.	637.	815.	835.	815.	687.	298.	49
50	Influenza (J09, J10–J11)	–	–	2.	1.	–	1.	3.	5.	6.	19.	13.	29.	40.	52.	55.	50.	17.	50
51	Pneumonie (J12–J18)	–	–	–	2.	3.	1.	1.	2.	9.	14.	18.	52.	93.	148.	210.	225.	122.	51
52	Chronische Krankheiten der unteren Atemwege (J40–J47)	–	–	–	–	1.	2.	11.	30.	87.	193.	279.	520.	601.	557.	457.	336.	118.	52
53	Asthma (J45–J46)	–	–	–	–	–	–	1.	2.	5.	2.	2.	8.	8.	13.	10.	19.	4.	53
54	Sonstige chronische Krankheiten der unteren Atemwege (J40–J44, J47)	–	–	–	–	1.	2.	10.	28.	82.	191.	277.	512.	593.	544.	447.	317.	114.	54
55	**Krankheiten der Verdauungsorgane (K00–K92)**	2.	2.	3.	14.	26.	58.	96.	173.	238.	273.	287.	319.	421.	422.	438.	310.	124.	55
56	Magen-, Duodenal- und Gastrojejunalgeschwür (K25–K28)	–	–	–	–	–	1.	2.	4.	7.	6.	17.	11.	27.	23.	28.	24.	11.	56
57	Chronische Leberkrankheit und -zirrhose (K70, K73–K74)	–	–	2.	8.	17.	45.	69.	130.	171.	209.	178.	187.	162.	116.	53.	21.	2.	57

(continued)

(continued)

Lfd. Nr.	Todesursache (Pos. Nr. ICD10)	Gestorbene im Alter von ... bis unter ... Jahren																	Lfd. Nr.
		15–20	20–25	25–30	30–35	35–40	40–45	45–50	50–55	55–60	60–65	65–70	70–75	75–80	80–85	85–90	90–95	95 und älter	
58	Krankheiten der Haut und der Unterhaut (L00–L99)	–	–	–	–	–	–	1.	–	2.	2.	3.	6.	5.	14.	11.	13.	6.	58
59	Krankheiten des Muskel-Skelett-Systems und des Bindegewebes (M00–M99)	–	1.	–	1.	1.	–	4.	6.	5.	4.	18.	27.	35.	58.	83.	68.	32.	59
60	Chronische Polyarthritis und Arthrose (M05–M06, M15–M19)	–	–	–	–	–	–	2.	–	–	1.	3.	4.	6.	10.	14.	10.	6.	60
61	Krankheiten des Urogenitalsystems (N00–N99)	–	1.	–	2.	1.	4.	2.	12.	19.	35.	69.	110.	246.	382.	554.	584.	274.	61
62	Krankheiten der Niere und des Ureters (N00–N29)	–	1.	–	1.	1.	3.	2.	10.	13.	23.	47.	70.	175.	269.	372.	440.	215.	62
63	Komplikationen in der Schwangerschaft, bei der Geburt und im Wochenbett (O00–O99)	–	–	1.	1.	–	–	–	–	–	–	–	–	–	–	–	–	–	63
64	Perinatale Affektionen (P00–P96)	–	–	–	–	–	–	–	–	–	–	–	–	–	–	–	–	–	64
65	Angeborene Fehlbildungen, Deformitäten und Chromosomenanomalien (Q00–Q99)	3.	10.	4.	9.	4.	7.	9.	21.	22.	28.	17.	5.	8.	5.	6.	2.	–	65
66	Symptome und schlecht bezeichnete Affektionen (R00–R99)	7.	12.	5.	19.	23.	26.	41.	83.	90.	178.	292.	347.	343.	315.	392.	423.	338.	66
67	Plötzlicher Kindstod (R95)	–	–	–	–	–	–	–	–	–	–	–	–	–	–	–	–	–	67
68	Ungenau bezeichnete und unbekannte Todesursachen (R96–R99)	7.	12.	5.	17.	23.	26.	41.	81.	89.	173.	285.	325.	291.	204.	134.	61.	19.	68
69	**Verletzungen und Vergiftungen (V01–Y89)**	68.	98.	112.	150.	125.	151.	171.	241.	280.	295.	251.	326.	487.	588.	678.	612.	244.	69
70	Unfälle (V01–X59, Y85–Y86)	29.	44.	45.	45.	33.	53.	54.	90.	122.	131.	125.	194.	330.	425.	574.	557.	228.	70
71	Transportmittelunfälle (V01–V99, Y85)	22.	32.	26.	15.	9.	24.	13.	35.	29.	29.	21.	35.	24.	29.	12.	6.	2.	71
72	Unfall durch Sturz (W00–W19)	1.	7.	10.	18.	9.	10.	17.	23.	31.	38.	40.	59.	113.	148.	209.	193.	58.	72
73	Unfall durch Ertrinken und Untergehen (W65–W74)	–	–	–	3.	2.	1.	–	1.	4.	4.	3.	1.	4.	3.	1.	1.	–	73
74	Unfälle durch Vergiftungen (X40–X49)	3.	1.	1.	3.	5.	1.	1.	2.	2.	4.	3.	–	2.	2.	–	1.	–	74

Nr.	Todesursache																		Nr.
75	Selbsttötung und Selbstbeschädigung (X60–X84, Y87.0)	19.	32.	28.	54.	47.	59.	65.	106.	111.	111.	84.	78.	96.	85.	62.	27.	5.	75
76	Mord, tätlicher Angriff (X85–Y09, Y87.1)	–	2.	5.	4.	3.	4.	–	2.	3.	2.	–	1.	2.	2.	1.	1.	–	76
77	Ereignisse, dessen nähere Umstände unbestimmt sind (Y10–Y34, Y87.2)	19.	18.	32.	44.	38.	28.	44.	30.	30.	30.	18.	13.	28.	33.	17.	8.	2.	77
78	**Sonstige Krankheiten (A00–B99, D00–H95, L00–R99, U07–U10)**	24.	42.	43.	77.	96.	130.	187.	424.	609.	945.	1.335	1.992	3.123	4.307	5.007	4.312	2.033	78
	männlich																		
1	**Alle Todesursachen (A00–Y89)**	77.	130.	134.	216.	248.	409.	610.	1.262	2.141	2.930	3.607	5.069	6.970	7.809	7.148	4.942	1.458	1
	Infektiöse und parasitäre Krankheiten (A00–B99, U07–U10)	1.	2.	2.	7.	12.	11.	23.	52.	96.	148.	250.	387.	625.	839.	724.	511.	148.	
2	Tuberkulose (A15–A19, B90)	–	1.	1.	–	–	1.	1.	–	3.	–	2.	4.	1.	3.	4.	3.	2.	2
3	AIDS (HIV-Krankheit) (B20–B24)	–	–	–	1.	1.	–	5.	2.	5.	2.	2.	2.	5.	1.	1.	–	1.	3
4	Virushepatitis (B15–B19,B94.2)	–	–	1.	3.	3.	1.	3.	3.	7.	11.	16.	6.	5.	5.	7.	7.	1.	4
5	COVID-19 (U07–U10)	1.	1.	1.	3.	5.	7.	10.	43.	75.	124.	210.	346.	559.	774.	663.	456.	125.	5
6	Neubildungen (C00–D48)	15.	8.	12.	22.	47.	76.	134.	390.	776.	1.066	1.326	1.736	2.142	1.863	1.312	661.	162.	6
7	**Bösartige Neubildungen (C00–C97)**	15.	8.	12.	21.	46.	75.	132.	389.	760.	1.052	1.302	1.690	2.090	1.776	1.226	616.	153.	7
8	Bösart.Neubild. der Lippe, der Mundhöhle und des Rachens (C00–C14)	–	–	–	–	2.	2.	9.	28.	49.	52.	67.	64.	54.	36.	14.	7.	5.	8
9	Bösart. Neubild. der Speiseröhre (C15)	–	–	–	–	–	4.	9.	19.	34.	53.	57.	52.	54.	31.	10.	8.	1.	9
10	Bösart. Neubild. des Magens (C16)	–	–	–	2.	3.	4.	10.	19.	28.	33.	38.	68.	81.	69.	57.	22.	4.	10
11	Bösart. Neubild. des Dünndarms (C17)	–	–	–	–	1.	–	–	6.	3.	3.	3.	7.	9.	7.	7.	3.	1.	11
12	Bösart. Neubild. des Colon, des Rektums und des Anus (C18–C21)	–	–	–	1.	4.	6.	17.	42.	63.	97.	134.	157.	253.	200.	168.	74.	19.	12
13	Bösart. Neubild. der Leber (C22)	–	–	–	–	2.	1.	2.	18.	47.	66.	95.	119.	134.	80.	41.	15.	2.	13
14	Bösart. Neubild. der Gallenblase und -wege (C23,C24)	–	–	–	–	–	–	6.	7.	5.	14.	20.	22.	28.	26.	16.	10.	–	14
15	Bösart. Neubild. der Bauchspeicheldrüse (C25)	–	–	–	–	4.	4.	7.	37.	74.	98.	120.	159.	173.	147.	76.	30.	5.	15
16	Bösart. Neubild. des Kehlkopfes (C32)	–	–	–	–	–	–	2.	5.	17.	16.	14.	22.	27.	11.	7.	4.	1.	16

(continued)

(continued)

Lfd. Nr.	Todesursache (Pos. Nr. ICD10)	Gestorbene im Alter von ... bis unter ... Jahren																	Lfd. Nr.
		15–20	20–25	25–30	30–35	35–40	40–45	45–50	50–55	55–60	60–65	65–70	70–75	75–80	80–85	85–90	90–95	95 und älter	
17	Bösart. Neubild. der Luftröhre, Bronchien und Lunge (C33–C34)	–	–	–	3.	4.	17.	23.	94.	217.	321.	364.	435.	434.	307.	142.	48.	3.	17
18	Bösartiges Melanom der Haut (C43)	–	1.	1.	1.	1.	4.	7.	7.	14.	24.	19.	27.	40.	35.	34.	19.	–	18
19	Bösart. Neubild. der Brustdrüse (C50)	–	–	–	–	–	–	1.	–	–	1.	–	2.	3.	4.	3.	2.	1.	19
20	Bösart. Neubild. der Zervix uteri (C53)	–	–	–	–	–	–	–	–	–	–	–	–	–	–	–	–	–	20
21	Bösart. Neubild. der anderen Teile der Gebärmutter (C54–C55)	–	–	–	–	–	–	–	–	–	–	–	–	–	–	–	–	–	21
22	Bösart. Neubild. des Ovariums (C56)	–	–	–	–	–	–	–	–	–	–	–	–	–	–	–	–	–	22
23	Bösart. Neubild. der Prostata (C61)	–	–	–	–	–	–	–	11.	23.	43.	95.	150.	267.	280.	304.	163.	62.	23
24	Bösart. Neubild. der Niere (C64)	–	–	–	–	2.	1.	2.	5.	13.	16.	27.	40.	35.	42.	22.	13.	1.	24
25	Bösart. Neubild. der Harnblase (C67)	–	–	–	–	–	1.	2.	2.	14.	21.	30.	51.	64.	66.	73.	44.	15.	25
26	Bösart. Neubild. des Gehirns und zentralen Nervensystems (C70–C72)	6.	2.	1.	2.	13.	15.	11.	33.	56.	36.	32.	45.	39.	28.	8.	3.	–	26
27	Bösart. Neubild. der Schilddrüse (C73)	–	–	–	–	–	–	–	1.	2.	3.	4.	3.	9.	9.	2.	3.	–	27
28	Morbus Hodgkin und Lymphome (C81–C85)	2.	–	4.	2.	1.	5.	6.	11.	15.	31.	29.	41.	63.	84.	43.	22.	2.	28
29	Leukämie (C91–C95)	1.	–	2.	–	1.	1.	3.	8.	13.	30.	34.	62.	100.	94.	57.	37.	9.	29
30	Sonstige bösartige Neubildungen des lymphatischen und blutbildenen Gewebes (C88, C90, C96)	–	–	–	–	–	–	1.	7.	5.	11.	22.	28.	48.	47.	26.	8.	2.	30
31	Sonst. bösartige Neubildungen (Rest von C00–C97)	6.	5.	4.	8.	8.	10.	14.	29.	70.	83.	98.	140.	175.	173.	116.	81.	21.	31
32	Neubildungen, ausgenommen bösartige (D00–D48)	–	–	–	1.	1.	1.	2.	1.	16.	14.	24.	46.	52.	87.	86.	45.	9.	32
33	Krankheiten des Blutes, der blutbildenden Organe sowie best. Störungen mit Beteiligung des Immunsystems (D50–D89)	–	–	–	–	1.	–	1.	1.	3.	4.	2.	6.	16.	17.	15.	11.	1.	33

No.	Krankheit (ICD-10)																		No.
34	Endokrine, Ernährungs- und Stoffwechselkrankheiten (E00–E90)	2.	4.	8.	7.	9.	13.	25.	53.	85.	150.	181.	232.	303.	335.	291.	160.	43.	34
35	Diabetes mellitus (E10–E14)	–	2.	2.	–	2.	2.	3.	14.	36.	86.	105.	173.	242.	295.	262.	143.	38.	35
36	Psychische Krankheiten (F01–F99)	1.	3.	6.	10.	16.	28.	25.	51.	78.	80.	77.	105.	156.	222.	255.	188.	56.	36
37	Demenz (F01,F03)	–	–	–	–	–	–	–	2.	5.	5.	30.	90.	188.	235.	183.	54.		37
38	Störungen durch Alkohol (F10)	–	–	–	2.	6.	14.	12.	40.	69.	70.	66.	73.	63.	28.	12.	2.	–	38
39	Drogenabhängigkeit, Toxikomanie (F11–F16, F18–F19)	1.	3.	6.	8.	9.	14.	13.	10.	7.	3.	6.	1.	2.	–	1.	–	–	39
40	Krankheiten des Nervensystems und der Sinnesorgane (G00–H95)	4.	6.	8.	6.	7.	10.	10.	36.	46.	70.	78.	148.	271.	339.	288.	177.	37.	40
41	Morbus Parkinson (G20)	–	–	–	–	–	–	–	–	2.	10.	22.	65.	126.	161.	145.	80.	15.	41
42	Alzheimer Krankheit (G30)	–	–	–	–	–	–	–	–	–	2.	3.	22.	54.	96.	100.	77.	17.	42
43	**Krankheiten des Herz-Kreislaufsystems (I00–I99)**	4.	6.	8.	15.	35.	82.	147.	261.	484.	689.	838.	1.334	2.090	2.788	2.951	2.325	705.	43
44	Ischämische Herzkrankheit (I20–I25)	–	–	1.	6.	17.	38.	92.	167.	295.	417.	509.	748.	1.066	1.373	1.305	951.	247.	44
45	Akuter Myokardinfarkt (I21–I22)	–	–	1.	4.	13.	29.	66.	120.	211.	257.	286.	353.	397.	450.	326.	216.	43.	45
46	Sonst. ischämische Herzkrankheiten (I20, I23–I25)	–	–	–	2.	4.	9.	26.	47.	84.	160.	223.	395.	669.	923.	979.	735.	204.	46
47	Andere Herzkrankheiten (I30–I33, I39–I51)	1.	3.	3.	5.	7.	18.	14.	31.	51.	77.	104.	171.	310.	481.	658.	563.	191.	47
48	Zerebrovaskuläre Krankheiten (I60–I69)	2.	2.	1.	2.	5.	13.	16.	26.	55.	82.	108.	219.	338.	423.	403.	297.	69.	48
49	**Krankheiten der Atmungsorgane (J00–J99)**	–	–	–	3.	4.	3.	11.	26.	73.	135.	198.	378.	479.	487.	408.	284.	92.	49
50	Influenza (J09, J10–J11)	–	–	–	1.	–	1.	2.	3.	4.	10.	6.	19.	27.	35.	24.	15.	7.	50
51	Pneumonie (J12–J18)	–	–	–	1.	–	–	–	2.	7.	9.	12.	30.	60.	81.	101.	84.	36.	51
52	Chronische Krankheiten der unteren Atemwege (J40–J47)	–	–	–	–	1.	2.	9.	19.	61.	108.	159.	303.	344.	325.	234.	155.	38.	52
53	Asthma (J45–J46)	–	–	–	–	–	–	1.	–	3.	1.	–	2.	3.	6.	4.	8.	2.	53
54	Sonstige chronische Krankheiten der unteren Atemwege (J40–J44, J47)	–	–	–	–	1.	2.	8.	19.	58.	107.	159.	301.	341.	319.	230.	147.	36.	54
55	**Krankheiten der Verdauungsorgane (K00–K92)**	–	–	–	6.	10.	43.	70.	140.	176.	199.	194.	193.	224.	208.	177.	83.	17.	55

(continued)

(continued)

Lfd. Nr.	Todesursache (Pos. Nr. ICD10)	Gestorbene im Alter von … bis unter … Jahren																	Lfd. Nr.
		15–20	20–25	25–30	30–35	35–40	40–45	45–50	50–55	55–60	60–65	65–70	70–75	75–80	80–85	85–90	90–95	95 und älter	
56	Magen-, Duodenal- und Gastrojejunalgeschwür (K25–K28)	–	–	–	–	–	–	2.	2.	6.	3.	9.	5.	16.	12.	12.	6.	–	56
57	Chronische Leberkrankheit und -zirrhose (K70, K73–K74)	–	–	–	3.	7.	34.	47.	108.	124.	164.	133.	129.	98.	81.	36.	11.	–	57
58	Krankheiten der Haut und der Unterhaut (L00–L99)	–	–	–	–	–	–	–	–	1.	2.	3.	5.	3.	5.	3.	1.	1.	58
59	Krankheiten des Muskel-Skelett-Systems und des Bindegewebes (M00–M99)	–	1.	–	1.	1.	–	2.	2.	4.	3.	11.	15.	15.	25.	26.	12.	3.	59
60	Chronische Polyarthritis und Arthrose (M05–M06, M15–M19)	–	–	–	–	–	–	–	–	–	1.	2.	3.	3.	4.	6.	3.	–	60
61	Krankheiten des Urogenitalsystems (N00–N99)	–	–	–	2.	1.	1.	–	8.	13.	16.	38.	65.	132.	183.	210.	165.	69.	61
62	Krankheiten der Niere und des Ureters (N00–N29)	–	–	–	1.	1.	1.	–	7.	8.	9.	30.	43.	99.	132.	146.	118.	52.	62
63	Komplikationen in der Schwangerschaft, bei der Geburt und im Wochenbett (O00–O99)	–	–	–	–	–	–	–	–	–	–	–	–	–	–	–	–	–	63
64	Perinatale Affektionen (P00–P96)	–	–	–	–	–	–	–	–	–	–	–	–	–	–	–	–	–	64
65	Angeborene Fehlbildungen, Deformitäten und Chromosomenanomalien (Q00–Q99)	1.	4.	3.	5.	2.	2.	2.	12.	11.	12.	12.	1.	2.	2.	2.	1.	–	65
66	Symptome und schlecht bezeichnete Affektionen (R00–R99)	5.	9.	5.	16.	14.	21.	34.	55.	67.	128.	213.	233.	211.	166.	162.	117.	66.	66
67	Plötzlicher Kindstod (R95)	–	–	–	–	–	–	–	–	–	–	–	–	–	–	–	–	–	67
68	Ungenau bezeichnete und unbekannte Todesursachen (R96–R99)	5.	9.	5.	15.	14.	21.	34.	54.	66.	125.	209.	223.	186.	116.	65.	27.	4.	68
69	**Verletzungen und Vergiftungen (V01–Y89)**	44.	87.	82.	116.	90.	119.	126.	175.	228.	228.	186.	231.	301.	330.	324.	246.	58.	69
70	Unfälle (V01–X59, Y85–Y86)	21.	39.	33.	36.	26.	44.	45.	69.	101.	104.	96.	132.	192.	223.	249.	213.	51.	70
71	Transportmittelunfälle (V01–V99, Y85)	17.	28.	17.	14.	7.	18.	12.	29.	26.	26.	17.	21.	18.	19.	4.	5.	–	71

72	Unfälle durch Sturz (W00–W19)	1.	6.	8.	13.	7.	10.	11.	15.	22.	28.	28.	47.	70.	83.	92.	78.	12.	72
73	Unfall durch Ertrinken und Untergehen (W65–W74)	–	–	–	2.	2.	1.	–	1.	4.	4.	2.	–	3.	1.	1.	1.	–	73
74	Unfälle durch Vergiftungen (X40–X49)	1.	1.	1.	1.	4.	1.	1.	2.	1.	3.	1.	–	–	1.	–	–	–	74
75	Selbsttötung und Selbstbeschädigung (X60–X84, Y87.0)	12.	29.	22.	41.	36.	46.	43.	74.	95.	87.	66.	59.	78.	69.	51.	24.	4.	75
76	Mord, tätlicher Angriff (X85–Y09, Y87.1)	–	2.	2.	1.	2.	2.	–	1.	–	–	–	1.	1.	–	1.	–	–	76
77	Ereignisse, dessen nähere Umstände unbestimmt sind (Y10–Y34, Y87.2)	11.	15.	24.	37.	24.	23.	33.	21.	21.	20.	9.	8.	16.	21.	10.	3.	2.	77
78	**Sonstige Krankheiten (A00–B99, D00–H95, L00–R99, U07–U10)**	14.	29.	32.	55.	63.	87.	124.	271.	420.	627.	889.	1.243	1.786	2.220	2.062	1.388	433.	78
		weiblich																	
	Alle Todesursachen (A00–Y89)	**45.**	**37.**	**62.**	**103.**	**158.**	**206.**	**353.**	**653.**	**1.084**	**1.562**	**2.082**	**3.215**	**5.265**	**7.072**	**9.504**	**9.461**	**5.200**	
1	Infektiöse und parasitäre Krankheiten (A00–B99, U07–U10)	1.	–	1.	1.	5.	6.	14.	29.	42.	70.	99.	227.	374.	633.	843.	803.	415.	1
2	Tuberkulose (A15–A19, B90)	–	–	–	–	–	–	1.	1.	1.	–	1.	3.	1.	3.	–	1.	2.	2
3	AIDS (HIV-Krankheit) (B20–B24)	–	–	–	–	2.	1.	2.	–	–	–	–	2.	1.	–	–	1.	–	3
4	Virushepatitis (B15–B19,B94.2)	–	–	–	–	2.	1.	–	3.	3.	3.	4.	3.	2.	4.	16.	11.	6.	4
5	COVID-19 (U07–U10)	–	–	1.	1.	1.	2.	9.	17.	30.	63.	80.	197.	321.	547.	750.	709.	360.	5
6	Neubildungen (C00–D48)	7.	9.	14.	33.	59.	81.	177.	304.	600.	764.	957.	1.286	1.649	1.629	1.286	805.	365.	6
7	**Bösartige Neubildungen (C00–C97)**	7.	9.	13.	31.	57.	81.	176.	299.	594.	751.	941.	1.238	1.590	1.544	1.192	725.	329.	7
8	Bösart.Neubild. der Lippe, der Mundhöhle und des Rachens (C00–C14)	–	–	–	–	–	–	1.	6.	11.	18.	14.	24.	19.	20.	13.	10.	4.	8
9	Bösart. Neubild. der Speiseröhre (C15)	–	–	–	–	–	2.	2.	2.	7.	12.	13.	12.	12.	15.	11.	4.	–	9
10	Bösart. Neubild. des Magens (C16)	–	1.	–	3.	3.	1.	6.	8.	22.	25.	30.	28.	47.	49.	48.	33.	11.	10
11	Bösart. Neubild. des Dünndarms (C17)	–	–	–	–	–	–	2.	2.	1.	2.	1.	3.	8.	5.	4.	1.	–	11
12	Bösart. Neubild. des Colon, des Rektums und des Anus (C18–C21)	–	–	2.	1.	2.	11.	15.	25.	34.	62.	61.	95.	141.	163.	134.	102.	48.	12
13	Bösart. Neubild. der Leber (C22)	–	1.	–	–	1.	1.	6.	3.	12.	20.	18.	30.	42.	47.	24.	9.	7.	13
14	Bösart. Neubild. der Gallenblase und-wege (C23,C24)	–	1.	–	–	1.	1.	–	7.	7.	4.	12.	19.	23.	22.	28.	12.	5.	14

(continued)

(continued)

Lfd. Nr.	Todesursache (Pos. Nr. ICD10)	Gestorbene im Alter von ... bis unter ... Jahren																	Lfd. Nr.
		15–20	20–25	25–30	30–35	35–40	40–45	45–50	50–55	55–60	60–65	65–70	70–75	75–80	80–85	85–90	90–95	95 und älter	
15	Bösart. Neubild. der Bauchspeicheldrüse (C25)	–	–	–	1.	2.	4.	8.	19.	62.	71.	100.	134.	181.	177.	107.	44.	19.	15
16	Bösart. Neubild. des Kehlkopfes (C32)	–	–	–	–	–	–	1.	–	–	3.	4.	1.	6.	3.	2.	–	–	16
17	Bösart. Neubild. der Luftröhre, Bronchien und Lunge (C33–C34)	–	–	–	1.	2.	3.	20.	58.	137.	205.	245.	305.	314.	189.	92.	51.	13.	17
18	Bösartiges Melanom der Haut (C43)	–	–	1.	1.	2.	2.	3.	7.	12.	8.	10.	15.	24.	28.	20.	13.	14.	18
19	Bösart. Neubild. der Brustdrüse (C50)	–	–	2.	9.	21.	31.	53.	63.	105.	101.	157.	182.	237.	245.	206.	143.	91.	19
20	Bösart. Neubild. der Zervix uteri (C53)	–	–	–	4.	6.	3.	9.	8.	16.	20.	17.	13.	6.	20.	9.	9.	3.	20
21	Bösart. Neubild. der anderen Teile der Gebärmutter (C54–C55)	–	–	1.	–	–	–	2.	6.	23.	21.	30.	38.	58.	49.	34.	23.	7.	21
22	Bösart. Neubild. des Ovariums (C56)	–	–	1.	–	7.	5.	9.	27.	36.	45.	47.	58.	85.	85.	56.	26.	5.	22
23	Bösart. Neubild. der Prostata (C61)	–	–	–	–	–	–	–	–	–	–	–	–	–	–	–	–	–	23
24	Bösart. Neubild. der Niere (C64)	1.	–	–	–	–	–	–	1.	4.	4.	3.	22.	36.	25.	36.	15.	8.	24
25	Bösart. Neubild. der Harnblase (C67)	–	–	–	–	–	1.	–	3.	7.	9.	7.	20.	22.	28.	39.	27.	9.	25
26	Bösart. Neubild. des Gehirns und zentralen Nervensystems (C70–C72)	3.	3.	2.	4.	8.	7.	15.	23.	26.	30.	31.	26.	44.	31.	20.	5.	–	26
27	Bösart. Neubild. der Schilddrüse (C73)	–	–	–	–	–	–	–	1.	2.	1.	1.	6.	10.	11.	9.	8.	–	27
28	Morbus Hodgkin und Lymphome (C81–C85)	–	–	1.	2.	1.	–	7.	5.	11.	11.	31.	30.	54.	65.	39.	26.	9.	28
29	Leukämie (C91–C95)	1.	–	–	2.	1.	2.	6.	2.	12.	13.	20.	48.	59.	91.	71.	41.	23.	29
30	Sonstige bösartige Neubildungen des lymphatischen und blutbildenen Gewebes (C88, C90, C96)	–	–	–	1.	–	–	1.	3.	7.	11.	7.	27.	34.	41.	27.	5.	2.	30
31	Sonst. bösartige Neubildungen (Rest von C00–C97)	2.	3.	3.	2.	–	7.	10.	20.	40.	55.	82.	102.	128.	135.	163.	118.	51.	31
32	Neubildungen, ausgenommen bösartige (D00–D48)	–	–	1.	2.	2.	–	1.	5.	6.	13.	16.	48.	59.	85.	94.	80.	36.	32

No.	19.	24.	27.	23.	15.	5.	3.	1.	1.	1.	2.	2.	2.						Disease	No.
33	–	–	–	–	–	–	–	–	–	–	–	–	–	–	–	–	–	–	Krankheiten des Blutes, der blutbildenden Organe sowie best. Störungen mit Beteiligung des Immunsystems (D50–D89)	33
34	157.	335.	439.	378.	214.	144.	105.	73.	45.	27.	4.	8.	5.	3.	1.	3.	–	–	Endokrine, Ernährungs- und Stoffwechselkrankheiten (E00–E90)	34
35	132.	294.	380.	304.	149.	91.	46.	28.	16.	4.	6.	–	1.	1.	2.	4.	3.	2.	Diabetes mellitus (E10–E14)	35
36	302.	527.	477.	242.	135.	43.	34.	19.	16.	13.	6.	–	6.	4.	6.	6.	–	2.	Psychische Krankheiten (F01–F99)	36
37	297.	513.	468.	230.	115.	25.	10.	2.	–	–	–	–	–	–	–	–	–	–	Demenz (F01,F03)	37
38	–	4.	7.	8.	17.	14.	20.	8.	13.	7.	5.	2.	1.	–	–	–	–	–	Störungen durch Alkohol (F10)	38
39	–	3.	1.	1.	–	2.	3.	9.	3.	5.	2.	5.	5.	3.	2.	1.	2.	–	Drogenabhängigkeit, Toxikomanie (F11–F16, F18–F19)	39
40	160.	361.	422.	333.	266.	106.	67.	56.	37.	34.	8.	13.	8.	4.	3.	3.	2.	2.	Krankheiten des Nervensystems und der Sinnesorgane (G00–H95)	40
41	40.	91.	106.	116.	82.	20.	7.	3.	2.	–	–	–	–	–	–	–	–	–	Morbus Parkinson (G20)	41
42	99.	221.	226.	133.	93.	25.	7.	4.	5.	–	–	–	–	–	–	–	–	–	Alzheimer Krankheit (G30)	42
43	2.772	4.816	4.345	2.621	1.619	748.	401.	251.	151.	86.	39.	33.	14.	7.	2.	4.	2.	2.	**Krankheiten des Herz-Kreislaufsystems (I00–I99)**	43
44	900.	1.546	1.494	914.	612.	322.	182.	112.	74.	27.	11.	15.	2.	–	1.	–	1.	–	Ischämische Herzkrankheit (I20–I25)	44
45	150.	339.	371.	317.	228.	136.	108.	76.	46.	22.	6.	10.	1.	–	–	–	2.	1.	Akuter Myokardinfarkt (I21–I22)	45
46	750.	1.207	1.123	597.	384.	186.	74.	36.	28.	5.	5.	5.	1.	–	1.	–	1.	–	Sonst. ischämische Herzkrankheiten (I20, I23–I25)	46
47	755.	1.232	970.	533.	302.	104.	57.	35.	16.	12.	10.	5.	3.	4.	1.	1.	1.	–	Andere Herzkrankheiten (I30–I33, I39–I51)	47
48	265.	612.	645.	485.	300.	139.	84.	54.	33.	32.	8.	8.	4.	3.	2.	2.	2.	2.	Zerebrovaskuläre Krankheiten (I60–I69)	48
49	206.	403.	407.	348.	336.	259.	136.	101.	36.	16.	4.	2.	3.	1.	–	3.	–	–	**Krankheiten der Atmungsorgane (J00–J99)**	49
50	10.	35.	31.	17.	13.	10.	7.	9.	2.	2.	1.	–	–	–	–	2.	–	–	Influenza (J09, J10–J11)	50
51	86.	141.	109.	67.	33.	22.	6.	5.	2.	–	1.	1.	1.	1.	–	–	–	–	Pneumonie (J12–J18)	51
52	80.	181.	223.	232.	257.	217.	120.	85.	26.	11.	2.	2.	–	1.	2.	1.	–	2.	Chronische Krankheiten der unteren Atemwege (J40–J47)	52
53	2.	11.	6.	7.	5.	6.	2.	1.	2.	2.	–	–	–	–	–	–	–	–	Asthma (J45–J46)	53
54	78.	170.	217.	225.	252.	211.	118.	84.	24.	9.	2.	9.	–	–	–	2.	–	2.	Sonstige chronische Krankheiten der unteren Atemwege (J40–J44, J47)	54

(continued)

(continued)

Gestorbene im Alter von ... bis unter ... Jahren

Lfd. Nr.	Todesursache (Pos. Nr. ICD10)	15–20	20–25	25–30	30–35	35–40	40–45	45–50	50–55	55–60	60–65	65–70	70–75	75–80	80–85	85–90	90–95	95 und älter	Lfd. Nr.
55	**Krankheiten der Verdauungsorgane (K00–K92)**	2.	–	3.	8.	16.	15.	26.	33.	62.	74.	93.	126.	197.	214.	261.	227.	107.	55
56	Magen-, Duodenal- und Gastrojejunalgeschwür (K25–K28)	–	–	–	–	–	1.	–	2.	1.	3.	8.	6.	11.	11.	16.	18.	11.	56
57	Chronische Leberkrankheit und -zirrhose (K70, K73–K74)	–	2.	2.	5.	10.	11.	22.	22.	47.	45.	45.	58.	64.	35.	17.	10.	2.	57
58	Krankheiten der Haut und der Unterhaut (L00–L99)	–	–	–	–	–	–	1.	–	1.	–	–	1.	2.	9.	8.	12.	5.	58
59	Krankheiten des Muskel-Skelett-Systems und des Bindegewebes (M00–M99)	–	–	–	–	1.	–	2.	4.	1.	1.	7.	12.	20.	33.	57.	56.	29.	59
60	Chronische Polyarthritis und Arthrose (M05–M06, M15–M19)	–	–	–	–	–	–	2.	–	–	–	1.	1.	3.	6.	8.	7.	6.	60
61	Krankheiten des Urogenitalsystems (N00–N99)	–	1.	–	–	–	3.	2.	4.	6.	19.	31.	45.	114.	199.	344.	419.	205.	61
62	Krankheiten der Niere und des Ureters (N00–N29)	–	1.	–	–	–	2.	2.	3.	5.	14.	17.	27.	76.	137.	226.	322.	163.	62
63	Komplikationen in der Schwangerschaft, bei der Geburt und im Wochenbett (O00–O99)	–	–	1.	1.	–	–	–	–	–	–	–	–	–	–	–	–	–	63
64	Perinatale Affektionen (P00–P96)	–	–	–	–	–	–	–	–	–	–	–	–	–	–	–	–	–	64
65	Angeborene Fehlbildungen, Deformitäten und Chromosomenanomalien (Q00–Q99)	2.	6.	1.	4.	2.	5.	7.	9.	11.	16.	5.	4.	6.	3.	4.	1.	–	65
66	Symptome und schlecht bezeichnete Affektionen (R00–R99)	2.	3.	–	3.	9.	5.	7.	28.	23.	50.	79.	114.	132.	149.	230.	306.	272.	66
67	Plötzlicher Kindstod (R95)	–	–	–	–	–	–	–	–	–	–	–	–	–	–	–	–	–	67
68	Ungenau bezeichnete und unbekannte Todesursachen (R96–R99)	2.	3.	–	2.	9.	5.	7.	27.	23.	48.	76.	102.	105.	88.	69.	34.	15.	68
69	**Verletzungen und Vergiftungen (V01–Y89)**	24.	11.	30.	34.	35.	32.	45.	66.	52.	67.	65.	95.	186.	258.	354.	366.	186.	69

	C1	C2	C3	C4	C5	C6	C7	C8	C9	C10	C11	C12	C13	C14	C15	C16	C17
70 Unfälle (V01–X59, Y85–Y86)	8.	5.	12.	9.	7.	9.	9.	21.	9.	21.	27.	29.	62.	138.	202.	325.	344.	177.
71 Transportmittelunfälle (V01–V99, Y85)	5.	4.	9.	1.	2.	6.	1.	6.	3.	3.	4.	14.	6.	10.	8.	1.	1.	2.
72 Unfälle durch Sturz (W00–W19)	–	1.	2.	5.	2.	–	6.	8.	9.	10.	12.	12.	43.	65.	117.	115.	46.
73 Unfall durch Ertrinken und Untergehen (W65–W74)	–	–	–	1.	–	–	–	–	–	–	1.	1.	1.	2.	–	–	–
74 Unfälle durch Vergiftungen (X40–X49)	2.	–	–	2.	1.	–	1.	–	1.	2.	2.	2.	1.	1.	1.	–	1.
75 Selbsttötung und Selbstbeschädigung (X60–X84, Y87.0)	7.	3.	6.	13.	11.	13.	22.	32.	16.	24.	18.	19.	18.	16.	11.	3.	1.
76 Mord, tätlicher Angriff (X85–Y09, Y87.1)	–	–	3.	3.	1.	2.	1.	1.	3.	2.	–	1.	1.	2.	1.	1.	–
77 Ereignisse, dessen nähere Umstände unbestimmt sind (Y10–Y34, Y87.2)	8.	3.	8.	7.	14.	5.	11.	9.	9.	10.	9.	5.	12.	12.	7.	5.	–
78 Sonstige Krankheiten (A00–B99, D00–H95, L00–R99, U07–U10)	10.	13.	11.	22.	33.	43.	63.	153.	189.	318.	446.	749.	1.337	2.087	2.945	2.924	1.600
ingesamt (im 1. Lebensjahr Gestorbene: auf 100.000 Lebendgeborene)																	
Alle Todesursachen (A00–Y89)	28.	32.6	32.8	52.3	66.2	108.6	156.5	271.6	465.5	778.1	1.249	2.067	3.397	5.617	11.687	21.798	37.519
1 Infektiöse und parasitäre Krankheiten (A00–B99, U07–U10)	.5	.4	.5	1.3	2.8	3.	6.	11.5	19.9	37.8	76.6	153.2	277.4	555.6	1.100	1.989	3.173
2 Tuberkulose (A15–A19, B90)	–	.2	.2	–	–	.2	.3	.1	.6	–	.7	1.7	.6	2.3	2.8	6.1	22.5
3 AIDS (HIV-Krankheit) (B20–B24)	–	–	–	–	.5	.2	1.1	.3	.7	.3	.4	.1	1.7	.4	.7	1.5	5.6
4 Virushepatitis (B15–B19,B94.2)	.3	–	.2	.2	.8	.4	.5	.9	1.4	2.4	4.4	2.2	1.9	3.4	16.1	27.2	39.4
5 COVID-19 (U07–U10)	.2	.2	.3	.7	1.	1.6	3.1	8.5	15.2	32.4	63.7	135.5	244.3	498.6	991.7	1.763	2.733
6 Neubildungen (C00–D48)	5.1	3.3	4.3	9.	17.3	27.7	50.5	98.4	198.6	317.	501.2	753.8	1.053	1.318	1.823	2.219	2.970
7 Bösartige Neubildungen (C00–C97)	5.1	3.3	4.2	8.5	16.8	27.5	50.1	97.6	195.5	312.3	492.4	730.4	1.022	1.253	1.697	2.030	2.716
8 Bösart.Neubild. der Lippe, der Mundhöhle und des Rachens (C00–C14)	–	–	–	–	.3	.4	1.6	4.8	8.7	12.1	17.8	22.	20.3	21.1	18.9	25.7	50.7
9 Bösart. Neubild. der Speiseröhre (C15)	–	–	–	–	–	1.1	1.8	3.	5.9	11.3	15.4	16.	18.3	17.4	14.7	18.2	5.6
10 Bösart. Neubild. des Magens (C16)	–	.2	–	.8	1.	.9	2.6	3.8	7.2	10.	14.9	23.9	35.5	44.5	73.7	83.2	84.5
11 Bösart. Neubild. des Dünndarms (C17)	–	–	–	–	.2	–	.3	1.1	.3	.9	.9	1.5	4.7	4.5	7.7	6.1	5.6
12 Bösart. Neubild. des Colon, des Rektums und des Anus (C18–C21)	–	.3	.3	.5	1.	3.	5.2	9.5	14.	27.5	42.8	62.9	109.4	137.	212.	266.4	377.6

(continued)

(continued)

Lfd. Nr.	Todesursache (Pos. Nr. ICD10)	Gestorbene im Alter von ... bis unter ... Jahren																	Lfd. Nr.
		15–20	20–25	25–30	30–35	35–40	40–45	45–50	50–55	55–60	60–65	65–70	70–75	75–80	80–85	85–90	90–95	95 und älter	
13	Bösart. Neubild. der Leber (C22)	–	.2	–	.2	.5	.4	1.3	3.	8.5	14.9	24.8	37.2	48.9	47.9	45.6	36.3	50.7	13
14	Bösart. Neubild. der Gallenblase und -wege (C23,C24)	–	.2	–	–	.2	.2	1.	2.	1.7	3.1	7.	10.2	14.2	18.1	30.9	33.3	28.2	14
15	Bösart. Neubild. der Bauchspeicheldrüse (C25)	–	–	–	.2	1.	1.4	2.4	7.9	19.6	29.3	48.3	73.1	98.3	122.3	128.4	112.	135.2	15
16	Bösart. Neubild. des Kehlkopfes (C32)	–	–	–	–	–	–	.5	.7	2.5	3.3	4.	5.7	9.2	5.3	6.3	6.1	5.6	16
17	Bösart. Neubild. der Luftröhre, Bronchien und Lunge (C33–C34)	–	–	–	.7	1.	3.5	7.	21.6	51.1	91.1	133.7	184.6	207.7	187.2	164.2	149.8	90.2	17
18	Bösartiges Melanom der Haut (C43)	–	.2	.3	.3	.5	1.1	1.6	2.	3.8	5.5	6.4	10.5	17.8	23.8	37.9	48.4	78.9	18
19	Bösart. Neubild. der Brustdrüse (C50)	–	–	.3	1.5	3.4	5.5	8.8	8.9	15.2	17.7	34.5	45.9	66.6	94.	146.7	219.4	518.4	19
20	Bösart. Neubild. der Zervix uteri (C53)	–	–	–	.7	1.	.5	1.5	1.1	2.3	3.5	3.7	3.2	1.7	7.5	6.3	13.6	16.9	20
21	Bösart. Neubild. der anderen Teile der Gebärmutter (C54–C55)	–	–	.2	–	–	–	.3	.9	3.3	3.6	6.6	9.5	16.1	18.5	23.9	34.8	39.4	21
22	Bösart. Neubild. des Ovariums (C56)	–	–	.2	–	1.1	.9	1.5	3.8	5.2	7.8	10.3	14.5	23.6	32.1	39.3	39.3	28.2	22
23	Bösart. Neubild. der Prostata (C61)	–	–	–	–	–	–	–	1.6	3.3	7.4	20.9	37.4	74.1	105.7	213.4	246.7	349.4	23
24	Bösart. Neubild. der Niere (C64)	.2	–	–	–	.3	.2	.3	.9	2.5	3.5	6.6	15.5	19.7	25.3	40.7	42.4	50.7	24
25	Bösart. Neubild. der Harnblase (C67)	–	–	–	–	–	.4	.3	.7	3.	5.2	8.1	17.7	23.9	35.5	78.6	107.5	135.2	25
26	Bösart. Neubild. des Gehirns und zentralen Nervensystems (C70–C72)	2.1	1.	.5	1.	3.4	3.9	4.2	7.9	11.8	11.4	13.8	17.7	23.	22.3	19.7	12.1	–	26
27	Bösart. Neubild. der Schilddrüse (C73)	–	–	–	–	–	–	–	.3	.6	.7	1.1	2.2	5.3	7.5	7.7	16.6	–	27
28	Morbus Hodgkin und Lymphome (C81–C85)	.5	–	.8	.7	.3	.9	2.1	2.3	3.8	7.3	13.2	17.7	32.5	56.2	57.6	72.6	62.	28
29	Leukämie (C91–C95)	.5	–	.3	.3	.3	.5	1.5	1.4	3.6	7.4	11.9	27.4	44.1	69.8	89.8	118.	180.3	29
30	Sonstige bösartige Neubildungen des lymphatischen und blutbildenen Gewebes (C88, C90, C96)	–	–	–	.2	–	–	.3	1.4	1.7	3.8	6.4	13.7	22.8	33.2	37.2	19.7	22.5	30

No.	Kategorie (ICD)																			No.
31	Sonst. bösartige Neubildungen (Rest von C00–C97)	1.8	1.6	1.6	1.2	1.6	1.3	3.	3.9	6.9	15.9	23.9	39.5	60.4	84.1	116.3	195.8	301.2	405.7	31
32	Neubildungen, ausgenommen bösartige (D00–D48)	–	–	–	–	.2	.5	.5	.2	.9	3.2	4.7	8.8	23.4	30.8	64.9	126.3	189.2	253.6	32
33	Krankheiten des Blutes, der blutbildenden Organe sowie best. Störungen mit Beteiligung des Immunsystems (D50–D89)	–	–	–	–	–	.2	.4	.5	.1	.6	.9	1.1	2.7	8.6	15.1	29.5	53.	112.7	33
34	Endokrine, Ernährungs- und Stoffwechselkrankheiten (E00–E90)	–	.5	.8	1.5	1.6	2.3	3.7	5.	11.3	18.8	38.6	62.8	93.8	143.6	269.1	512.3	749.1	1.127	34
35	Diabetes mellitus (E10–E14)	–	–	.4	.3	.2	.5	1.1	.5	2.6	7.5	19.7	33.1	65.9	108.6	226.1	450.6	661.4	958.	35
36	Psychische Krankheiten (F01–F99)	–	.9	1.	1.5	2.3	3.6	6.	5.4	9.1	13.6	17.1	24.4	36.9	80.8	175.1	513.7	1.082	2.017	36
37	Demenz (F01,F03)	–	–	–	–	–	–	–	–	–	.3	1.2	3.3	13.7	56.9	157.8	493.4	1.053	1.978	37
38	Störungen durch Alkohol (F10)	–	–	–	–	.5	1.1	2.8	2.8	6.7	11.8	13.5	18.9	21.7	22.2	13.6	13.3	9.1	–	38
39	Drogenabhängigkeit, Toxikomanie (F11–F16, F18–F19)	–	–	.8	1.3	1.8	2.3	3.2	2.4	2.1	1.4	2.1	2.	.7	.6	.4	1.4	4.5	–	39
40	Krankheiten des Nervensystems und der Sinnesorgane (G00–H95)	–	1.4	1.4	1.8	1.6	1.6	3.2	3.7	9.9	12.	21.8	31.8	63.4	149.1	253.7	498.3	814.2	1.110	40
41	Morbus Parkinson (G20)	–	–	–	–	–	–	–	–	.6	.6	2.3	6.4	21.2	57.8	104.6	176.2	258.8	309.9	41
42	Alzheimer Krankheit (G30)	–	–	–	–	–	–	–	–	.7	.7	1.	2.2	11.7	40.8	86.4	228.8	451.	653.7	42
43	**Krankheiten des Herz-Kreislaufsystems (I00–I99)**	–	1.4	2.	3.6	1.7	8.	20.3	30.2	49.2	91.7	162.8	272.	519.4	1.030	2.042	5.121	10.807	19.594	43
44	Ischämische Herzkrankheit (I20–I25)	–	–	.2	.2	1.	3.1	9.4	16.7	27.5	53.3	91.6	151.7	266.9	465.9	863.3	1.965	3.779	6.464	44
45	Akuter Myokardinfarkt (I21–I22)	–	–	–	.7	.2	2.3	6.9	11.7	20.1	37.1	57.7	86.5	122.	173.5	289.5	489.2	840.	1.088	45
46	Sonst. ischämische Herzkrankheiten (I20, I23–I25)	–	–	–	.2	.3	.8	2.5	5.	7.4	16.2	33.9	65.2	144.9	292.4	573.8	1.475	2.939	5.376	46
47	Andere Herzkrankheiten (I30–I33, I39–I51)	–	.2	.8	1.5	.5	1.6	4.1	3.9	6.1	9.7	19.4	35.3	68.6	169.9	382.8	1.143	2.717	5.331	47
48	Zerebrovaskuläre Krankheiten (I60–I69)	–	–	.5	.8	.3	1.5	3.7	3.9	8.2	12.7	23.6	42.1	89.3	177.1	342.7	735.5	1.376	1.882	48
49	**Krankheiten der Atmungsorgane (J00–J99)**	–	–	–	–	.5	1.1	.9	2.4	6.	15.7	40.9	73.3	158.9	226.3	315.2	572.	1.040	1.679	49
50	Influenza (J09, J10–J11)	–	–	–	–	.2	.3	.2	.5	.7	.9	3.3	2.9	7.2	11.1	19.6	38.6	75.7	95.8	50
51	Pneumonie (J12–J18)	–	–	–	.3	.5	.3	.2	.2	.3	1.3	2.4	4.	13.	25.8	55.9	147.4	340.5	687.5	51

(continued)

(continued)

Lfd. Nr.	Todesursache (Pos. Nr. ICD10)	Gestorbene im Alter von ... bis unter ... Jahren																	Lfd. Nr.
		15–20	20–25	25–30	30–35	35–40	40–45	45–50	50–55	55–60	60–65	65–70	70–75	75–80	80–85	85–90	90–95	95 und älter	
52	Chronische Krankheiten der unteren Atemwege (J40–J47)	–	–	–	–	.2	.4	1.8	4.3	12.6	33.4	61.2	129.7	166.9	210.3	320.7	508.5	665.	52
53	Asthma (J45–J46)	–	–	–	–	–	–	.2	.3	.7	.3	.4	2.	2.2	4.9	7.	28.8	22.5	53
54	Sonstige chronische Krankheiten der unteren Atemwege (J40–J44, J47)	–	–	–	–	.2	.4	1.6	4.	11.8	33.1	60.8	127.7	164.7	205.3	313.7	479.8	642.4	54
55	**Krankheiten der Verdauungsorgane (K00–K92)**	.5	–	.5	2.3	4.2	10.2	15.6	24.5	34.4	47.3	63.	79.6	116.9	159.3	307.4	469.2	698.8	55
56	Magen-, Duodenal- und Gastrojejunalgeschwür (K25–K28)	–	–	–	–	–	.2	.3	.6	1.	1.	3.7	2.7	7.5	8.7	19.7	36.3	62.	56
57	Chronische Leberkrankheit und -zirrhose (K70, K73–K74)	–	–	.3	1.3	2.8	7.9	11.2	18.4	24.7	36.2	39.1	46.6	45.	43.8	37.2	31.8	11.3	57
58	Krankheiten der Haut und der Unterhaut (L00–L99)	–	–	–	–	.2	–	.2	–	.3	.3	.7	1.5	1.4	5.3	7.7	19.7	33.8	58
59	Krankheiten des Muskel-Skelett-Systems und des Bindegewebes (M00–M99)	–	.2	–	.2	.2	.7	.7	.9	.7	.7	4.	6.7	9.7	21.9	58.3	102.9	180.3	59
60	Chronische Polyarthritis und Arthrose (M05–M06, M15–M19)	–	–	–	–	–	–	.3	–	–	.2	.7	1.	1.7	3.8	9.8	15.1	33.8	60
61	Krankheiten des Urogenitalsystems (N00–N99)	–	.2	–	.3	.2	.7	.3	1.7	2.7	6.1	15.1	27.4	68.3	144.2	388.8	883.8	1.544	61
62	Krankheiten der Niere und des Ureters (N00–N29)	–	.2	–	.2	.2	.5	.3	1.4	1.9	4.	10.3	17.5	48.6	101.5	261.1	665.9	1.212	62
63	Komplikationen in der Schwangerschaft, bei der Geburt und im Wochenbett (O00–O99)	–	–	.2	.2	–	–	–	–	–	–	–	–	–	–	–	–	–	63
64	Perinatale Affektionen (P00–P96)	–	–	–	–	–	–	–	–	–	–	–	–	–	–	–	–	–	64
65	Angeborene Fehlbildungen, Deformitäten und Chromosomenanomalien (Q00–Q99)	.7	2.	.7	1.5	.7	1.2	1.5	3.	3.2	4.8	3.7	1.2	2.2	1.9	4.2	3.	–	65
66	Symptome und schlecht bezeichnete Affektionen (R00–R99)	1.6	2.3	.8	3.1	3.8	4.6	6.7	11.8	13.	30.8	64.1	86.6	95.2	118.9	275.1	640.2	1.905	66

No.	Todesursache																			No.
67	Plötzlicher Kindstod (R95)	—	—	—	—	—	—	—	—	—	—	—	—	—	—	—	—	—	67
68	Ungenau bezeichnete und unbekannte Todesursachen (R96–R99)	1.6	2.3	.8	3.8	2.8	4.6	6.7	11.5	12.8	30.	62.6	81.1	80.8	77.	94.	92.3	107.1	68
69	**Verletzungen und Vergiftungen (V01–Y89)**	15.6	19.1	18.7	24.6	20.4	26.7	27.8	34.2	40.4	51.1	55.1	81.3	135.2	222.	475.8	926.2	1.375	69
70	Unfälle (V01–X59, Y85–Y86)	6.7	8.6	7.5	7.4	5.4	9.4	8.8	12.8	17.6	22.7	27.4	48.4	91.6	160.4	402.9	843.	1.285	70
71	Transportmittelunfälle (V01–V99, Y85)	5.1	6.3	4.3	2.5	1.5	4.2	2.1	5.	4.2	5.	4.6	8.7	6.7	10.9	8.4	9.1	11.3	71
72	Unfälle durch Sturz (W00–W19)	.2	1.4	1.7	2.9	1.5	1.8	2.8	3.3	4.5	6.6	8.8	14.7	31.4	55.9	146.7	292.1	326.8	72
73	Unfall durch Ertrinken und Untergehen (W65–W74)	—	—	—	.5	.3	.2	—	.1	.6	.7	.7	.2	1.1	1.1	.7	1.5	—	73
74	Unfälle durch Vergiftungen (X40–X49)	.7	.2	.2	.5	.8	.2	.2	.3	.3	.7	.7	—	.6	.8	—	1.5	—	74
75	Selbsttötung und Selbstbeschädigung (X60–X84, Y87.0)	4.4	6.3	4.7	8.8	7.7	10.4	10.6	15.	16.	19.2	18.4	19.5	26.7	32.1	43.5	40.9	28.2	75
76	Mord, tätlicher Angriff (X85–Y09, Y87.1)	—	.4	.8	.7	.5	.7	—	.3	.4	.3	—	.2	.6	.8	.7	1.5	—	76
77	Ereignisse, dessen nähere Umstände unbestimmt sind (Y10–Y34, Y87.2)	4.4	3.5	5.4	7.2	6.2	4.9	7.2	4.3	4.3	5.2	4.	3.2	7.8	12.5	11.9	12.1	11.3	77
78	**Sonstige Krankheiten (A00–B99, D00–H95, L00–R99, U07–U10)**	5.5	8.2	7.2	12.6	15.7	23.	30.4	60.1	87.9	163.7	293.1	496.9	867.1	1.626	3.514	6.526	11.456	78
	männlich (im 1. Lebensjahr Gestorbene: auf 100.000 Lebendgeborene)																		
	Alle Todesursachen (A00–Y89)	34.3	49.4	43.7	69.5	80.2	143.7	200.1	357.	618.8	1.038	1.672	2.749	4.412	7.083	13.876	24.991	40.562	
1	Infektiöse und parasitäre Krankheiten (A00–B99, U07–U10)	.4	.8	.7	2.3	3.9	3.9	7.5	14.7	27.7	52.5	115.9	209.9	395.7	761.	1.406	2.584	4.117	1
2	Tuberkulose (A15–A19, B90)	—	.4	.3	—	—	.4	.3	—	.9	—	.9	2.2	.6	2.7	7.8	15.2	55.6	2
3	AIDS (HIV-Krankheit) (B20–B24)	—	—	—	—	.3	—	1.6	.6	1.4	.7	.9	1.1	3.2	.9	1.9	—	27.8	3
4	Virushepatitis (B15–B19,B94.2)	—	—	—	.3	1.	.4	1.	.8	2.	3.9	7.4	3.3	3.2	4.5	13.6	35.4	27.8	4
5	COVID-19 (U07–U10)	.4	.4	.3	1.	1.6	2.5	3.3	12.2	21.7	43.9	97.3	187.6	353.9	702.	1.287	2.306	3.478	5
6	Neubildungen (C00–D48)	6.7	3.	3.9	7.1	15.2	26.7	44.	110.3	224.3	377.8	614.5	941.4	1.356	1.690	2.547	3.343	4.507	6
7	**Bösartige Neubildungen (C00–C97)**	6.7	3.	3.9	6.8	14.9	26.3	43.3	110.	219.7	372.8	603.4	916.5	1.323	1.611	2.380	3.115	4.257	7
8	Bösart.Neubild. der Lippe, der Mundhöhle und des Rachens (C00–C14)	—	—	—	—	.6	.7	3.	7.9	14.2	18.4	31.	34.7	34.2	32.7	27.2	35.4	139.1	8
9	Bösart. Neubild. der Speiseröhre (C15)	—	—	—	—	—	1.4	3.	5.4	9.8	18.8	26.4	28.2	34.2	28.1	19.4	40.5	27.8	9

(continued)

(continued)

Lfd. Nr.	Todesursache (Pos. Nr. ICD10)	Gestorbene im Alter von ... bis unter ... Jahren																	Lfd. Nr.
		15–20	20–25	25–30	30–35	35–40	40–45	45–50	50–55	55–60	60–65	65–70	70–75	75–80	80–85	85–90	90–95	95 und älter	
10	Bösart. Neubild. des Magens (C16)	–	–	–	.6	1.	1.4	3.3	5.4	8.1	11.7	17.6	36.9	51.3	62.6	110.7	111.2	111.3	10
11	Bösart. Neubild. des Dünndarms (C17)	–	–	–	–	.3	–	–	1.7	.3	1.1	1.4	1.6	5.7	6.3	13.6	15.2	–	11
12	Bösart. Neubild. des Colon, des Rektums und des Anus (C18–C21)	–	–	–	.6	1.3	2.1	5.6	11.9	18.2	34.4	62.1	85.1	160.2	181.4	326.1	374.2	528.6	12
13	Bösart. Neubild. der Leber (C22)	–	–	–	.3	.6	.4	.7	5.1	13.6	23.4	44.	64.5	84.8	72.6	79.6	75.9	55.6	13
14	Bösart. Neubild. der Gallenblase und -wege (C23,C24)	–	–	–	–	–	–	2.	2.	1.4	5.	9.3	11.9	17.7	23.6	31.1	50.6	–	14
15	Bösart. Neubild. der Bauchspeicheldrüse (C25)	–	–	–	–	1.3	1.4	2.3	10.5	21.4	34.7	55.6	86.2	109.5	133.3	147.5	151.7	139.1	15
16	Bösart. Neubild. des Kehlkopfes (C32)	–	–	–	–	–	–	.7	1.4	4.9	5.7	6.5	11.9	17.1	10.	13.6	20.2	27.8	16
17	Bösart. Neubild. der Luftröhre, Bronchien und Lunge (C33–C34)	–	–	–	1.	1.3	6.	7.5	26.6	62.7	113.8	168.7	235.9	274.7	278.5	275.7	242.7	83.5	17
18	Bösartiges Melanom der Haut (C43)	–	.4	.3	.3	.3	1.4	2.3	2.	4.	8.5	8.8	14.6	25.3	31.7	66.	96.1	–	18
19	Bösart. Neubild. der Brustdrüse (C50)	–	–	–	–	–	–	.3	–	–	.4	–	1.1	1.9	3.6	5.8	10.1	27.8	19
20	Bösart. Neubild. der Zervix uteri (C53)	–	–	–	–	–	–	–	–	–	–	–	–	–	–	–	–	–	20
21	Bösart. Neubild. der anderen Teile der Gebärmutter (C54–C55)	–	–	–	–	–	–	–	–	–	–	–	–	–	–	–	–	–	21
22	Bösart. Neubild. des Ovariums (C56)	–	–	–	–	–	–	–	–	–	–	–	–	–	–	–	–	–	22
23	Bösart. Neubild. der Prostata (C61)	–	–	–	–	–	–	–	3.1	6.6	15.2	44.	81.3	169.	254.	590.1	824.3	1.725	23
24	Bösart. Neubild. der Niere (C64)	–	–	–	–	.6	.4	.7	1.4	3.8	5.7	12.5	21.7	22.2	38.1	42.7	65.7	27.8	24
25	Bösart. Neubild. der Harnblase (C67)	–	–	–	–	–	–	–	.6	4.	7.4	13.9	27.7	40.5	59.9	141.7	222.5	417.3	25
26	Bösart. Neubild. des Gehirns und zentralen Nervensystems (C70–C72)	2.7	.8	.3	.6	4.2	5.3	3.6	9.3	16.2	12.8	14.8	24.4	24.7	25.4	15.5	15.2	–	26
27	Bösart. Neubild. der Schilddrüse (C73)	–	–	–	–	–	–	–	.3	.6	1.1	1.9	1.6	5.7	8.2	3.9	15.2	–	27
28	Morbus Hodgkin und Lymphome (C81–C85)	.9	–	1.3	.6	.3	1.8	2.	3.1	4.3	11.	13.4	22.2	39.9	76.2	83.5	111.2	55.6	28

#																		
29	Leukämie (C91–C95)	.4	—	.7	—	.3	.4	1.	2.3	3.8	10.6	15.8	33.6	63.3	85.3	110.7	187.1	250.4
30	Sonstige bösartige Neubildungen des lymphatischen und blutbildenen Gewebes (C88, C90, C96)	—	—	—	—	—	—	.3	2.	1.4	3.9	10.2	15.2	30.4	42.6	50.5	40.5	55.6
31	Sonst. bösartige Neubildungen (Rest von C00–C97)	2.7	1.9	1.3	2.6	2.6	3.5	4.6	8.2	20.2	29.4	45.4	75.9	110.8	156.9	225.2	409.6	584.2
32	Neubildungen, ausgenommen bösartige (D00–D48)	—	—	—	.3	.3	.4	.7	.3	4.6	5.	11.1	24.9	32.9	78.9	166.9	227.6	250.4
33	Krankheiten des Blutes, der blutbildenden Organe sowie best. Störungen mit Beteiligung des Immunsystems (D50–D89)	—	—	—	—	.3	—	.3	.3	.9	1.4	.9	3.3	10.1	15.4	29.1	55.6	27.8
34	Endokrine, Ernährungs- und Stoffwechselkrankheiten (E00–E90)	.9	1.5	2.6	2.3	2.9	4.6	8.2	15.	24.6	53.2	83.9	125.8	191.8	303.8	564.9	809.1	1.196
35	Diabetes mellitus (E10–E14)	—	.8	.7	—	.6	.7	1.	4.	10.4	30.5	48.7	93.8	153.2	267.6	508.6	723.1	1.057
36	Psychische Krankheiten (F01–F99)	.4	1.1	2.	3.2	5.2	9.8	8.2	14.4	22.5	28.4	35.7	56.9	98.8	201.4	495.	950.7	1.558
37	Demenz (F01,F03)	—	—	—	—	—	—	—	—	.6	1.8	2.3	16.3	57.	170.5	456.2	925.4	1.502
38	Störungen durch Alkohol (F10)	—	—	—	.6	1.9	4.9	3.9	11.3	19.9	24.8	30.6	39.6	39.9	25.4	23.3	10.1	—
39	Drogenabhängigkeit, Toxikomanie (F11–F16, F18–F19)	.4	1.1	2.	2.6	2.9	4.9	4.3	2.8	2.	1.1	2.8	.5	1.3	—	1.9	—	1.029
40	Krankheiten des Nervensystems und der Sinnesorgane (G00–H95)	1.8	2.3	2.6	1.9	2.3	3.5	3.3	10.2	13.3	24.8	36.1	80.3	171.6	307.5	559.1	895.	—
41	Morbus Parkinson (G20)	—	—	—	—	—	—	—	—	.6	3.5	10.2	35.2	79.8	146.	281.5	404.5	417.3
42	Alzheimer Krankheit (G30)	—	—	—	—	—	—	—	—	—	.7	1.4	11.9	34.2	87.1	194.1	389.4	472.9
43	**Krankheiten des Herz-Kreislaufsystems (I00–I99)**	1.8	2.3	2.6	4.8	11.3	28.8	48.2	73.8	139.9	244.2	388.4	723.4	1.323	2.529	5.729	11.757	19.613
44	Ischämische Herzkrankheit (I20–I25)	—	—	.3	1.9	5.5	13.3	30.2	47.2	85.3	147.8	235.9	405.6	674.8	1.245	2.533	4.809	6.872
45	Akuter Myokardinfarkt (I21–I22)	—	—	.3	1.3	4.2	10.2	21.7	33.9	61.	91.1	132.5	191.4	251.3	408.2	632.8	1.092	1.196
46	Sonst. ischämische Herzkrankheiten (I20, I23–I25)	—	—	—	.6	1.3	3.2	8.5	13.3	24.3	56.7	103.3	214.2	423.5	837.2	1.901	3.717	5.675
47	Andere Herzkrankheiten (I30–I33, I39–I51)	.4	1.1	1.	1.6	2.3	6.3	4.6	8.8	14.7	27.3	48.2	92.7	196.2	436.3	1.277	2.847	5.314

(continued)

(continued)

Lfd. Nr.	Todesursache (Pos. Nr. ICD10)	Gestorbene im Alter von ... bis unter ... Jahren																	Lfd. Nr.
		15–20	20–25	25–30	30–35	35–40	40–45	45–50	50–55	55–60	60–65	65–70	70–75	75–80	80–85	85–90	90–95	95 und älter	
48	Zerebrovaskuläre Krankheiten (I60–I69)	.9	.8	.3	.6	1.6	4.6	5.2	7.4	15.9	29.1	50.1	118.8	214.	383.7	782.3	1.502	1.920	48
49	**Krankheiten der Atmungsorgane (J00–J99)**	–	–	–	1.	1.3	1.1	3.6	7.4	21.1	47.8	91.8	205.	303.2	441.7	792.	1.436	2.560	49
50	Influenza (J09, J10–J11)	–	–	–	.3	–	.4	.7	.8	1.2	3.5	2.8	10.3	17.1	31.7	46.6	75.9	194.7	50
51	Pneumonie (J12–J18)	–	–	–	.3	.6	–	–	.6	2.	3.2	5.6	16.3	38.	73.5	196.1	424.8	1.002	51
52	Chronische Krankheiten der unteren Atemwege (J40–J47)	–	–	–	–	.3	.7	3.	5.4	17.6	38.3	73.7	164.3	217.8	294.8	454.3	783.8	1.057	52
53	Asthma (J45–J46)	–	–	–	–	–	–	.3	–	.9	.4	–	1.1	1.9	5.4	7.8	40.5	55.6	53
54	Sonstige chronische Krankheiten der unteren Atemwege (J40–J44, J47)	–	–	–	–	.3	.7	2.6	5.4	16.8	37.9	73.7	163.2	215.9	289.3	446.5	743.3	1.002	54
55	**Krankheiten der Verdauungsorgane (K00–K92)**	–	–	–	1.9	3.2	15.1	23.	39.6	50.9	70.5	89.9	104.7	141.8	188.7	343.6	419.7	472.9	55
56	Magen-, Duodenal- und Gastrojejunalgeschwür (K25–K28)	–	–	–	–	–	–	.7	.6	1.7	1.1	4.2	2.7	10.1	10.9	23.3	30.3	–	56
57	Chronische Leberkrankheit und -zirrhose (K70, K73–K74)	–	–	–	1.	2.3	11.9	15.4	30.6	35.8	58.1	61.6	70.	62.	73.5	69.9	55.6	–	57
58	Krankheiten der Haut und der Unterhaut (L00–L99)	–	–	–	–	–	–	–	–	.3	.7	1.4	2.7	1.9	4.5	5.8	5.1	27.8	58
59	Krankheiten des Muskel-Skelett-Systems und des Bindegewebes (M00–M99)	–	.4	–	.3	–	–	.7	.6	1.2	1.1	5.1	8.1	9.5	22.7	50.5	60.7	83.5	59
60	Chronische Polyarthritis und Arthrose (M05–M06, M15–M19)	–	–	–	–	–	–	–	–	–	.4	.9	1.6	1.9	3.6	11.6	15.2	–	60
61	Krankheiten des Urogenitalsystems (N00–N99)	–	–	–	.6	.3	.4	–	2.3	3.8	5.7	17.6	35.2	83.6	166.	407.7	834.4	1.920	61
62	Krankheiten der Niere und des Ureters (N00–N29)	–	–	–	.3	.3	.4	–	2.	2.3	3.2	13.9	23.3	62.7	119.7	283.4	596.7	1.447	62
63	Komplikationen in der Schwangerschaft, bei der Geburt und im Wochenbett (O00–O99)	–	–	–	–	–	–	–	–	–	–	–	–	–	–	–	–	–	63

Nr	Todesursache (ICD-10)																		Nr
64	Perinatale Affektionen (P00–P96)	–	–	–	–	–	–	–	–	–	–	–	–	–	–	–	–	–	64
65	Angeborene Fehlbildungen, Deformitäten und Chromosomenanomalien (Q00–Q99)	–	.4	1.5	1.	1.6	.6	.7	.7	3.4	3.2	4.3	5.6	.5	1.3	1.8	3.9	5.1	65
66	Symptome und schlecht bezeichnete Affektionen (R00–R99)	2.2	3.4	1.6	5.2	4.5	7.4	11.2	15.6	19.4	45.4	98.7	126.4	133.6	150.6	314.5	591.6	1.836	66
67	Plötzlicher Kindstod (R95)	–	–	–	–	–	–	–	–	–	–	–	–	–	–	–	–	–	67
68	Ungenau bezeichnete und unbekannte Todesursachen (R96–R99)	2.2	3.4	1.6	4.8	4.5	7.4	11.2	15.3	19.1	44.3	96.9	120.9	117.7	105.2	126.2	136.5	111.3	68
69	Verletzungen und Vergiftungen (V01–Y89)	19.6	33.	26.7	37.3	29.1	41.8	41.3	49.5	65.9	80.8	86.2	125.3	190.5	299.3	629.	1.244	1.614	69
70	Unfälle (V01–X59, Y85–Y86)	9.4	14.8	10.8	11.6	8.4	15.5	14.8	19.5	29.2	36.9	44.5	71.6	121.5	202.3	483.4	1.077	1.419	70
71	Transportmittelunfälle (V01–V99, Y85)	7.6	10.6	5.5	4.5	2.3	6.3	3.9	8.2	7.5	9.2	7.9	11.4	11.4	17.2	7.8	25.3	25.3	71
72	Unfälle durch Sturz (W00–W19)	.4	2.3	2.6	4.2	2.3	3.5	3.6	4.2	6.4	9.9	13.	25.5	44.3	75.3	178.6	394.4	333.8	72
73	Unfall durch Ertrinken und Untergehen (W65–W74)	–	–	–	–	.6	.4	–	.3	1.2	1.4	.9	–	1.9	.9	1.9	5.1	–	73
74	Unfälle durch Vergiftungen (X40–X49)	.4	.4	.3	1.3	.6	.4	.3	.6	.3	1.1	.5	–	.9	.9	–	–	–	74
75	Selbsttötung und Selbstbeschädigung (X60–X84, Y87.0)	5.4	11.	7.2	13.2	11.6	16.2	14.1	20.9	27.5	30.8	30.6	32.	49.4	62.6	99.	121.4	111.3	75
76	Mord, tätlicher Angriff (X85–Y09, Y87.1)	–	.8	.7	.6	.7	.3	.3	.3	–	.5	.6	–	1.9	1.9	–	–	–	76
77	Ereignisse, dessen nähere Umstände unbestimmt sind (Y10–Y34, Y87.2)	4.9	5.7	7.8	11.9	7.8	8.1	10.8	5.9	6.1	7.1	4.2	4.3	10.1	19.	19.4	15.2	55.6	77
78	Sonstige Krankheiten (A00–B99, D00–H95, I00–R99, U07–U10)	6.2	11.	10.4	17.7	20.4	30.6	40.7	76.7	121.4	222.2	412.	674.1	1.131	2.014	4.003	7.019	12.046	78
	weiblich (im 1. Lebensjahr Gestorbene: auf 100.000 Lebendgeborene)																		
	Alle Todesursachen (A00–Y89)	**21.3**	**14.9**	**21.3**	**34.4**	**52.**	**73.1**	**113.7**	**185.7**	**312.6**	**529.2**	**868.4**	**1.485**	**2.604**	**4.572**	**10.448**	**20.434**	**36.747**	
1	Infektiöse und parasitäre Krankheiten (A00–B99, U07–U10)	.5	–	.3	.3	1.6	2.1	4.5	8.2	12.1	23.7	41.3	104.9	185.	409.3	926.7	1.734	2.933	1
2	Tuberkulose (A15–A19, B90)	–	–	–	–	–	–	.3	.3	.3	.4	–	1.4	.5	1.9	–	2.2	14.1	2
3	AIDS (HIV-Krankheit) (B20–B24)	–	–	–	–	.7	.4	.4	.6	–	–	1.7	.9	.5	–	–	2.2	–	3
4	Virushepatitis (B15–B19, B94.2)	–	–	–	–	.7	.4	–	.9	.9	1.	1.7	1.4	1.	2.6	17.6	23.8	42.4	4
5	COVID-19 (U07–U10)	–	.3	.3	.7	.3	.7	2.9	4.8	8.7	21.3	33.4	91.	158.8	353.7	824.5	1.531	2.544	5

(continued)

(continued)

Lfd. Nr.	Todesursache (Pos. Nr. ICD10)	Gestorbene im Alter von ... bis unter ... Jahren																95 und älter	Lfd. Nr.
		15–20	20–25	25–30	30–35	35–40	40–45	45–50	50–55	55–60	60–65	65–70	70–75	75–80	80–85	85–90	90–95		
6	Neubildungen (C00–D48)	3.3	3.6	4.8	11.	19.4	28.7	57.	86.4	173.	258.8	399.2	594.1	815.6	1.053	1.414	1.739	2.579	6
7	**Bösartige Neubildungen (C00–C97)**	3.3	3.6	4.5	10.3	18.8	28.7	56.7	85.	171.9	254.4	392.5	571.9	786.4	998.3	1.310	1.566	2.325	7
8	Bösart.Neubild. der Lippe, der Mundhöhle und des Rachens (C00–C14)	–	–	–	–	–	–	.3	1.7	3.2	6.1	5.8	11.1	9.4	12.9	14.3	21.6	28.3	8
9	Bösart. Neubild. der Speiseröhre (C15)	–	–	–	–	–	.7	.6	.6	2.	4.1	5.4	5.5	5.9	9.7	12.1	8.6	–	9
10	Bösart. Neubild. des Magens (C16)	–	.4	–	1.	1.	.4	1.9	2.3	6.3	8.5	12.5	12.9	23.2	31.7	52.8	71.3	77.7	10
11	Bösart. Neubild. des Dünndarms (C17)	–	–	–	–	–	–	.6	.6	.3	.7	.4	1.4	4.	3.2	4.4	2.2	–	11
12	Bösart. Neubild. des Colon, des Rektums und des Anus (C18–C21)	–	–	.7	.3	.7	3.9	4.8	7.1	9.8	21.	25.4	43.9	69.7	105.4	147.3	220.3	339.2	12
13	Bösart. Neubild. der Leber (C22)	–	.4	–	–	.3	.4	1.9	.9	3.5	6.8	7.5	13.9	20.8	30.4	26.4	19.4	49.5	13
14	Bösart. Neubild. der Gallenblase und-wege (C23,C24)	–	.4	–	–	.3	.4	–	2.	2.	1.4	5.	8.8	11.4	14.2	30.8	25.9	35.3	14
15	Bösart. Neubild. der Bauchspeicheldrüse (C25)	–	–	–	.3	.7	1.4	2.6	5.4	17.9	24.1	41.7	61.9	89.5	114.4	117.6	95.	134.3	15
16	Bösart. Neubild. des Kehlkopfes (C32)	–	–	–	–	–	–	.3	–	–	1.	1.7	.5	3.	1.9	2.2	–	–	16
17	Bösart. Neubild. der Luftröhre, Bronchien und Lunge (C33–C34)	–	–	–	.3	.7	1.1	6.4	16.5	39.5	69.5	102.2	140.9	155.3	122.2	101.1	110.2	91.9	17
18	Bösartiges Melanom der Haut (C43)	–	–	.3	.3	.7	.7	1.	2.	3.5	2.7	4.2	6.9	11.9	18.1	22.	28.1	98.9	18
19	Bösart. Neubild. der Brustdrüse (C50)	–	–	.7	3.	6.9	11.	17.1	17.9	30.3	34.2	65.5	84.1	117.2	158.4	226.5	308.9	643.1	19
20	Bösart. Neubild. der Zervix uteri (C53)	–	–	–	1.3	2.	1.1	2.9	2.3	4.6	6.8	7.1	6.	3.	12.9	9.9	19.4	21.2	20
21	Bösart. Neubild. der anderen Teile der Gebärmutter (C54-C55)	–	–	.3	–	–	–	.6	1.7	6.6	7.1	12.5	17.6	28.7	31.7	37.4	49.7	49.5	21
22	Bösart. Neubild. des Ovariums (C56)	–	–	.3	–	2.3	1.8	2.9	7.7	10.4	15.2	19.6	26.8	42.	55.	61.6	56.2	35.3	22
23	Bösart. Neubild. der Prostata (C61)	–	–	–	–	–	–	–	–	–	–	–	–	–	–	–	–	–	23
24	Bösart. Neubild. der Niere (C64)	.5	–	–	–	–	–	–	.3	1.2	1.4	1.3	10.2	17.8	16.2	39.6	32.4	56.5	24
25	Bösart. Neubild. der Harnblase (C67)	–	–	–	–	–	.4	–	.9	2.	3.	2.9	9.2	10.9	18.1	42.9	58.3	63.6	25

#		1.4	1.2	.7	1.3	2.6	2.5	4.8	6.5	7.5	10.2	12.9	12.	21.8	20.	22.	10.8		#
26	Bösart. Neubild. des Gehirns und zentralen Nervensystems (C70–C72)	1.4	–	.7	–	–	–	–	–	–	–	–	–	–	–	–	10.8	–	26
27	Bösart. Neubild. der Schilddrüse (C73)	–	–	–	–	–	–	–	.3	.6	.3	.4	2.8	4.9	7.1	9.9	17.3	–	27
28	Morbus Hodgkin und Lymphome (C81–C85)	–	–	.3	.7	.3	–	2.3	1.4	3.2	3.7	12.9	13.9	26.7	42.	42.9	56.2	63.6	28
29	Leukämie (C91–C95)	.5	–	–	.7	.3	.7	1.9	.6	3.5	4.4	8.3	22.2	29.2	58.8	78.	88.6	162.5	29
30	Sonstige bösartige Neubildungen des lymphatischen und blutbildenden Gewebes (C88, C90, C96)	–	–	–	.3	–	–	.3	.9	2.	3.7	2.9	12.5	16.8	26.5	29.7	10.8	14.1	30
31	Sonst. bösartige Neubildungen (Rest von C00–C97)	.9	1.2	1.	.7	–	2.5	3.2	5.7	11.5	18.6	34.2	47.1	63.3	87.3	179.2	254.9	360.4	31
32	Neubildungen, ausgenommen bösartige (D00–D48)	–	–	.3	.7	.7	–	.3	1.4	1.7	4.4	6.7	22.2	29.2	55.	103.3	172.8	254.4	32
33	Krankheiten des Blutes, der blutbildenden Organe sowie best. Störungen mit Beteiligung des Immunsystems (D50–D89)	–	–	–	–	–	.7	.6	–	.3	.3	1.3	2.3	7.4	14.9	29.7	51.8	134.3	33
34	Endokrine, Ernährungs- und Stoffwechselkrankheiten (E00–E90)	–	–	.3	1.	1.6	2.8	1.9	7.7	13.	24.7	43.8	66.5	105.8	244.4	482.6	723.5	1.110	34
35	Diabetes mellitus (E10–E14)	–	–	–	.3	.3	1.4	–	1.1	4.6	9.5	19.2	42.	73.7	196.6	417.7	635.	932.8	35
36	Psychische Krankheiten (F01–F99)	1.4	.8	1.	1.3	2.	2.1	2.6	3.7	4.6	6.4	14.2	19.9	66.8	156.5	524.4	1.138	2.134	36
37	Demenz (F01,F03)	–	–	–	–	–	–	–	–	–	.7	4.2	11.5	56.9	148.7	514.5	1.108	2.099	37
38	Störungen durch Alkohol (F10)	–	–	–	.3	.3	.7	1.6	2.	3.7	2.7	8.3	6.5	8.4	5.2	7.7	8.6	–	38
39	Drogenabhängigkeit, Toxikomanie (F11–F16, F18–F19)	.9	.4	.7	1.	1.6	1.4	.6	1.4	.9	3.	1.3	.9	–	.6	1.1	6.5	–	39
40	Krankheiten des Nervensystems und der Sinnesorgane (G00–H95)	.9	.4	1.	1.3	1.	2.8	4.2	9.7	10.7	19.	27.9	49.	131.6	215.3	463.9	779.7	1.131	40
41	Morbus Parkinson (G20)	–	–	–	–	–	–	–	–	.6	1.	2.9	9.2	40.6	75.	116.5	196.5	282.7	41
42	Alzheimer Krankheit (G30)	–	–	–	–	–	–	–	–	1.4	1.4	2.9	11.5	46.	86.	248.4	477.3	699.6	42
43	**Krankheiten des Herz-Kreislaufsystems (I00–I99)**	.9	1.6	.7	2.3	4.6	11.7	12.6	24.5	43.5	85.	167.3	345.5	800.8	1.695	4.776	10.402	19.589	43
44	Ischämische Herzkrankheit (I20–I25)	–	.4	–	–	.7	5.3	3.5	7.7	21.3	37.9	75.9	148.7	302.7	590.9	1.642	3.339	6.360	44

(continued)

(continued)

Lfd. Nr.	Todesursache (Pos. Nr. ICD10)	Gestorbene im Alter von ... bis unter ... Jahren																	Lfd. Nr.
		15–20	20–25	25–30	30–35	35–40	40–45	45–50	50–55	55–60	60–65	65–70	70–75	75–80	80–85	85–90	90–95	95 und älter	
45	Akuter Myokardinfarkt (I21–I22)	–	–	–	–	.3	3.5	1.9	6.3	13.3	25.7	45.	62.8	112.8	205.	407.8	732.2	1.060	45
46	Sonst. ischämische Herzkrankheiten (I20, I23–I25)	–	.4	–	–	.3	1.8	1.6	1.4	8.1	12.2	30.9	85.9	189.9	386.	1.235	2.607	5.300	46
47	Andere Herzkrankheiten (I30–I33, I39–I51)	–	.4	–	1.3	1.	1.8	3.2	3.4	4.6	11.9	23.8	48.	149.4	344.6	1.066	2.661	5.335	47
48	Zerebrovaskuläre Krankheiten (I60–I69)	–	.8	.3	1.	1.3	2.8	2.6	9.1	9.5	18.3	35.	64.2	148.4	313.6	709.	1.322	1.873	48
49	**Krankheiten der Atmungsorgane (J00–J99)**	–	–	1.	.3	1.	.7	1.3	4.5	10.4	34.2	56.7	119.6	166.2	225.	447.4	870.4	1.456	49
50	Influenza (J09, J10–J11)	–	–	.7	–	–	–	.3	.6	.6	3.	2.9	4.6	6.4	11.	34.1	75.6	70.7	50
51	Pneumonie (J12–J18)	–	–	–	.3	.3	.4	.3	–	.6	1.7	2.5	10.2	16.3	43.3	119.8	304.5	607.7	51
52	Chronische Krankheiten der unteren Atemwege (J40–J47)	–	–	–	–	–	–	.6	3.1	7.5	28.8	50.1	100.2	127.1	150.	245.1	390.9	565.3	52
53	Asthma (J45–J46)	–	–	–	–	–	–	–	.6	.6	.3	.8	2.8	2.5	4.5	6.6	23.8	14.1	53
54	Sonstige chronische Krankheiten der unteren Atemwege (J40–J44, J47)	–	–	–	–	–	–	.6	2.6	6.9	28.5	49.2	97.5	124.6	145.5	238.5	367.2	551.2	54
55	**Krankheiten der Verdauungsorgane (K00–K92)**	.9	–	1.	2.7	5.3	5.3	8.4	9.4	17.9	25.1	38.8	58.2	97.4	138.4	286.9	490.3	756.1	55
56	Magen-, Duodenal- und Gastrojejunalgeschwür (K25–K28)	–	–	–	–	–	.4	–	.6	.3	1.	3.3	2.8	5.4	7.1	17.6	38.9	77.7	56
57	Chronische Leberkrankheit und -zirrhose (K70, K73–K74)	–	.7	.7	1.7	3.3	3.9	7.1	6.3	13.6	15.2	18.8	26.8	31.7	22.6	18.7	21.6	14.1	57
58	Krankheiten der Haut und der Unterhaut (L00–L99)	–	–	–	–	–	–	.3	–	.3	–	–	.5	1.	5.8	8.8	25.9	35.3	58
59	Krankheiten des Muskel-Skelett-Systems und des Bindegewebes (M00–M99)	–	–	–	–	.3	–	.6	1.1	.3	.3	2.9	5.5	9.9	21.3	62.7	121.	204.9	59
60	Chronische Polyarthritis und Arthrose (M05–M06, M15–M19)	–	–	–	–	–	–	.6	–	–	–	.4	.5	1.5	3.9	8.8	15.1	42.4	60
61	Krankheiten des Urogenitalsystems (N00–N99)	–	.4	–	–	1.1	1.1	.6	1.1	1.7	6.4	12.9	20.8	56.4	128.7	378.2	905.	1.449	61

No.	Todesursache (ICD)																			
62	Krankheiten der Niere und des Ureters (N00–N29)	–	.4	–	–	–	–	–	.7	.6	.9	1.4	4.7	7.1	12.5	37.6	88.6	248.4	695.5	1.152
63	Komplikationen in der Schwangerschaft, bei der Geburt und im Wochenbett (O00–O99)	–	–	.3	–	.3	–	–	–	–	–	–	–	–	–	–	–	–	–	–
64	Perinatale Affektionen (P00–P96)	–	–	–	–	–	–	–	–	–	–	–	–	–	–	–	–	–	–	–
65	Angeborene Fehlbildungen, Deformitäten und Chromosomenanomalien (Q00–Q99)	.9	2.4	.3	1.3	.7	1.8	2.3	2.6	3.2	5.4	2.1	1.8	3.	1.9	–	4.4	2.2	–	–
66	Symptome und schlecht bezeichnete Affektionen (R00–R99)	.9	1.2	–	1.	3.	.7	1.8	1.8	2.3	8.	6.6	16.9	33.	52.7	65.3	96.3	252.8	660.9	1.922
67	Plötzlicher Kindstod (R95)	–	–	–	–	–	–	–	–	–	–	–	–	–	–	–	–	–	–	–
68	Ungenau bezeichnete und unbekannte Todesursachen (R96–R99)	.9	1.2	–	.7	3.	3.	1.8	2.3	2.3	7.7	6.6	16.3	31.7	47.1	51.9	56.9	75.8	73.4	106.
69	**Verletzungen und Vergiftungen (V01–Y89)**	11.4	4.4	10.3	11.3	11.5	11.4	14.5	18.8	15.	22.7	27.1	43.9	92.	166.8	389.1	790.5	1.314	–	–
70	Unfälle (V01–X59, Y85–Y86)	3.8	2.	3.	3.	2.3	3.2	2.9	6.	6.1	9.1	12.1	28.6	68.3	130.6	357.3	743.	1.251	–	–
71	Transportmittelunfälle (V01–V99, Y85)	2.4	1.6	3.1	.3	.7	2.1	.3	1.7	.9	1.	1.7	6.5	3.	6.5	8.8	2.2	14.1	–	–
72	Unfälle durch Sturz (W00–W19)	–	.4	.7	1.7	.7	–	1.9	2.3	2.6	3.4	5.	5.5	21.3	42.	128.6	248.4	325.1	–	–
73	Unfall durch Ertrinken und Untergehen (W65–W74)	–	–	–	.3	–	–	–	–	–	–	.4	.5	.5	1.3	–	–	–	–	–
74	Unfälle durch Vergiftung (X40–X49)	.9	–	–	.7	.3	–	–	–	.3	.3	.8	–	.6	2.2	–	2.2	–	–	–
75	Selbsttötung und Selbstbeschädigung (X60–X84, Y87.0)	3.3	1.2	2.1	4.3	3.6	4.6	7.1	9.1	4.6	8.1	7.5	8.9	10.3	12.1	6.5	7.1	–	–	–
76	Mord, tätlicher Angriff (X85–Y09, Y87.1)	–	–	1.	1.	.7	–	.3	.3	.9	.7	–	.5	1.3	–	2.2	–	–	–	–
77	Ereignisse, dessen nähere Umstände unbestimmt sind (Y10–Y34, Y87.2)	3.8	1.2	2.7	2.3	4.6	1.8	3.5	2.6	2.6	3.4	3.8	5.9	7.8	10.8	–	–	–	–	–
78	**Sonstige Krankheiten (A00–B99, D00–H95, L00–R99, U07–U10)**	4.7	5.2	3.8	7.3	10.9	15.3	20.3	43.5	54.5	107.7	186.	346.	661.3	1.349	3.237	6.315	11.307	–	–

Q: STATISTIK AUSTRIA. Todesursachenstatistik. Erstellt am 28.06.2021. Ab 2002 Klassifikation der Todesursachen auf entsprechend ICD10 Version 2013 und altersstandardisierte Raten auf Basis der Eurostat-Standardbevölkerung 2013. Ab 2009 vollzähligere Erfassung von im Ausland verstorbenen Personen mit Wohnsitz in Österreich

References

Aeskulap Praxis. (2022, June). Integrative Medizin, Schwerpunkt Long Covid Therapie. Retrieved June 4, 2022, from https://www.aeskulap-praxis.ch/schwerpunkte/long-covid-therapie

Agence France Press in Vienna. (2022, March 9). Austria suspends mandatory COVID vaccination law. The Guardian. Retrieved June 17, 2022, from https://www.theguardian.com/world/2022/mar/09/austria-suspends-mandatory-covid-vaccination-law

Alvarez, M. C. (2020, August). The EU recovery plan: Funding arrangements and their impact. Funcas. Retrieved June 22, 2022, from https://www.funcas.es/articulos/the-eu-recovery-plan-funding-arrangements-and-their-impacts/

Anderson, M. (2017). No disease can exist in an alkaline environment. Natural Food Pantry. Retrieved August 06, 2021, from https://naturalfoodpantry.ca/blogs/mind-body/no-disease-can-exist-in-an-alkaline-environment

APA. (2020, May 29), Austria Presse Agentur, Das Milieu macht's: Coronaviren mögen es nicht basisch. Retrieved June 11, 2022, from 2https://www.ots.at/presseaussendung/OTS_20200529_OTS0115/das-milieu-machts-coronaviren-moegen-es-nicht-basisch

Bardosh K., de Figueiredo A., Gur-Arie R., Jamrozik, E., Doidge, J., Lemmens, T., Keshavjee, S., Graham, J. E., & Baral, S. (2022, May). The unintended consequences of COVID-19 vaccine policy: Why mandates, passports and restrictions may cause more harm than good. *BJM Global Health 7*, e008684. Retrieved April 4, 2022, from https://gh.bmj.com/content/bmjgh/7/5/e008684.full.pdf.

BMSGPK. (2022, June 1). Coronavirus: Aktuelle Maßnahmen. Bundesministerium Soziales, Gesundheit, Pflege und Konsumentenschutz. Retrieved June 20, 2022, from https://www.sozialministerium.at/Informationen-zum-Coronavirus/Coronavirus%2D%2D-Aktuelle-Ma%C3%9Fnahmen.html

Brown, F.Z., et al. (2020, April 6). How will the coronavirus reshape democracy and governance globally? Carnegie Endowment for International Peace. Retrieved June 5, 2022, from https://carnegieendowment.org/2020/04/06/how-will-coronavirus-reshape-democracy-and-governance-globally-pub-81470

BVZ. (2021, March 10). Rund neun Prozent der Sterbefälle 2021 aufgrund von COVID-19. Statistik Austria. Retrieved June 5, 2022, from https://m.bvz.at/in-ausland/statistik-austria-rund-neun-prozent-der-sterbefaelle-2021-aufgrund-von-covid-19-demografie-epidemie-statistik-austria-viruserkrankung-wien-oesterreich-314967874

Campra, P. (2021, November). Detection of Graphene in COVID 19 vaccines. Retrieved June 11, 2022, from https://www.researchgate.net/publication/355979001_DETECTION_OF_GRAPHENE_IN_COVID19_VACCINES

Cattel F, Giordano S, Bertiond C, Lupia T, Corcione S, Scaldaferri M, Angelone L, De Rosa FG (2021, January 2). Ozone therapy in COVID-19: A narrative review. *Virus Research, 291*, 198207. https://doi.org/10.1016/j.virusres.2020.198207. Retrieved June 4, 2022, from https://pubmed.ncbi.nlm.nih.gov/33115670/.

Cenci, A., Macchia, I., La Sorsa, V., Sbarigia, C., Di Donna, V., & Pietraforte, D. (2022, April). Mechanisms of action of ozone therapy in emerging viral diseases: Immunomodulatory effects and therapeutic advantages with reference to SARS-CoV-2. Journal. *Frontiers in Microbiology, 13*. Retrieved June 5, 2022, from https://www.frontiersin.org/articles/10.3389/fmicb.2022.871645/full

Cochrane, Ö. (2021, November). Ozontherapie bei COVID-19: Fehlende Hinweise auf Wirksamkeit. Retrieved June 4, 2022, from https://www.medizin-transparent.at/ozontherapie-covid-19/

ECFIN. (2021). Austria Forecast Winter 2021. European Commission. Retrieved June 22, 2022, from https://ec.europa.eu/economy_finance/forecasts/2021/winter/ecfin_forecast_winter_2021_at_en.pdf.

EEAS. (2021, March 13). EU statement on WHO-led Covid-19 origins study. Retrieved May 29, 2022, from https://www.eeas.europa.eu/eeas/eu-statement-who-led-covid-19-origins-study_en

European Commission. (2020, March 19). COVID-19: Commission creates first ever rescue stockpile for medical equipment. Retrieved June 20, 2022, from https://ec.europa.eu/commission/presscorner/detail/en/ip_20_476,

European Commission. (2022a, June). Coronavirus response. Retrieved June 21, 2022, from https://ec.europa.eu/info/live-work-travel-eu/coronavirus-response/timeline-eu-action_en

European Commission. (2022b). Questions and answers on COVID-19 vaccination in the EU. Retrieved June 24, 2022, from https://ec.europa.eu/info/live-work-travel-eu/coronavirus-response/safe-covid-19-vaccines-europeans/questions-and-answers-covid-19-vaccination-eu_en#vaccination

European Council. (2022, June 17). A recovery plan for Europe. Retrieved June 20, 2022, from https://www.consilium.europa.eu/en/policies/eu-recovery-plan/#:~:text=EU%20leaders%20agreed%20a%20deal,EU%20budget%20for%202021%2D2027.

European Parliament. (2022, January 24). Time for the truth on the presence of graphene in COVID-19 vaccines. Parliamentary Questions. Retrieved June 12, 2022, from https://www.europarl.europa.eu/doceo/document/P-9-2022-000303_EN.html

Gaub, F., & Boswinkel, L. (2020, May) ISS, European Institute for Security Studies, Who first wins? International crisis response to Covid-19. Retrieved May 29, 2022, from https://www.iss.europa.eu/content/who%E2%80%99s-first-wins-international-crisis-response-covid-19

Heneghan, C., & One, J. (2020, August 12). Public Health England has changed its definition of deaths: here's what it means. The Center for Evidence-Based Medicine. Retrieved June 20, 2022, from https://www.cebm.net/covid-19/public-health-england-death-data-revised/

Himmel, M., & Frey, S. (2022, February 3). SARS-CoV-2: International Investigation under the WHO or BWC. Policy Brief article, Front Public Health. Retrieved May 29, 2022, from https://www.frontiersin.org/articles/10.3389/fpubh.2021.636679/full

International Scientific Commiitee, ISCO3. (2020, March). Mögliche Verwendung von Ozon bei SARS-CoV-2/COVID-19. Madrid 2020, Internationales Wissenschaftliches Komitee für Ozontherapie, www.isco3.org. Retrieved June 6, 2022, from https://aepromo.org/coronavirus/pdfs_doc_ISCO3/Covid19_de.pdf

IPEV, Institut für Prävention un Ernährung/IPEV. (2020, May 29). Das Milieu macht's: Coronaviren mögen es nicht basisch - Wissenschaft weist Einfluß von pH-Wert auf Virenaktivität nach. Retrieved June 11, 2022, from https://www.presseportal.de/pm/143476/4609518

Jerving, S. (2022, May 5). WHO: Excess deaths from COVID-19 pandemic 3 times more than reported. Devex. Retrieved June 10, 2022, from https://www.devex.com/news/who-excess-deaths-from-covid-19-pandemic-3-times-more-than-reported-103157

John, G. (2021, September 13). Zu wenige Intensivbetten: Hat die Politik den Ausbau verschlafen? Der Standard, Corona. Retrieved June 22, 2022, from https://www.derstandard.at/story/2000129529461/zu-wenige-intensivbetten-hat-die-politik-den-ausbau-verschlafen

John Hopkins. (2022). Corona Virus Resource Center. Cases and mortality. Retrieved June 12, 2022, from https://coronavirus.jhu.edu/data/mortality

Karp P., & Davidson H. (2020, April 29). China bristles at Australia's call for investigation into coronavirus origin. The Guardian. Retrieved May 30, 2022, from https://www.theguardian.com/world/2020/apr/29/australia-defends-plan-to-investigate-china-over-covid-19-outbreak-as-row-deepens

Khandar, J., et al. (2021, December). Fighting COVID-19 with the help of alkaline diet. Journal of Pharmaceutical Research International. Retrieved August 6, 2022, from https://journaljpri.com/index.php/JPRI/article/view/33743/63551

Kopp, E. (2022, March 28). Lab accident is "most likely but least probed COVID-origin, State Dept. memo says. US RTK, US right to know. Retrieved June 19, 2022, from https://usrtk.org/biohazards-blog/lab-accident-is-most-likely-but-least-probed/

Kuhbandner, C. (2021, February 21). Corona-Todesfälle: Die Mär von den 10 verlorenen Lebensjahren. Telepolis/Wissenschaft. Retrieved June 10, 2022, from https://www.heise.de/tp/features/Corona-Todesfaelle-Die-Maer-von-den-zehn-verlorenen-Lebensjahren-5060636.html?seite=all

Kupferschmidt, K. (2021, January 5). Science, Viral mutations may cause another very very bad covid-19 wave scientists warn. Retrieved June 11, 2022, from https://www.science.org/content/article/viral-mutations-may-cause-another-very-very-bad-covid-19-wave-scientists-warn

Lang, H. (2022, January 31). Ein "starker" Impfstoff, Leserbrief zu den Corona Impfstoffen. Retrieved June 12, 2022, from https://kraichgau.news/bretten/c-politik-wirtschaft/ein-starker-impfstoff_a87521

Loe, M. (2022, February 4). Yes, the CDC did change the definition of a vaccine. Verify. Retrieved June 1, 2022, from https://www.king5.com/article/news/verify/coronavirus-verify/cdc-changed-vaccine-definition-more-transparent/536-03ce7891-2604-4090-b548-b1618d286834

Macrotrends. (2022, June). World death rate 150–2022. Retrieved June 10, 2022, from https://www.macrotrends.net/countries/WLD/world/death-rate

Marini, S, Maggiorotti, M, Dardes, N., Bonetti, M., Martinelli, M., Re, L., Carinci, F., & Tavera, C. (2020, May–June). NUOVA F.I.O. (Italian Oxygen-Ozone Federation). Oxygen-ozone therapy as adjuvant in the current emergency in SARS-COV-2 infection: A clinical study. *Journal of Biological Regulators and Homeostatic Agents, 34*(3), 757–766. https://doi.org/10.23812/20-250-e-56. Retrieved June 4, 2022, from https://europepmc.org/article/med/32462858.

Martinez-Sanchez, G. (2022, April). *Ozone therapy for prevention and treatment of COVID-19*. XHP Publishing. Retrieved June 4, 2022, from https://www.xiahcpublishing.com/2572-5505/JERP-2022-00015.

Mazumdaru, S. (2020, April 17). DW, What influence does China have over the WHO? Retrieved May 30, 2022, from https://www.dw.com/en/what-influence-does-china-have-over-the-who/a-53161220

Müller, W., & Tomaselli, E. (2021, September 27). Erdrutschsieg der Kommunisten in Graz: Woher rührt der Erfolg, und wo will die KPÖ hin? Der Standard. Retrieved June 4, 2022, from https://www.derstandard.at/story/2000129973927/erdrutschsieg-der-kommunisten-in-grazwoher-ruehrt-der-erfolg-und-wo

Newton, J. (2020, August 12). Behind the headlines: Counting COVID-19 deaths. UK Health Security Agency, Coronavirus (COVID-19) data blog. Retrieved June 20, 2022, from https://ukhsa.blog.gov.uk/2020/08/12/behind-the-headlines-counting-covid-19-deaths/

Ni V., & Burger J. (2021, August). All theories on origins of Covid-19 outbreak still, on the table', says WHO. The Guardian. Retrieved August 5, 2022, from https://www.theguardian.com/world/2021/aug/25/us-intelligence-biden-inconclusive-report-covid-origins-wuhan-lab-animals

NPR. (2021, July 22), China has rejected a WHO plan for further investigation into the origins of COVID-19, Coronavirus update. Retrieved May 29, 2022, from https://www.npr.org/sections/coronavirus-live-updates/2021/07/22/1019244601/china-who-coronavirus-lab-leak-theory

OECD. (2022, January 21). OECD policy responses to coronavirus (COVID-19). First lessons from government evaluations of COVID-19 responses: A synthesis. Retrieved May 29, 2022, from https://www.oecd.org/coronavirus/policy-responses/first-lessons-from-government-evaluations-of-covid-19-responses-a-synthesis-483507d6/

ORF. (2022a, January 30). Waidhofen: Mit freundlichen Grüßen zieht in Gemeinderat ein. Retrieved June 22, 2022, from https://noe.orf.at/stories/3140962/

ORF. (2022b, February 23). In aller Munde: Der CT Wert und das Familiencluster. Retrieved June 5, 2022, from https://orf.at/stories/3247548/

ORF. (2022c, June 24). Aus für Impfpflicht: Opposition ortet Regierungsscheitern. Retrieved June 24, 2022, from https://orf.at/stories/3272713/

ORF Science. (2010, January 1). Influenza Epidemie: Bis zu 6000 Tote in Österreich. Retrieved May 31, 2022, from https://sciencev1.orf.at/science/news/96097

ÖRH/Rechnungshof Österreich. (2022a, June 3). COVID-19-Pandemie: Herausforderungen des Krisenmanagements bislang ungelöst. Retrieved June 6, 2022, from https://www.rechnungshof. gv.at/rh/home/news/news/aktuelles/COVID-19-Pandemie-_Herausforderungen_ungeloest.html

ÖRH/Rechnungshof Österreich. (2022b, April 8). Covid-19 HIltsmassnahmen: Rechnungshofe veröffentlicht aktualisierte Daten auf Bundes- und Landesebene. Presseinformation. Retrieved June 6, 2022, from https://www.rechnungshof.gv.at/rh/home/home_1/fragen-medien/ Presseinfo_8.4._COVID_Datenaktualisierung_BF.pdf

Palmai, J. (2021, July 11). Risikofaktor Ernährung: "Erstaunlich, dass der Zusammenhang in der Pandemie kaum thematisiert wurde". Der Standard. Retrieved June 4, 2022, from https://www. derstandard.de/story/2000127724434/risikofaktor-ernaehrung-erstaunlich-dass-der-zusammenhang-in-der-pandemie-kaum

Paredes, C. F. (2022, January 1). Transmissibility of SARS-CoV-2 among fully vaccinated individuals. The Lancet, Infectious Diseases. Retrieved June 20, 2022, from https://www. thelancet.com/journals/laninf/article/PIIS1473-3099(21)00768-4/fulltext

Pezenik, S. (2022, February 3). Scientists demand new investigation of covid-19 origins ahead of Beijing Olympics. ABC News. Retrieved May 29, 2022, from https://abcnews.go.com/ International/scientists-demand-investigation-covid-19-origins-ahead-beijing/story?id=826503 83

Prairy Naturals. (2017, December). The PH connections – Colds & Flu. Nature's Fare Markets. Retrieved August 6, 2022, from https://www.naturesfare.com/health/ph-connection-colds-flu/

Raynolde, P. (2021, August 21). The world needs a proper investigation into how Covid-19 started. The Economist. Retrieved May 29, 2022, from https://www.economist.com/interna tional/2021/08/21/the-world-needs-a-proper-investigation-into-how-covid-19-started

Regan, H. (2022, June). CNN, Scientists are trying to understand how the COVID pandemic began. Here's what they know. 7NEWS. Retrieved August 7, 2022, from https://7news.com.au/news/ coronavirus/scientists-are-trying-to-understand-how-the-covid-pandemic-began-heres-what-they-know-c-7121200

Republic of Kenya/Ministry of Health. (2020, March). Guidance for Nutrition management of COVID-19 for health Workers in COVID-19 Treatment and Isolation Centers. Retrieved June 11, 2022, from https://www.health.go.ke/wp-content/uploads/2020/03/FINAL-GUIDANCE-FOR-COVID-19-NUTRITION-MANAGEMENT.pdf

Riedel, M. (2020, April 9). Sweden goes it alone: The EU#s coronavirus exception. European Council of Foreign Relations. Retrieved June 7, 2022, from https://ecfr.eu/article/commentary_ sweden_goes_it_alone_the_eus_coronavirus_exception/

RT. (2021, December 22). Entlassener Professor Sönnichsen geht wohl gegen Kündigung der MedUni Wien vor. Retrieved June 22, 2022, from https://de.rt.com/oesterreich/128835-entlassener-professor-sonnichsen-geht-wohl/

Sager, F., & Mavrot, C. (2020, October). Switzerland's COVID-19 policy response: Consociational crisis management and Neo-corporatist reopening. European Policy Analysis. Retrieved June 14, 2022, from https://www.researchgate.net/publication/345349491_Switzerland's_COVID-19_policy_response_Consociational_crisis_management_and_neo-corporatist_reopening.

Salomon, L. (2021, July 3). Laut israelischen und europäischen Experten könnten Anti-Covid-Impfstoffe die Quelle für Varianten sein. Qactus. Retrieved June 11, 2022, from https://qactus. fr/2021/07/05/q-scoop-les-vaccins-anti-covid-19-pourraient-etre-a-lorigine-des-variants-selon-des-experts-israeliens-et-europeens/

Schmidt-Vierthaler, R. (2022, April 21). Covid-Tote in der "ZIB2": Ich verstehe es auch nicht wirklich. Die Presse, TV Notiz. Retrieved June 5, 2022, from https://www.diepresse.com/612 8687/covid-tote-in-der-zib-2-ich-verstehe-es-auch-nicht-wirklich

Schwallenberg, G.K. (2012, October). The Alkaline diet: Is there evidence that an Alkaline Ph Diet benefits health? *Journal of environmental Public Health, Natural Library of Medicine, National*

Center of Biotechnology Information. Retrieved August 7, 2022, from https://www.ncbi.nlm. nih.gov/pmc/articles/PMC3195546/

Seneff, S., & Nigh, G. (2021, May 10). Worse than disease? Reviewing some possible unintended consequences of the mRNA vaccines against COVID-19. *International Journal of Vaccine Theory, Practice, and Research.* Retrieved June 11, 2022, from https://dpbh.nv.gov/ uploadedFiles/dpbhnvgov/content/Boards/BOH/Meetings/2021/SENEFF~1.PDF

Setti, L., Passarini, F., De Gennaro, G., Barbieri, P., Perrone, M. G., Borelli, M., Palmissani, J., Di Gilio, A., Piscitelli, P., & Miani, A. (2020, April). Airborne transmission route of COVID-19: Why 2 meters/6 feet of interpersonal distance could not be enough. *International Journal of Environmental Research and Public Health..* Retrieved June 20, 2022, from https://www.ncbi. nlm.nih.gov/pmc/articles/PMC7215485/

Sprenger, M. (2021a, October 21). Die Corona Pandemie aus der Public Health Perspective, interdisziplinäre Ringvorlesung, Universität Wien. Retrieved June 15, 2022, from https:// www.youtube.com/watch?v=TXt7MAS9HzY

Sprenger, M. (2021b, December). Stellungnahme zum Impflichtsgesetz. Retrieved June 17, 2022, from https://www.parlament.gv.at/PtWeb/api/s3serv/file/96066910-5fa4-499a-98a3-be1 9bccacb7b

Stacher, P. (2021, September 28), OÖ-Wahl: Mit freundlichen Grüßen in 27 Gemeinderäten vertreten. Kurier. Retrieved June 18, 2022, from https://kurier.at/chronik/oberoesterreich-wahl-2021/ooe-wahl-mfg-in-27-gemeinderaeten-vertreten/401750259

Statista. (2022a). Number of deaths due to the Coronavirus (COVID-19) in 2022 in Switzerland by age group. Retrieved June 20, 2022, from https://www.statista.com/statistics/1110092/ coronavirus-covid-19-deaths-age-group-switzerland/

Statista. (2022b, February 16). Bevölkerung Österreich 2012 bis 2022. Retrieved June 6, 2022, from https://de.statista.com/statistik/daten/studie/19292/umfrage/gesamtbevoelkerung-in-oesterreich/

Statista. (2022c, June 9). Auslastungsgrad der für Corona-Patienten zur Verfügung gestellten Normal- und Intensivbetten in Österreich seit April 2020. Retrieved June 22, 2022, from https://de.statista.com/statistik/daten/studie/1155556/umfrage/auslastungsgrad-von-normal-und-intensivbetten-durch-corona-patienten-in-oesterreich/

Statista. (2022d, June). Number of tests and confirmed illness cases for the coronavirus (COVID-19) in Austria 2022. Retrieved June 22, 2022, from https://www.statista.com/statistics/1109136/ coronavirus-covid-19-tests-and-confirmed-cases-austria/

Statistik Austria. (2022a, May 16). COVID-19 Todesursachen nach Alter. Retrieved June 20, 2022, from https://www.statistik.at/suche?tx_solr%5Bq%5D=covid+19+todesursachen+nach+alter

Statistik Austria. (2022b, May). Todesursachen 2020. Retrieved May 30, 2022, from http://www. statistik.at/web_de/statistiken/menschen_und_gesellschaft/gesundheit/todesursachen/index. html

Swissinfo. (2022a, April 26). Swiss Covid-19 crisis management gets thumbs from experts. Retrieved June 8, 2022, from https://www.swissinfo.ch/eng/swiss-covid-crisis-management-deemed%2D%2Dbasically-good-/47545774

Swissinfo. (2022b, June). COVID-19 Coronavirus: The situation in Switzerland. Retrieved June 20, 2022, from https://www.swissinfo.ch/eng/covid-19_coronavirus%2D%2Dthe-situation-in-switzerland/45592192

Tech2. (2021, April). WHO COVID-19 Origins Report: What are WHO's four theories on the emergence of SARS-CoV-2? Retrieved August 5, 2022, from https://www.firstpost.com/tech/ science/who-covid-19-origins-report-what-are-whos-four-theories-on-the-emergence-of-sars-cov-2-9487721.html

The AP-HP/Universities/Inserm COVID-19 Research Collaboration. (2021, October 27). Predictive usefulness of RT-PCR testing in different patterns of Covid-10 symptomatology: analysis of a French cohort of 12 810 outpatients. The Nature. Retrieved June 15, 2022, from https:// www.nature.com/articles/s41598-021-99991-6#citeas

The Local. (2021, April 5). What are Vienna's coronavirus "gurgle tests"? Retrieved June 22, 2022, from https://www.thelocal.at/

Times of India. (2021, May 18). COVID-19: The role of alkaline foods in recovery and prevention. Retrieved June 11, 2022, from https://timesofindia.indiatimes.com/life-style/food-news/covid-1 9-the-role-of-alkaline-foods-in-recovery-and-prevention/photostory/82729987.cms?picid=82 730068

Varesi, A., Chirumbolo, S. & Ricevuti, G. (2022, March). Oxygen-ozone treatment and COVID-19: Antioxidants targeting endothelia lead the scenery. *International Emergency Medicine, 17*, 593–596. Retrieved June 5, 2022, from https://link.springer.com/article/10.1007/s11739-021-02865-y#citeas.

Vavrik, U. (2020a, April 6). Covid-19 Peak überschritten: Ein neuer Maßnahmenmix ist nun gefragt. Die Presse. Retrieved June 11, 2022, from https://www.diepresse.com/5796398/ covid-19-peak-ueberschritten-ein-neuer-massnahmenmix-ist-nun-gefragt

Vavrik, U. (2020b, April 23). Gastkommnentar Covid-19: Improvise, adapt, overcome - Auf der Suche nach der optimalen Exit-Strategie, in Österreich, in Europa, weltweit. International Institute for Peace Vienna. Retrieved May 29, 2022, from www.iipvienna.com, https://static1. squarespace.com/static/58a2c691b3db2b3c6990193a/t/5ea2b0300793ed73ac20dd60/1 587720241135/Gastkommentar+Covid+16042020+REV1.pdf

Velot, C. (2021, June 9). Ne faisons pas un remed pire que le mal: l'entretien essentiel avec Christian Velot France Soir. Retrieved June 11, 2022, from https://www.francesoir.fr/videos-lentretien-essentiel-videos-ne-pas-manquer/ne-faisons-pas-un-remede-pire-que-le-mal, https:// www.youtube.com/watch?v=UVS4Tm7X1Wo

VKI/Verein für Konsumenteninformation. (2021, December). Faktencheck Medizin: Ozontherapie bei COVID-19. Retrieved June 5, 2022, from https://konsument.at/faktencheck-medizin-ozontherapie-bei-covid-19/6435633

Wadman, M. (2020, September 8). Why Covid-19 is more deadly in people with obesity – even if they are young. *Science*. Retrieved June 4, 2022, from https://www.science.org/content/article/ why-covid-19-more-deadly-people-obesity-even-if-theyre-young

WFOT/World Federation of Ozone-Therapy. (2020, May). Romania fights against COVID-19 using medical ozone. Retrieved June 4, 2022, from https://www.wfoot.org/romania-fights-against-covid-19-using-medical-ozone/

WHO. (2020a, August). Estimating mortality from Covid. Retrieved May 31, 2022, from https:// www.who.int/news-room/commentaries/detail/estimating-mortality-from-covid-19

WHO. (2020b). Food and nutrition tips during self-quarantine. Retrieved June 11, 2022, from https://www.euro.who.int/en/health-topics/health-emergencies/coronavirus-covid-19/publica tions-and-technical-guidance/food-and-nutrition-tips-during-self-quarantine

WHO. (2021, May 5). 14.9 million excess deaths associated with the COVID-19 pandemic in 2020 and 2021. Retrieved June 10, 2022, from https://www.who.int/news/item/05-05-2022-14.9-million-excess-deaths-were-associated-with-the-covid-19-pandemic-in-2020-and-2021

WHO. (2022a). #HealthyAtHome: Healthy Diet. Retrieved June 11, 2022, from https://www.who. int/campaigns/connecting-the-world-to-combat-coronavirus/healthyathome/healthyathome%2 D%2D-healthy-diet#:~:text=While%20no%20foods%20or%20dietary,some%20types%20of %20cancer

WHO. (2022b, April 13). COVID-19 advice for the public: Getting vaccinated. Retrieved May 31, 2022, from https://www.who.int/emergencies/diseases/novel-coronavirus-2019/covid-19-vaccines/advice

Williamson, A., Forman, R., Azzopardi-Muscat, N., Battista, R., Colombo, F., Glassman, A., et al. (2022, May 16). Effective post-pandemic governance must focus on shared challenges. *The Lancet*. Retrieved June 1, 2022, from https://www.thelancet.com/journals/lancet/article/PIIS0140-6736(22)00891-1/fulltext

Worldometer. (2020, May 13). Age, sex, existing conditions of COVID-19 cases and death. Retrieved June 12, 2022, from https://www.worldometers.info/coronavirus/coronavirus-age-sex-demographics/

Worldometer. (2022a, May). Countries. Retrieved May 31, 2022, from https://www.worldometers.info/coronavirus/#countries

Worldometer. (2022b, June). Countries. Retrieved June 14, 2022, from https://www.worldometers.info/coronavirus/#countries

Ursula A. Vavrik holds a master's degree and a PhD (1990) in Economic and Social Sciences; her doctoral thesis was awarded with the Rudolf Sallinger Prize of the Austrian Chamber of Commerce for outstanding research. She pursued an international career having worked for multinational organizations (UN, OECD), EU institutions, academia, the private sector, and NGOs. Areas of specific expertise encompass international and EU policy, particularly in the fields of environment, development, and sustainable development. She served during all three Austrian EU Council Presidencies, 2006 as senior advisor and 2018 also at the Austrian Ministry of Foreign Affairs. As international consultant, she facilitated the Water Sector Review in Uganda, advised the European Environment and Sustainable Development Advisory Councils on EU Budget Reform and the CAP, and acted as EU and OSCE election observer. She held numerous positions as guest professor and lecturer (BOKU/Vienna, Sciences-Po/Paris, ITAM/Mexico, Diplomatic Academy Vienna, etc.), launched the first seminar cycle on Sustainable Development at the Vienna University of Economics and Business Administration, and published in Europe, USA, and Mexico. Since 2009, she leads the NEW WAYS Center for Sustainable Development. In 2018, she initiated the European Award "Excellence in the Implementation of the UN SDGs" and awarded 50 companies. More recently, the focus of her activities lies in research and conferences on planetary boundaries concepts, sustainable and organic agriculture, and COVID-19 pandemic management.

Corporate Social Responsibility Initiatives and Programs in the Health System of Greece due to the Pandemic of COVID-19

Φ. Ioannis Panagiotopoulos

1 The Pandemic Era in Greece

Humanity is confronting an unprecedented phenomenon with the severe respiratory syndrome coronavirus 2 (SARS-CoV-2 or COVID-19), which has been a real global pandemic that has shocked the entire world with its intensity and duration having caused a tremendous death toll that reminds human loses due to war. The two previous coronaviruses of twenty-first century which had as main effect respiratory disease outbreaks and finally death, SARS in China in 2002 and MERS in KSA in 2012, had not these global, massive, and time-persistent characteristics that COVID-19 has. Both SARS and MERS had higher fatality rates than COVID-19, but the last has much easier dispersion that leads to greater number of cases and deaths. World Health Organization (WHO) declared COVID-19 as a pandemic on March 11, 2020 (Hewings-Martin, 2020). The pandemic has caused more than 500 M cases and more than 6 M deaths globally up to May 2022. Greece has approximately 3.5 M cases and 30 K deaths in the same period (John Hopkins University, 2022).

Greece started the organization of its defense line against the pandemic when the first COVID-19 case was detected on February 26, 2020 (Sypsa et al., 2021). The dramatic situation in neighboring Italy was a kind of first alert for Greece. The knowledge that the National Health System (NHS) had weak points, especially after more than 10 years of deep financial depression (2009–2020) and underfunding from the government in combination with the limited number of intensive care units (ICUs), left no ground for complacency. The element of older population of Greece was one more negative factor, which intensified the concerns since COVID-19 causes more health issues to elderly people and people with comorbidity.

Most of the countries have applied several types of social distancing and lockdown to their societies and markets, respectively, aimed at limiting person-to-person

Φ. I. Panagiotopoulos (✉)
Department of Business Administration, University of the Aegean, Chios, Greece

© The Author(s), under exclusive license to Springer Nature Switzerland AG 2023
S. O. Idowu et al. (eds.), *Corporate Social Responsibility in the Health Sector*, CSR, Sustainability, Ethics & Governance, https://doi.org/10.1007/978-3-031-23261-9_3

contact. The first general lockdown across Greece started on March 23, 2020, limiting any unnecessary exit of citizens out of their homes. Other relevant actions like remote working, social distancing, travel-related limitations, and case-based interventions followed (Sypsa et al., 2021).

Greece had in total of 540 intensive care units(ICUs) in the beginning of 2020. For Greece, the proper number of ICUs is 3500 based on the population and the ideal health coverage. However, taking into consideration the European average, a number of 1250 ICUs suits to Greece based on its population and the european standards. Having only 540 ICUs, Greece was put in the bottom of European statistics with less than 6 ICUs per 100,000 citizens in comparison with the European average that indicates 11.5 ICUs per 100,000 citizens. This lag highlighted that Greece should have as first target to double its number of ICUs achieving to cover almost one out of three of the ideal number of ICUs for its citizens (Rhodes et al., 2012). Greece has managed to increase vertically the number of ICUs in a year since March 2020 up to March 2021 to 951 ICUs. This 76% increase within a year has been mainly the result of donations from private companies as part of their critical CSR programs for the confrontation of the pandemic.

Although coronavirus is a health enemy, it strikes equally strongly the economy since the applied precautionary measures affect economy with multiple ways. Such side effects of coronavirus are the decrease of production in primary sector like agriculture and in the secondary sector like car industry since social distancing affects the occupation of workforce. The same applies for the tertiary sector with the example of the crisis in transportation, supply chain, tourism, and aviation due to the same reasons. The pandemic of COVID-19 has symmetrical characteristics because it affects adversely and evenly: (a) the demand for goods and services and (b) the demand and the offer of them (Panagiotopoulos, 2021).

Corporations must deal with the threat of coronavirus and the consequent recession by adjusting their activities under the new pandemic conditions keeping their personnel, customers, suppliers, and rest of the stakeholders safe, protecting their profitability on the same time. This complex problem consists of an urgent call for healthcare organizations, to express their social sensitivity and implement critical CSR programs for their employees, their patients and customers, and society in general (Panagiotopoulos, 2021).

2   Methodology

The current research is based on an unobtrusive research, a qualitative and comparative content analysis (Babbie, 2011) in the field of CSR programs implemented during the period of the pandemic by healthcare companies in Greece. The unit of analysis is the CSR policy of a healthcare firm, and the units of observations are all these firms that constitute the health system of Greece and have expressed their corporate social responsibility (CSR) amid the pandemic. The research is based in sources in Greek and English language, corporate websites, governmental documents, business press and public newspapers, and scientific articles.

3   What Is a Healthcare Organization?

The health system of Greece is constituted by any healthcare institution or organization that is activated in the country. Thus a healthcare organization embraces any facility that is designed to provide in-patient or out-patient treatment, diagnostic or therapeutic interventions, nursing, rehabilitative, palliative, convalescent, preventive, or other health services for public and private use by medical practitioners (Republic of Kenya, 2019). Furthermore, healthcare organizations could be recognized as those that provide health-related services or relevant products, too. Concluding, as healthcare institutions can be considered indicatively but not exclusively, public, and private hospitals, pharmaceutical companies, clinics, paramedic firms, industries that manufacture high-tech medical equipment, rehabilitation centers, biotechnology firms, etc. The healthcare sector encompasses several professions like doctors and professional nurses, several health-allied professions like medical engineers, chemists, biologists, and managers of the healthcare organizations, and nonmedical staff like cleaners and drivers for ambulances.

Health systems' task is the provision of healthcare services meaning services that upgrade, retain, or restore the health of people. This task includes the usual care provided by public and private hospitals and clinics, family doctors, other activities like the prevention and control of communicable disease, health promotion, health workforce planning, and improving the social, economic, or environmental conditions in which people live. Health systems should take care of the equal approach to these services by all citizens that these services are responsive to individual needs and vulnerabilities and that they do not impact a high cost on citizens and their families (World Health Organization Europe, n.d.-b). *This scope of work is very close to the concept of sustainability and corporate social responsibility (CSR) and implies that healthcare organizations have obtained familiarization with these terms and possibly could realize easier the idea of corporate citizenship than other companies.*

4   Critical CSR and Sustainability

Corporate social responsibility constitutes the concept that urges firms to take into consideration social and environmental parameters in their corporate decision processes (European Commission, 2001). Other terms that are used to describe exactly the same concept are corporate citizenship, social responsibility, social responsiveness, and corporate responsibility (Panagiotopoulos, 2020). All these terms represent the continuous commitment of companies for moral behavior and positive contribution to the quality of life of their employees and their families, the local communities, and society in general. Companies have started to abandon the sole pursuit of profit in favor of a triple bottom line that consists of financial profit along with social and environmental profit (Kaptein & Van Tulder, 2003). This balance among economy, society, and environment puts a spotlight on the concept of sustainability from where CSR is sourced. Sustainability secures that not only the company and its shareholders but also all the stakeholders could develop themselves fruitfully in parallel and assures the proper conditions for the further development and well-being of future generations (Strange & Bayley, 2008). There is internationally a common understanding by people, politicians and scientists that the financial development cannot automatically resolves all the issues that burden society has led in the recognition of three main pillars of sustainability: economy, society, environment which could be served equally in order current and future generations to attain and maintain a high level of living. As private companies (including healthcare companies and especially pharmaceutical companies) gain more and more financial growth, they concentrate power and could compete with the power of medium- and small-sized countries. Public society expects that these companies will realize their corporate citizenship and their global potential to assist in environmental and societal issues.

The idea of CSR could concern any kind of company, and thus, healthcare organizations could also develop CSR strategies. There is a very interesting element that maybe no other business sector has regarding the special attention that healthcare organizations pay for human life since this is exactly their core business. CSR is based on the corporate understanding that each company should operate as a responsible member of its market and within the local community where it is active. It refers to programs that support or at least do not harm societal and environmental interests. Thus, CSR is being built on the strict compliance with the legislation and is unfolded beyond this level. Therefore, CSR keeps a strong voluntary nature since it starts to be creative and dynamic when goes beyond the requirements of legislation (Panagiotopoulos, 2021).

A crucial element in the theoretical complex of CRS is the concept of stakeholders. Stakeholders are considered those groups without whose support and consent a company cannot survive. From another point of view, stakeholder of a company is any group or individual who can affect or is affected by the operation of a company. This complete consideration of stakeholders indicates that the usual groups of them are managers, employees, customers, investors, suppliers, shareholders, government, society in general, and local community. For the sake of clarity and based on sustainability principles, we could consider stakeholders for a firm as the dimensions of natural environment and future, too. It is expected that an individual could be a multiple stakeholder for a firm. For example, an employee of a health administration could be in parallel a member of the local community and a potential patient (Crowther & Aras, 2008: 25–30).

In general, a company develops either strategic or tactical CSR programs (Bansal et al., 2015). Indicatively strategic CSR implies (a) medium to long term regarding time, (b) big human resource commitment either regarding employees needed to plan and implement a CSR activity or regarding the number of employees involved in general, (c) medium to big size of investment, (d) clear correlation with core activities, and (e) clear impact on organization structure or everyday activities of firm. Proportionally, tactical CSR implies (a) short to medium term regarding time, (b) small to medium human resource commitment either regarding employees needed to plan and implement a CSR activity or regarding the number of employees involved in general, (c) small to medium size of investment, (d) no correlation with core activities, and (e) no impact on organization structure or everyday activities of firm (Panagiotopoulos, 2021).

It is noticed that amid the pandemic, companies have managed to develop a hybrid type of CSR that shares characteristics from both already established and well-recognized types. The CSR reply to the current pandemic looks tactical since it is unexpected, flexible, short term, and fast-track. At the same time, it looks strategic since it is robust regarding its resources and significance, and powerful concerning its possibilities for building trust and social capital (Sacconi & Degli Antoni, 2011). Thus, during this pandemic that is considered simultaneously a global recession, a new CSR approach has been raised combining characteristics from both tactical and strategic CSR. A window of threat and chance is come up, and therefore, urgent CSR programs are designed under fast-track conditions which spring from a solid CSR strategy. This hybrid CSR activity maintains external characteristics which are tactical, while its internal structural elements are strategic. This global brand-new type of CSR is called critical CSR, and it is developed as a plan against the pandemic and the implied financial recession. Critical CSR inevitably has strong ontological character since the threat to public health and the market consequently threats human resources and customers. Without these two players, customers and employees, the financial game cannot be maintained (Panagiotopoulos, 2021).

5   Health Value as a Global Sustainable Goal

In accordance with World Health Organization (WHO), health is defined as the situation within which an individual enjoys physical, psychic, and social well-being and not just the absence of sickness of infirmity. It demands the application of various interventions like prevention, diagnosis, care, and rehabilitation (Gulluscio, 2013). The concept of health value has been recognized as a Sustainable Goal by United Nations (UN). In 2015, all the countries in the UN adopted the 2030 Agenda for Sustainable Development. It sets out 17 goals, which include 169 targets. These wide-ranging and ambitious goals interconnect. Goal 3 is to ensure healthy lives and promote well-being for all at all ages (World Health Organization Europe, n.d.-c). Furthermore, UN promotes the concept of universal health coverage (UHC). An effective health system governance should ensure UHC for the whole population in each country. In the framework of the Sustainable Development Goals agenda, WHO works to support countries to exercise effective health system governance, focused on strengthening the capacity of governments to develop and implement strategies toward achieving UHC by 2030 (World Health Organization Europe, n.d.-a).

Healthcare value is described as the coverage of expressed and concealed needs of each individual and of society in general related to health issues. The satisfaction of health needs should be achieved in a smooth manner without waste of resources, lack of planning, and budgeting. On the same time, the satisfaction of health needs would support the further development of medical knowledge, the improvement of offered services, and a mutual positive experience for health professionals, patients, and their relatives (Gulluscio, 2013). The social relevance of this activity is very wide because it safeguards a basic need of individuals and of the whole administered community: health. Depending on the national context, the tasks connected to the supply of these services can be attributed to public administrations, private healthcare organizations, or both. Business administration theory has initially discriminated the ideas of CSR related to social responsibility of for-profit organizations, and public sector social responsibility (PSSR), concerning public administrations. The latter kind of organization is an ordinary example of nonprofit organization. There have been developed recently broader definitions of CSR that embrace both private organizations (profit and nonprofit) and public administrations. In that case, social responsibility considers any possible relation of all organizations with all stakeholders of reference (citizens, workers, customers, community in general, etc.). Finally, a total consideration of CSR supports the idea that all organizations whether public or private, for-profit, or not-for-profit set the same common goal: value production and its fair distribution and delivery among the various stakeholders. The concept of value of healthcare includes (Gulluscio, 2013) the following:

- Quality of care
- Costs
- Hotel amenities
- Communication and exchange of documents, booking of examinations
- Medical explanation
- Psychological support for the patient and the family

Regarding healthcare administrations either public or private, profit or nonprofit and analyzing them on the axis of CSR, it is apparent that the pursuit of profit, the creation of value only for management or equity holders or funders, or even the sole satisfaction of the patients as the principal stakeholders, as separate elements cannot operate toward the sustainability of the organization. The value of health care should be distributed equally to all stakeholders, and this is in line with CSR and sustainability principles. Healthcare industry should be approached with a multidimensional concept. Derivative idea of the equal distribution of healthcare value is the proper allocation of funds for clinical care to different parts of a region or country in a manner that considers special needs and local characteristics and maximizes value for the whole population (Gray, 2011).

Another important concept is the allocation of funds to different patient groups by decision-making that is equitable and maximizes value for the whole population.

Thus, this implies that even during the pandemic healthcare administrations should not forget their healing role for all diseases, and thus, they cannot put aside other critical groups of patients that need necessarily their services like surgeries and chemotherapies. Conclusively, healthcare institutions should prioritize their tasks and recognize the vulnerable group of patients including but not exclusively COVID-19 patients and they should target business continuity amid the pandemic (Gray, 2011).

It is not unusual in CSR that discrete stakeholders' group have different interests or at least conflicted expectations concerning the operation of an organization or the results of a CSR program. In the specific case of hospitals, patients wish to receive the best possible health assistance with the most hi-tech equipment and enjoy the maximum available comfort. On the contrary, the management of a hospital expects to provide the highest possible medical care taking into consideration that the available resources are limited and not all patients' expectations can be fulfilled (Gray, 2011).

The main stakeholders of healthcare organizations could be as follows:

- Citizens
- Health and nonhealthy personnel
- Subjects operating according to agreements with the health organization
- Governmental bodies and authorities that supervise the NHS
- Suppliers
- Scientific institutions like universities and research centers
- Trade unions
- Professional associations
- Volunteers' associations
- Media

However, some stakeholders are more directly interdepended to the healthcare organizations, and therefore, they are recognized as privileged stakeholders:

- Citizens both as patients and possible funders of a health organization
- Governance bodies that supervise the operation of healthcare organizations like Ministry of Health
- Human resources of a health administration

The development of the concepts of governance and leadership in the management of healthcare organizations could empower it delivering a more efficient legal frame and regulation system, coalition building, cooperation between public and private sectors, tighten bonds between research and development (R&D) and clinical practice, attention to system design, and accountability. The three main types of stakeholders that affect more the formation and operation of a health system are (World Health Organization Europe, n.d.-a) as follows:

- The government (public institutions in national and regional levels)
- The health organizations either profit or nonprofit, public, or private, medical, paramedical, and nonclinical institutions, trade unions and other associations of health workers, and networks of care of services
- Population with the form of patients' or citizens' representatives, clubs for protecting vulnerable groups of people who are potentially health service users

Healthcare value concerns all health administrations and includes all possible stakeholders even though some of them could be recognized as more important among others. Especially, amid the pandemic, the interrelation of planned CSR programs, CSR activities implemented, and consequent outputs provide plenty of information for the allocation of healthcare value among the stakeholders that allow the measurement and record of critical CSR footprint of health organizations.

6 Critical CSR Programs in Greece during the Pandemic

Healthcare companies must compete inside a complicated frame with demanding regulations, needs for employees of high expertise, increased cost for hi-tech medical equipment, strict quality requirements, and a hefty relationship with the community (Hossain et al., 2019). Generally, CSR policies of healthcare organizations could include medicine supply facilities, nursing facilities, food supply facilities, ambulance service, medical assistance, and diagnostic medical examinations (Rahman et al., 2010). Greek healthcare institutions have supported through their critical CSR programs Greek State in flattening the curve of COVID-19. This support is expressed with multiple ways. Greek healthcare organizations have donated money to NHS, medicines, antiseptics, disinfectants, protective equipment, and medical appliances for ICUs and even completed brand-new ICUs. They have tried to keep their organizations running successfully despite the pandemic and continue providing to anyone their services and products as they used to do prior to the pandemic. At the same time, they protect their employees as much as possible with several actions like free COVID-19 tests, provision of COVID-19 medical, diagnostic, and protective equipment, and work-for-home programs when this is possible. They continue working at full steam so that they can supply patients, consumers, doctors, and hospitals with the necessary equipment, medicines, and consumables for COVID-19 and other diseases. Business continuity should not be taken as a fact under the unprecedented conditions of a pandemic and a global recession.

A representative grouping of critical CSR actions of companies that constitute the NHS of Greece is presented in Table 1.

The abovementioned critical CSR activities designed by healthcare organizations including public and private hospitals, clinics, and pharmaceutical companies aim for population's health. This wide variety of critical CSR programs could be categorized as follows (Droppert & Bennett, 2015): (a) donations, (b) health system

Table 1 Brief recording of critical CSR programs amid the pandemic by Greek healthcare companies

Organization	CSR action	Public/private
Public hospitals	Cooperation with private doctors with specializations related to the pandemic like pneumonologists to increase the capacity to deal with COVID and non-COVID cases daily	Public and private
Public hospitals	Medical assistance at home for COVID patients	Public
Psychological Clinic of Athens University Medical School	Operation of telephone center in national level for COVID and non-COVID patients and their families in cooperation with nongovernmental organizations and philanthropic institutions	Public
General Hospital of Volos	Cooperation with University of Thessaly for a patent of reusable protection shields for COVID-19 in cooperation with private sector (TED 3D) and funding from private company (Energean)	Public and private
IASO Hospital	• Provision of beds for non-COVID patients and consequent release of beds for COVID patients in public hospitals • Provision of medical equipment for ICU in public hospitals	Private
Hellenic Healthcare Group (Hygeia Metropolitan)	Provision of beds for COVID cases in its premises	Private
Henry Dunant Hospital Center	Provision of ICUs and simple beds for the hospitalization of COVID cases in isolated areas in its premises	Private
Euroclinic	• Provision of beds in their premises for the hospitalization of non-COVID cases and consequent release of beds for COVID patients in public hospitals • Free access to emergency department for non-COVID cases	Private
Athens Medical Group	Provision of a whole hospital unit for COVID cases with more than 200 health professionals	Private
Uni-PHARMA SA (OFET Pharmaceutical Group)	• Production of antiseptics in cooperation with government for public hospitals • Donation of medicine Unikinon with the drastic substance chloroquine to public hospitals • Organization and funding of medical clinic research about chloroquine with public hospitals and universities	Private
ELPEN Pharmaceutical Company	• Donation of medical consumables, PPE, medical equipment to public sector • Funding of medical clinic research GRECCO-19	Private
Vianex Pharmaceutical Company	• Donation of oxygen generation for COVID and non-COVID patients to the public hospital of Kos island • Donation for the full refurbishment of the unique pharmacy in Kastellorizo Island • Continuation of already designed CSR program in parallel with COVID critical CSR actions	Private

(continued)

Table 1 (continued)

Organization	CSR action	Public/ private
Euromedica Private Hospitals	Provision of simple beds, ICUs, and beds in artificial kidney unit for non-COVID patients in Makedonia (north Greece where less ICUs are allocated in comparison with the capital in central Greece)	Private
Panhellenic Association of Pharmaceutical Industries	• Donation of medicines, antiseptics, beds for ICUs, medical equipment, PPE, respirators, monitors, etc., to public hospitals, Hellenic Police, and other organizations of public sector • Maintain stock for critical medicines for Greek patients for 4 months • Increase in shifts—24-h production for medicines, pharmaceutical and para-pharmaceutical products	Private

strengthening, and (c) investments in R&D in cooperation with the private sector. The concerned organizations put measures in place to protect their product supply, thereby ensuring that their customers and patients will continue having access to any essential medication they need. They prioritize their product range and at first secure the needed resources to create products related either to COVID-19 or other serious illnesses and characterize other products like cosmetics as of less priority. In general, healthcare organizations have oriented their urgent CSR programs toward (a) the health of their employees, (b) the security of business continuity for their customers and patients, (c) and the welfare of society through donations of money, healthcare services, and products mainly medical equipment, medicines, and personal protective equipment (PPE) to confront COVID-19. Their critical CSR programs target both their internal environment (employees) and their external environment (suppliers, customers, public health).

7 Incentives for Critical CSR Policies in the Healthcare Sector

The application of a CSR policy theoretically could reciprocate various tangible and intangible benefits back to the firm. Thus, there are additional reasons for the design and implementation of a CSR strategy further to ethical motives. These profits could be of mid- or long term and possibly difficult to be measured at least in short term. However, careful consideration of them is of high importance for the better understanding of CSR philosophy and how CSR enhances the sustainability of a firm. The rational consideration of pros and cons would offer the proper reply to the Operational Ethical Dilemma that in brief condenses all these questions raised regarding the usefulness, the advantages and disadvantages, and the sustainability of a firm's

Table 2 Groups of incentives behind CSR activities of healthcare organizations

Type	Incentives
Reputational advantage	• Perception of organization by patients and their relatives • Perception of organization by potential investors • Relationship and trust building with potential partners and stakeholders • Employee recruitment, satisfaction, engagement, and innovation • Patients' satisfaction • Ranking on stock market indices • Remedying of public relations and repair of compromised reputation
Competitive advantage	• Entering new markets • Expanding consumer base • Intelligence gathering on new markets • Anticipated long-term financial gain • Increased cost-effectiveness of interventions and programs in developing countries • Improved efficiencies • Opportunity for innovations • Special access to local government officials and decision-making • Customer and employee loyalty
Philanthropy and health impact	• Ethical responsibility as a healthcare company • Improved population health impact • Increasing patient access to necessary medications/health services • Greater patient involvement and compliance with therapeutical process

CSR strategy (Panagiotopoulos, 2022). Especially within the current framework of the pandemic, healthcare companies are seeking ways to respond, react, and interact with their stakeholders and assist this multifactor complex of their organization to stay protected.

Several potential advantages could be sourced from the critical CSR strategy of a healthcare organization, and thus, there are various motives that are presented in Table 2 (Coţiu et al., 2014; Droppert & Bennett, 2015; Hossain et al., 2019). They are grouped into three main categories: (a) incentives that strengthen the reputation of the organization, (b) incentives that heighten and differentiate the organization among its competitors, and (c) incentives that are interwoven with the benevolent dimension of CSR and business ethics.

Possibly, the biggest inception for a CSR policy and at sure a concern of the highest priority for the management team of a healthcare organization is patient satisfaction. As a recognition of that, patient satisfaction is the most popular examined parameter in relevant surveys (York & McCarthy, 2011). Patient satisfaction is closely related to various benefits like greater patient involvement and compliance with therapeutical process, high staff morale, positive word of mouth, or a minimization in patients' litigation intent. Participants highly appreciate positive interactions with the medical staff, expect a correct diagnosis and treatment, and seem to be less affected by administrative procedures or accommodation facilities (Coţiu et al.,

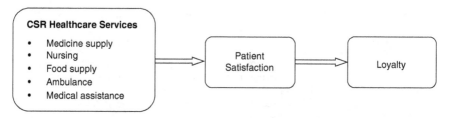

Fig. 1 Example of positive outcomes from a CSR healthcare service program (Hossain et al., 2019)

2014). Thus, there is a statistically significant link between satisfaction and loyalty (Kessler & Mylod, 2011). Patient satisfaction could cause customer loyalty that could be translated into stable profits for an organization (Newsome & Wright, 1999). Healthcare service providers in advanced economies consider patient satisfaction as a critical element for their long-term viability and profitability (Deber, 1994; Thompson et al., 2012). Pascoe (1983) supports that *"patient satisfaction may be one of the desired outcomes of healthcare information and patient satisfaction should be as indispensable to assessments of quality as to the design and management of healthcare systems."* Patient satisfaction and loyalty are based on total patients' attitude toward a service provider (Elstad & Eide, 2009) and an emotional counteraction to the possible discrepancy between patient expectations and what they really receive (Hankins et al., 2007). The creation of patient satisfaction is gradual, and it is aggregated each time there is interaction between the healthcare organization and the patient. Both patient satisfaction and loyalty could be measured as the consecutive assessment of the service or the product offered by a specific CSR healthcare program, and its interaction is positive (Hossain et al., 2019). CSR healthcare services cultivate patient satisfaction and consequently the loyalty of patients as presented in Fig. 1. The higher the quality level of CSR healthcare services offered to the patients, the more they are satisfied with their interaction with the organization and the biggest is the loyalty that is developed from patients' side (Hossain et al., 2019).

8 Criticism about the Response to Coronavirus in Greece

Greek researchers have tried to investigate the response of NHS of Greece during the pandemic regarding its efficiency in examining parameters like geographic information, number of cases, number of deaths, and available ICUs. A research by Lytras and Tsiodras (2021b) has provided useful conclusions about any possible regional disparities regarding the healthcare quality on COVID-19 patients in Greece and any other misfunctions in the NHS that should be taken into consideration by all stakeholders. Especially in Greece which has faced an economic recession since 2009 after the global crisis of 2007–2009, up to the outbreak of the pandemic at the beginning of 2020, the weaknesses of the Greek NHS are in the spotlight from the very beginning of the pandemic.

As the pandemic is developing in Greece and globally, the healthcare services have increased and the same applies to the number of ICUs. However, it seems that the quality of care has not followed the same surge course. The researchers have found that in-hospital mortality of severely ill COVID-19 patients has been negatively affected by high patient load even without exceeding capacity of ICUs. The tremendous increase in the number of ICUs in Greece due to the big endeavors of private firms through their critical CSR programs has led to almost doubling of ICUs. However, this intervention has taken place without proportional improvements in the care of COVID-19 and non-COVID-19 patients as especially the number of COVID-19 deaths indicates during 2021 and up to the end of May 2022 by government's end. The research detects a compromise in quality even though the capacity of ICUs is not exceeded. The chronic underfunding due to the recession and poor pandemic preparedness did not allow the doubling of ICUs to increase vertically the real performance of NHS to confront the pandemic. The current total number of intubated COVID-19 patients at each point was used as an indicator of healthcare system stress during this research. There was a significant association between mortality and total intubated patients above 400 even though the capacity was not exceeded.

The same research underlines that there are regional disparities regarding the quality of health care in the metropolitan regions of Attica and Thessaloniki, in comparison with the rest of Greece. Being hospitalized outside the capital region of Attica is associated with increased in-hospital mortality. These outcomes indicate that the further enhancement of NHS should pay extra attention on equity and quality of care besides just expanding capacity (Lytras & Tsiodras, 2021b). This could be the result of chronically uneven regional distribution of healthcare resources in the country, with beds, equipment, and trained healthcare personnel to be disproportionally allocated close to the capital. Thus, this rural–urban discrimination has caused a higher fatality ratio in remote and rural areas despite higher exposure to coronavirus in capital.

Another finding of the research by Lytras and Tsiodras (2021b) is that being intubated outside an ICU is strongly associated with mortality. Recent documents from the National Public Health Organization unfortunately reveal that 16,619 deaths out of 25,914 up to March 1, 2022, concern patients that had not been introduced in ICU at all. Furthermore, the average time for intubation outside an ICU independently of the aftermath course of the patient like introduction into ICU, death, or cure is estimated in 4.7 days (Greek National Public Health Organization, 2022). This fact reveals that only 36% of the victims of COVID-19 had the chance to receive the maximum healthcare service that could be provided to them in an ICU by the NHS.

9 Discussion and Conclusions

The abovementioned facts reveal that the number of ICUs is not an absolute factor that sufficiently supports higher level of quality and effectiveness in health care. The creation of more ICUs in a short period of time due to critical CSR initiatives as happened in Greece without more and well-trained medical and paramedical personnel cannot protect patients. The increased needs for healthcare workers due to the pandemic have been faced with staff redeployments, short-term hiring, and requisitioning of private practitioners' services. However, it should be mentioned that an important percentage of medical and nonmedical staff in public hospitals (initially was around 20% and later was limited approximately to 5% in 2022) is put in suspension since the government has established the compulsory vaccination of health workers (Kathimerini, 2021). The Greek government did not manage to replace sufficiently and on time the staff loses due to the mandatory vaccination and relevant work suspension measures, and consequently, it has never achieved to staff the health organizations with the proper number of experienced personnel to face the pandemic although it has hired 15,700 health workers with short contracts including 1940 doctors and 8700 nurses among them (Bouloutza, 2022).

Furthermore, there is an interconnection among the number of health professionals in a specific place and the percentage of coverage of health services and health outcomes: health professionals improve health and save lives (Anyangwe & Mtonga, 2007). Any expected improvements in the result of healthcare processes cannot be disconnected from the impediments in the access to necessary health care. These challenges cannot be overcome without sufficient numbers of health professionals with proper skills and experience to provide these interventions, especially under critical circumstances like a pandemic (Haskins et al., 2017). Additionally, people in rural and remote regions, which are usually underdeveloped financially, have increased health needs. The inexistence of policies that provide motivation to skilled health workers to move to such areas instead of metropolitan areas is an explanation for the inequity in the allocation of health workers and consequently medical machinery and healthcare premises (Haskins et al., 2017).

The measures of social distancing and especially in the form of lockdowns have been characterized as controversial because even though they are in favor of public health by controlling the transmission of coronavirus they operate against economy (Ioannidis, 2020). Lockdown was introduced initially as a measure of protection when there were no available vaccines and effective medicines (Chin et al., 2021).

The intention was to prevent the massive introduction of COVID-19 cases in hospitals and ICUs, especially in countries with weak NHS, limited number of ICUs, and intense social inequalities (Lytras & Tsiodras, 2021a). However, the negative side effects of lockdowns are not limited to the field of economy. There are negative effects on people's health like obstacles to access healthcare centers for patients with chronic diseases and long-term treatment programs and mental and physical implications like depression and anxiety as a result of social distancing and lack of social and sports activities. The problematic information and communication between the state and its citizens not only in Greece have exacerbated these symptoms. The absence of kids from school could be considered as another source of health issues for them. Furthermore, the economic regression due to lockdowns increases unemployment and social inequalities, thus burdening public health indirectly (Lytras & Tsiodras, 2021a).

The pandemic scene is multidimensional because it is characterized by polyphony regarding medical and political instructions through mass and social media. The scientific uncertainty, the skepticism for vaccination programs and the predominance of an absolute directive for even compulsory vaccination, the inertia of politicians like in the case of EU where each country tried to deal with COVID-19 competitively with other countries of the same coalition especially during the first year of the pandemic, the expression of conflicted opinions among doctors, politicians, journalists, and even the commercialization of the crisis, have created a cacophony of confused messages, fake news, political games, and suspiciousness (Tsiodras, 2021). Exactly at this point of uncertainty, with the memory of global financial crisis of 2007–2009, companies in general and healthcare organizations especially should stand out with their initiatives for critical CSR programs. They should take care of COVID-19 and non-COVID-19 patients, support people who suffer from COVID-19-related issues, and finally build a relationship of trust with their stakeholders. In parallel, coronavirus has shed light on a) the weaknesses of NHS, b) the value of doctors and nurses and even of nonmedical staff of healthcare organizations, c) the value of cooperation between researchers from medical and nonmedical fields, and d) the importance of cooperation between public and private sectors.

Thus, current and older governmental decisions, the aggregated problems of chronic underfunding, and understaffing of public healthcare services due to the over 10 years of long economic depression have led Greece to the lower positions regarding the confrontation of COVID-19 after 2 and half years of pandemic even though Greece had one of the best performance regarding cases and deaths during the first year of the pandemic (John Hopkins University, 2022; Panagiotopoulos, 2021). It is crucial government to implement long-term investments for the enhancement of NHS in personnel, equipment, and premises for the post-COVID era, assuring equitable access to high quality for all citizens.

References

Anyangwe, S., & Mtonga, C. (2007). Inequities in the global health workforce: The greatest impediment to health in Sub-Saharan Africa. *International Journal of Environmental Research and Public Health, 4*(2), 93–100. https://doi.org/10.3390/ijerph2007040002

Babbie, E. (2011). *Introduction to social research* (5th ed.). Wadsworth.

Bansal, P., Jiang, G. F., & Jung, J. C. (2015). Managing responsibility in tough economic times: Strategic and tactical CSR during the 2008-2009 global recession. *Long Range Planning, 48*(2), 69–79. https://doi.org/10.1016/j.lrp.2014.07.002

Bouloutza, P. (2022, March 1). Time for decisions for the unvaccinated health workers of national health system. *Kathimerini*. https://www.kathimerini.gr/society/561738589/ora-ton-apofaseon-gia-toys-anemvoliastoys-toy-esy/

Chin, V., Ioannidis, J. P. A., Tanner, M. A., & Cripps, S. (2021). Effect estimates of COVID-19 non-pharmaceutical interventions are non-robust and highly model-dependent. *Journal of Clinical Epidemiology, 136*, 96–132. https://doi.org/10.1016/j.jclinepi.2021.03.014

Coțiu, M. A., Crișan, I. M., & Catană, G. A. (2014). Patient satisfaction with healthcare—A focus group exploratory study. In S. Vlad & R. V. Ciupa (Eds.), *International conference on advancements of medicine and health care through technology; 5th–7th June 2014, Cluj-Napoca, Romania* (Vol. 44, pp. 119–124). Springer International Publishing. https://doi.org/10.1007/978-3-319-07653-9_24

Crowther, D., & Aras, G. (2008). *Corporate social responsibility*. Ventus Publishing ApS.

Deber, R. B. (1994). Physicians in health care management: 7. The patient-physician partnership: Changing roles and the desire information. *Canadian Medical Association Journal, 151*, 171.

Droppert, H., & Bennett, S. (2015). Corporate social responsibility in global health: An exploratory study of multinational pharmaceutical firms. *Globalization and Health, 11*(1), 15. https://doi.org/10.1186/s12992-015-0100-5

Elstad, T. A., & Eide, A. H. (2009). User participation in community mental health services: Exploring the experiences of users and professionals. *Scandinavian Journal of Caring Sciences, 23*(4), 674–681. https://doi.org/10.1111/j.1471-6712.2008.00660.x

European Commission. (2001). *Green paper: Promoting a European framework for corporate social responsibility*. Commission of the European Communities. https://ec.europa.eu/commission/presscorner/detail/en/DOC_01_9

Gray, J. A. M. (2011). *How to get better value healthcare* (2nd ed.). Offox Press.

Greek National Public Health Organization. (2022). *Reply to relevant questioning in parliament 6/3/15-11-2021*. Greek National Public Health Organization.

Gulluscio, C. (2013). Value and CSR in public health sector. In S. O. Idowu, N. Capaldi, L. Zu, & A. D. Gupta (Eds.), *Encyclopedia of corporate social responsibility* (pp. 1–8). Springer. https://doi.org/10.1007/978-3-642-28036-8

Hankins, M., Fraser, A., Hodson, A., Hooley, C., & Smith, H. (2007). Measuring patient satisfaction for the quality and outcomes framework. *The Journal of the Royal College of General Practitioners, 57*(542), 737–740.

Haskins, J. L., Phakathi, S. A., Grant, M., & Horwood, C. M. (2017). Factors influencing recruitment and retention of professional nurses, doctors and allied health professionals in rural hospitals in KwaZulu Natal. *Health SA Gesondheid, 22*, 174–183. https://doi.org/10.1016/j.hsag.2016.11.002

Hewings-Martin, Y. (2020, April 10). How do SARS and MERS compare with COVID-19? *Medical News Today*. https://www.medicalnewstoday.com/articles/how-do-sars-and-mers-compare-with-covid-19

Hossain, M. S., Yahya, S. B., & Khan, M. J. (2019). The effect of corporate social responsibility (CSR) health-care services on patients' satisfaction and loyalty – A case of Bangladesh. *Social Responsibility Journal, 16*(2), 145–158. https://doi.org/10.1108/SRJ-01-2018-0016

Ioannidis, J. P. A. (2020). Coronavirus disease 2019: The harms of exaggerated information and non-evidence-based measures. *European Journal of Clinical Investigation, 50*(4), e13222. https://doi.org/10.1111/eci.13222

John Hopkins University. (2022). *COVID-19 Dashboard by the Center for Systems Science and Engineering (CSSE) at Johns Hopkins University (JHU)*. https://coronavirus.jhu.edu/map.html

Kaptein, M., & Van Tulder, R. (2003). Toward effective stakeholder dialogue. *Business and Society Review, 108*(2), 203–224. https://doi.org/10.1111/1467-8594.00161

Kathimerini. (2021, August 26). National health system and unvaccinated health workers in Greece. *Kathimerini*. https://www.kathimerini.gr/society/561476071/esy-21-664-oi-anemvoliastoi-ergazomenoi-ta-pososta-ana-katigoria-grafima/

Kessler, D. P., & Mylod, D. (2011). Does patient satisfaction affect patient loyalty? *International Journal of Health Care Quality Assurance, 24*(4), 266–273. https://doi.org/10.1108/09526861111125570

Lins, K. V., Servaes, H., & Tamayo, A. (2017). Social capital, trust, and firm performance: The value of corporate social responsibility during the financial crisis. *The Journal of the American Finance Association, LXXII*(4), 1785–1824. https://doi.org/10.1111/jofi.12505

Lytras, T., & Tsiodras, S. (2021a). Lockdowns and the COVID-19 pandemic: What is the endgame? *Scandinavian Journal of Public Health, 49*(1), 37–40. https://doi.org/10.1177/1403494820961293

Lytras, T., & Tsiodras, S. (2021b). Total patient load, regional disparities and in-hospital mortality of intubated COVID-19 patients in Greece, from September 2020 to May 2021. *Scandinavian Journal of Public Health, 50*(6), 671–675. https://doi.org/10.1177/14034948211059968

Newsome, P., & Wright, G. (1999). A review of patient satisfaction: Concepts of satisfaction. *British Dental Journal, 186*, 161–165.

Panagiotopoulos, F. I. (2020). *Energy management and corporate social responsibility: The case of air-transportation.* University of the Aegean.

Panagiotopoulos, I. Φ. (2021). Novel CSR & novel coronavirus: Corporate social responsibility inside the frame of coronavirus pandemic in Greece. *International Journal of Corporate Social Responsibility, 6*(1), 10. https://doi.org/10.1186/s40991-021-00065-7

Panagiotopoulos, Φ. I. (2022). Operational ethical dilemma. In S. Idowu, R. Schmidpeter, N. Capaldi, L. Zu, M. Del Baldo, & R. Abreu (Eds.), *Encyclopedia of sustainable management* (pp. 1–7). Springer International Publishing. https://doi.org/10.1007/978-3-030-02006-4_1144-1

Pascoe, G. C. (1983). Patient satisfaction in primary health care: A literature review and analysis. *Evaluation and Program Planning, 6*(3/4), 185–210. https://doi.org/10.1016/0149-7189(83)90002-2

Rahman, S., Jahan, S., & McDonald, N. (2010). CSR by Islami Bank in healthcare & stakeholders perception. *Bangladesh Journal of Medical Science, 9*(4), 208–215.

Republic of Kenya. (2019). *Medical practitioners and dentists act chapter 253.* National Council for Law Reporting. http://kmpdc.go.ke/resources/Medical%20Practitioners%20and%20Dentists%20Act%20(2019%20Revision).pdf

Rhodes, A., Ferdinande, P., Flaatten, H., Guidet, B., Metnitz, P. G., & Moreno, R. P. (2012). The variability of critical care bed numbers in Europe. *Intensive Care Medicine, 2012*(38), 1647–1653. https://doi.org/10.1007/s00134-012-2627-8

Sacconi, L., & Degli Antoni, G. (2011). Modeling cognitive social capital and corporate social responsibility (CSR) as preconditions for sustainable networks of relations. *SSRN Electronic Journal.* https://doi.org/10.2139/ssrn.1778102

Strange, T., & Bayley, A. (2008). *Sustainable development: Linking economy, society, environment.* OECD. https://doi.org/10.1787/9789264055742-en

Sypsa, V., Roussos, S., Paraskevis, D., Lytras, T., Tsiodras, S., & Hatzakis, A. (2021). Effects of social distancing measures during the first epidemic wave of severe acute respiratory syndrome infection, Greece. *Emerging Infectious Diseases, 27*(2), 452–462. https://doi.org/10.3201/eid2702.203412

Thompson, J. M., Buscbinder, S. B., & Shanks, N. H. (2012). An overview of healthcare management. In *Introduction to health care management* (pp. 1–16). Jones & Bartlett Learning, LLC. https://samples.jbpub.com/9780763790868/90868_ch01_final_withoutcropmark.pdf

Tsiodras, S. (2021). COVID-19 research and science in the service of public health: The example of Greece. *Nature Immunology, 22*(5), 531–532. https://doi.org/10.1038/s41590-021-00919-z

World Health Organization Europe. (n.d.-a). *Health system governance* [WHO]. https://www.who.int/health-topics/health-systems-governance#tab=tab_1

World Health Organization Europe. (n.d.-b). *Health systems* [WHO Europe]. https://www.euro.who.int/en/health-topics/Health-systems/pages/health-systems

World Health Organization Europe. (n.d.-c). *Sustainable development goals* [International Organization]. Retrieved April 17, 2022, from https://www.euro.who.int/en/health-topics/health-policy/sustainable-development-goals

York, A. S., & McCarthy, K. A. (2011). Patient, staff and physician satisfaction: A new model, instrument and their implications. *International Journal of Health Care Quality Assurance, 24*(2), 178–191. https://doi.org/10.1108/09526861111105121

Φ. Ioannis Panagiotopoulos is a Greek Electrical and Electronic Engineer who holds a Bachelor in Electrical Engineering mainly in Energy from Greece (Technological Educational Institute of Larissa), a Bachelor in Electrical & Electronic Engineering mainly in Telecommunications from England (University of Bradford), a MBA with specialization in Technology and Innovation Management from France (Grenoble Ecole de Management), a Master in Lighting Design & Multimedia from Greece (Hellenic Open University), and a PhD in Business Administration mainly in Energy Management of Air Transportation & CSR from Greece (University of the Aegean). He is working as a Sustainable/ESG Supply Chain/Procurement Category Manager for international companies in construction, power, and oil and gas sector. In parallel, he is an independent researcher in the field of Business Administration mainly about Sustainability, CSR, Energy Management, Sustainable Supply Chain and furthermore in Lighting Design, Air Transportation, and Education. He has more than 30 publications with double-blind peer review in conference proceedings and scientific journals. He is also co-founder of the School Life Museum of Thisvi in Greece.

CSR Manifestations in Health Care Facilities in Poland During the COVID-19 Pandemic

Anna Cierniak-Emerych ⓘ, Ewa Mazur-Wierzbicka ⓘ, Piotr Napora, and Sylwia Szromba

1 Introduction

1.1 CSR: The Key Issues

In addition to producing high-quality products and services with the principle of economic rationality, contemporary challenges faced by entrepreneurs in the second decade of the twenty-first century increasingly include taking responsibility for the company's stakeholders (Sokołowska, 2013, p. 9). For many years, the literature has stressed that when making decisions concerning the enterprise, one should not forget about the necessity of reckoning with social realities, including the legitimacy of listening to what people say (stakeholders), as they are the most important for the enterprise (Crozier, 1993, p. 20). It is emphasized that the results of this listening should be seen especially in the interest of enterprise managers in the concept of management based on the ethical conduct of business, which is an expression of corporate social responsibility (CSR) (Gableta, 2012, p. 20).

In theoretical terms, the concept of corporate social responsibility is not a new issue. Nevertheless, the conscious and purposive application of its principles to specific areas of business operation continues to raise discussions, and sometimes even doubts, mainly among part of the representatives of business executives. It

A. Cierniak-Emerych (✉)
Wrocław University of Economics and Business, Wrocław, Poland
e-mail: anna.cierniak-emerych@ue.wroc.pl

E. Mazur-Wierzbicka
University of Szczecin, Szczecin, Poland

P. Napora · S. Szromba
Clinical Research Center in Wrocław, Wrocław, Poland
e-mail: napora.piotr@cbk.wroc.pl; szromba.sylwia@cbk.wroc.pl

© The Author(s), under exclusive license to Springer Nature Switzerland AG 2023
S. O. Idowu et al. (eds.), *Corporate Social Responsibility in the Health Sector*, CSR, Sustainability, Ethics & Governance, https://doi.org/10.1007/978-3-031-23261-9_4

seems that one of the causes of this situation is the fact that the focus of studies presented in scientific publications is often on the attempts to define this concept.

These attempts have been made both by scientists and international organizations, sectoral institutions, and even the enterprises that implement the policy of socially responsible activities within their structures. The first of these attempts go back as far as the 1950s. However, the development of an unambiguous definition still involves numerous methodological problems. This is mainly due to different interpretations of the meaning of corporate social responsibility that we are interested in here. Consequently, it is relatively difficult to identify unambiguous indicators of social responsibility that would determine under what conditions a given enterprise should be considered socially responsible. This is due, among other things, to the fact that social responsibility is a multi-faceted issue and encompasses many fields that are both different and complementary or even overlapping. As a result, attempts to describe its semantics are being made in various scientific disciplines. The multidimensionality of social responsibility is also justified by the fact that the main factor determining its understanding is the context of the enterprise that declares using CSR.

Examples of defining social responsibility are included in Table 1.

The authors of the definitions of corporate social responsibility enumerated in Table 1 emphasize certain aspects of social responsibility differently. In each of these definitions, one can see statements referring to respecting social requirements, i.e., accepting the presence of various interest groups inside the enterprise (organization) and in its surroundings. This indicates the need to go beyond the existing formal and legal framework in the direction of taking into account the social requirements in the operations of the organization. In many definitions, one can also find references to stakeholder theory and the principle of voluntariness and respect for the environment (Stefańska, 2013, p. 204).

At this point, it is worth stressing that these definitional approaches often assume that organizations have some degree of responsibility to society as a whole. This gives a specific meaning to the role of the enterprise (organization) as a member of society (participant in socioeconomic life), whose decisions affect the entities and people with whom the organization has relations. This includes people, e.g., employees (internal stakeholders), but also people to whom these organizations provide services (external stakeholders). This perception of CSR seems particularly relevant when one considers social responsibility in the context of its manifestations in such complex circumstances as the COVID-19 pandemic. With this perspective, social responsibility refers to the mutual relationship understood as the role of organizations (enterprises) in society, and society's expectations toward these organizations (ISO Advisory Group on Social Responsibility, 2004).

The definition of the concept of social responsibility proposed by the European Commission in 2001 fits well with this perception of CSR. It combines social responsibility with voluntary consideration of social and environmental issues by companies. It should be noted that this responsibility should make it possible to manage relations with the various stakeholder groups that have an impact on the functioning of the company (European Commission, 2001). This concerns both

Table 1 Selected definitions of corporate social responsibility

No.	Author/source	Definition
1	J. McGuire	The idea of social responsibility assumes that enterprises have not only economic and legal duties but also many other obligations toward society, far exceeding the above-mentioned obligations arising from the law or rules of the market economy.
2	K. Davis R. Blomstrom	Social responsibility means that enterprises are not only economically and legally responsible, but they are obliged to take actions that contribute to the protection and multiplication of social welfare.
3	A. B. Carroll	Social responsibility consists of different types of responsibility (economic, legal, ethical, philanthropic), which are affected by social demands to varying degrees
4	J. Backman	Social responsibility usually refers to the goals and motives that a business should take into account as an additional activity in pursuing its economic endeavors. Pursuing these goals and motives is supported by the so-called "umbrella of social responsibility" understood as the design of appropriate programs that contribute to improving the quality of life by, e.g., improving medical care or living conditions of the population.
5	A. Sokołowska	Social responsibility means an economic, legal, ethical, and philanthropic obligation of the company with respect to internal and external social groups (and individuals) and it can be the subject of deliberate, rational, and institutionalized action, which can become a source of competitive advantage.
6	M. Rojek-Nowosielska	Conscious actions taken in a given organization, inspired by the results of social dialogue and directed toward the creation of socially relevant values that meet the expectations of stakeholders

Source: Author's own study based on: (Rojek-Nowosielska, 2017, pp. 76–81; Sokołowska, 2013, pp. 31–32)

external stakeholders (suppliers, customers, representatives of financial and insurance institutions, local governments, etc.), and internal stakeholders (owners, hired managers of various management levels, employees performing executive tasks). At this point, it is also worth noting that in 2011, the European Commission proposed a new definition of CSR, which states that social responsibility is "the responsibility of enterprises for their impact on society" (European Commission, 2011). Therefore, it has to be emphasized that the European Commission, by proposing the new definition of CSR in 2011, strongly emphasized its social character. Social responsibility should therefore be considered both in terms of the interests of individuals as members of a given community, including the community of a given country, region, or world external to CSR. On the other hand, in narrower terms, considered within the organization (the internal nature of CSR), it should be related to, among others, employees. CSR should be expressed in meeting the expectations of employees concerning work, especially concerning appropriate physical and non-physical working conditions.

Working conditions are mostly considered as factors that are present in the environment, connected with the nature of the work and the environment where the work is performed (Pocztowski, 1998, p. 261 et seq.). These conditions include tangible elements, such as workstation and room equipment, lighting, microclimate, noise, and intangible elements, such as working time, social activities, and labor relations, including interpersonal relations. A specific connecting factor between the above-mentioned elements of working conditions is occupational safety and health, understood as the entirety of legal standards and research, organizational and technological resources which are aimed to provide employees with such conditions that they are able to perform their work productively, without unnecessary risk of accidents or occupational diseases and excessive physical and mental load (cf. Koradecka, 2000, p. 21).

The working conditions and the related occupational safety and health in the enterprises are given specific importance in international norms and standards, which provide a specific "guideline" for the implementation of the concept of social responsibility in organizations. Among them, noteworthy are the following documents: PN-ISO 26000:2012 *Guidelines for social responsibility* (in most of the seven areas distinguished in ISO 26000 standard, it is possible to identify activities that can be implemented in OSH management systems; this applies in particular to the core subject "Labor practices": subclause Health and safety at work (Mazur-Wierzbicka, 2021)), occupational safety and health management system compliant with PN-N- 18001 standard, or SAI SA8000:2008 *Social responsibility—Requirements*.

One of the most important challenges taken up today within the social responsibility of organizations should therefore be the concern for the health and life of a person as a member of the community and an employee (cf. Cierniak-Emerych & Mazur-Wierzbicka, 2022, pp. 131–133). This issue takes on particular importance with the onset of the COVID-19 pandemic, which has been ongoing for 2 years, as indicated above. As it seems, this concern taken by organizations can take different forms depending on many factors. Among them, it seems worth pointing out the specificity of the organization (enterprise) in relation to which one can consider the issues discussed here (in this study, these organizations will be public health care institutions operating in Poland, but also depending on the development of CSR, in a particular country).

1.2 Development of CSR in Poland

The origins of the idea of social responsibility can be traced back to the nineteenth century. Referring here primarily to the Polish realities of CSR development, it should be indicated that already in the middle of the nineteenth century such practices were observed in the territory of today's Poland among entrepreneurs, which certainly demonstrate their social involvement as a manifestation of activities consistent with the CSR concept. It is worth pointing out, for example, the figure of the famous Polish capitalist and philanthropist Hipolit Cegielski (1813–1868). Cegielski was a man of versatile abilities, combining the talents of a teacher, philologist, doctor of philosophy, publisher, social activist, and industrialist. He combined knowledge and impeccable manners with humanistic and liberal ideals and a deep sensitivity to material matters and the economic development of Polish lands, which were an important element of organic work supported and developed by him in Poznan (Cierniak-Emerych et al., 2021, pp. 287–310).

Activities undertaken by organizations operating in Poland that could be associated with social responsibility were in turn strongly marginalized in the post-war period (after World War II), during the times of a centrally planned economy. This situation lasted for a relatively long time and began to change slowly in 1989 as part of the transformation of the Polish economy toward a market economy. It is worth adding here that Poland's integration with European Union structures was particularly important for establishing social trust and thus for the development of CSR.

Considering the above, it can be concluded that CSR is a relatively new concept in Poland, as it emerged in Poland when international corporations entered the Polish market. These corporations, operating based on specific norms and standards, implemented them in their branches in Poland, also in terms of socially responsible activities. Attention should also be paid to the phenomenon of double standards of social responsibility, observed especially among corporations (a discrepancy between what companies declare and what they actually do in terms of responsible business), or to the Polanyi's Paradox (1944) (how is it possible for the market to function effectively without the simultaneous disintegration of society), associated with the development of corporations on the global market (Durning & Ryan, 1997; Robbins, 2006, pp. 177–182; Cierniak-Emerych et al., 2021, p. 287–310).

When considering the development of CSR in Poland in relation to the perspective of CSR manifestations during the COVID-19 pandemic, it should be noted that a direct reference to the idea of corporate social responsibility can be found in the Polish constitution. Although the Constitution of the Republic of Poland of 1997 does not contain provisions defining corporate social responsibility, it creates norms that are a direct source of values for CSR, especially in the area of individual rights and freedoms (Chap. II of the Constitution contains a catalog of freedoms, rights, and duties of man and citizen, which is a source of responsibility of entrepreneurs also in the field of CSR) (Konstytucja Rzeczypospolitej Polskiej, 1997).

It also refers, among other things, to the legal basis of the social market economy, set out in Article 20 of the Constitution of the Republic of Poland, an important foundation for the concept of socially responsible business. According to the assumptions of the social market economy, the economic policy of the state should promote meeting social objectives and ensure the right to economic freedom. The goal of a social market economy is social security and social justice, including a social welfare system, and a guarantee of full employment. The concept targets private property, the free market, and the establishment of rules of competitiveness. It is conceived as a social order. The economy is part of social life, hence its social effect. Article 32 of the Constitution indicates that everyone is equal before the law and everyone has the right to equal treatment by public authorities (Konstytucja Rzeczypospolitej Polskiej, 1997), which is very important in the context of the idea of CSR in terms of conducting business activity and respecting human rights.

In addition to the provisions of the Constitution, there are several other provisions concerning activities directly and indirectly related to CSR contained in other legal acts (e.g., Labour Law, Trade Union Act, Environmental Protection Law). However, it is important to be aware that corporate social responsibility should consist in the voluntary implementation of additional practices that go beyond the obligations arising directly from legal regulations. Despite the fact that direct references to CSR are present in many legal acts, there is still no specific and practically implemented national action plan for socially responsible business and human rights developed by the Polish authorities (cf. Cierniak-Emerych et al., 2021, pp. 287–310).

One can also point to government initiatives in Poland that foster the development of CSR. As part of the implementation of international standards for CSR in Poland, the Minister of Economy issued the ordinance of July 9, 2014, on the establishment of the Team for Corporate Social Responsibility, took the initiative to build a dialogue and exchange of experiences between public administration, business, social organizations, and academic institutions in the field of corporate social responsibility (Zarządzenie Ministra Gospodarki, 2014). Furthermore, in 2016, the Team for Sustainable Development and Corporate Social Responsibility was established by the Ordinance of the Minister of Development as of September 21, 2016. As an auxiliary body of the Minister of Development and Finance, the team supports the Minister in actions for corporate social responsibility. The Responsible Development Plan, the Responsible Development Strategy, and the 2030 United Nations Agenda for Sustainable Development should also be mentioned (Cierniak-Emerych et al., 2021, pp. 287–310).

Furthermore, it is worth pointing out the support for the development of CSR in Poland through the PARP Programme "Increasing the competitiveness of regions through corporate social responsibility (CSR)." The Polish Agency for Enterprise Development (PARP) has been supporting the development of micro-, small- and medium-sized enterprises for over 15 years, setting trends in the development of innovation and responding to the needs of entrepreneurs.

The program "Increasing the competitiveness of regions through corporate social responsibility (CSR)" was one of the most important projects implemented in Poland as part of the popularization of CSR in enterprises operating in Poland. It was implemented within the framework of the Swiss-Polish Cooperation Programme. Its purpose was to promote socially responsible entrepreneurs and encourage them to implement CSR concepts through pilot projects. The program consisted of main parts, i.e., the part for enhancing CSR competence and the part for directly supporting micro-, small-, and medium-sized enterprises in implementing CSR. Within the framework of contests announced in 2013–2014, entrepreneurs could obtain up to PLN 100,000 of non-returnable financial support for the implementation of a CSR project. The co-financed projects concerned ecological and employee aspects, including those related to shaping safe and healthy working conditions, and the social involvement of entrepreneurs (PARP, 2015).

The development of CSR in Poland is supported by various entities, including international and non-governmental organizations. Special support is offered by the

World Bank, UNDP, and International Business Leaders Forum. In Poland, supporting entities include in particular: the Academy for the Development of Philanthropy, the Business Ethics Centre, and the Responsible Business Forum (FOB).

As indicated earlier in this chapter, one of the most important challenges of corporate social responsibility and the development of CSR in Poland today should be the care for the health and lives of humans, viewed both as members of the community and employees. Taking such targeted CSR activities has become especially important since March 2020, when societies started to live in an environment of uncertainty related to the emergence of new waves of the COVID-19 pandemic. It seems that health care institutions play a special role in CSR activities aimed at taking care of the health and life of people, society, and at the same time the health, life, and occupational safety of employees. These include hospitals, but also public and non-public health care facilities, commonly referred to as clinics. In the following sections of this chapter, attention will be focused on CSR issues from the perspective of the operation of non-public health care institutions in Poland (*niepubliczny zakład opieki zdrowotnej*, NZOZs). Their role seems to be particularly important, as it is the physicians and nursing staff of the NZOZ that the patient contacts first when COVID-19 is suspected.

2 Non-public Health Care Facilities in Poland During the COVID-19 Pandemic: A CSR Perspective

An essential part of the health care system in Poland is primary health care (*podstawowa opieka zdrowotna, POZ*), which provides comprehensive health care services at the place of residence to all eligible persons residing or staying on the territory of Poland. Primary health care services are provided in an outpatient setting (in a doctor's surgery, clinic, or outpatient clinic), and in medically justified cases, also at the patient's home (also in a nursing home). They also include preventive care for children and adolescents provided by a nurse/hygienist in teaching and educational settings. The functioning of *POZ* (with the exception of preventive care services for children and adolescents in the teaching and educational environment, and night and holiday care services in POZ) is based on the right to a personal choice of a doctor, nurse, and midwife of primary health care (NFZ, 2022).

Primary health care in Poland includes public health care units created by such bodies as a minister, a voivode, or local government. The second group consists of non-public health care institutions (NZOZ), which in their legal and organizational form are similar to traditional enterprises. NZOZs are established as, e.g., limited liability companies, civil law partnerships, or professional partnerships. They are founded by legal and natural persons. Currently, with reference to the current legislation, they are more often referred to as *zakłady lecznicze* (medical establishments) or *przedsiębiorstwa podmiotów leczniczych* (health care entity enterprises), and sometimes the term *przychodnia prywatna* (private clinic) is also used.

Due to the kind of close organizational and legal nature of these entities to traditional businesses, considerations of CSR during the COVID-19 pandemic will focus on manifestations of their activities in Poland.

Analysis of the manifestations of CSR in non-public health care institutions during the COVID-19 pandemic should take into account that these entities, on the one hand, provide medical services to the public, whereas on the other hand, they provide employment for doctors, nurses, and non-medical staff. Therefore, there are two dimensions of CSR:

- patient-oriented (social dimension: external stakeholder),
- oriented toward creating safe and healthy working conditions for employees (social dimension: internal stakeholder).

In the first of the indicated dimensions, CSR should be viewed in particular from the perspective of activities undertaken by family doctors and nursing staff for the benefit of society, i.e., patients. This applies to patients with suspected COVID-19 as well as those with confirmed disease. There are also medical services provided to other patients, i.e., those without COVID-19. Furthermore, preventive measures, including COVID-19 vaccination, should also be looked at more closely.

The diagnosis of the first case of COVID-19 in Poland in early March 2020 led to major changes in the way family practitioners provide services to patients. The basic manifestation of socially responsible actions aimed at reducing public exposure to COVID-19 during contact with family doctors and other patients in outpatient clinics was the introduction of solutions related to the contact of patients with doctors and nursing staff that had not been used previously in Poland. In order to minimize the risk of infection, the health care system has actually changed the delivery of services from inpatient largely to remote.

The patients, regardless of the condition with which he or she was seeking medical advice, first had to use telephone consultation with the clinic. It should be noted that in situations where the doctor, after an initial interview, considered it appropriate to conduct a patient visit in the traditional form, such actions were taken. In addition, it is worth pointing out that the doctor could not refuse to provide consultation in a conventional manner when, among other things, the patient's health status worsened. The indicated manner of providing medical services as a basis for actions taken by physicians providing advice in outpatient clinics has been recommended for nearly 2 years since the first case of COVID-19 was confirmed in Poland.

The legal acts that are the basis for the provision of telephone consultation include Act of August 27, 2004, on health care services financed from public funds (Journal of Laws of 2021, item 1285, as amended), Ordinance of the Minister of Health of August 12, 2020, on the organizational standard of telephone consultation in primary health care (Journal of Laws of 2020, item 1395), and Ordinance of the Minister of Health of October 8, 2020, on the organizational standard of health care for patients suspected of being infected or infected with SARS-CoV-2 virus (Journal of Laws, items 1749, 1873 and 2043). Under the provisions of the indicated legislation, telephone consultation should be provided no later than on the first working day after the day the patient consulted a doctor at the facility via an information and communication technology system or in person, or at a later date if it is determined in consultation with the patient or the patient's legal guardian (Ustawa z dnia27sierpnia, 2004; Rozporządzenie Ministra Zdrowia z dnia 12 sierpnia, 2020; Rozporządzenia Ministra Zdrowia z dnia 8 października, 2020).

Analogous to the above-mentioned legal bases, the general principles of telephone consultation were defined, which can be synthesized in the form of the following stages:

1. Confirmation of the patient's identity based on the data provided by the patient or the data indicated in the medical records or the patient's declaration of choice, presented at the video consultation (e.g., identity card) or indicated in the Patient's Internet Account.
2. Interview with the patient and analysis of patient medical records.
3. Determination of disease entity with the recommendation of medical management.
4. Inform the patient of the possible need for an inpatient visit when the physician determines that a medical problem prevents the provision of services in the form of telephone consultation.

Referring directly to the provision of services to patients with suspected COVID-19, the physician is obliged to have such patients tested for SARS-CoV-2. The patient is quarantined at the time of test referral. Depending on the test results, further action is taken. If the test is positive, the patient is placed in home isolation or referred to the hospital depending on the severity of the disease.

Information about the COVID-19 test result is provided to the patient who can read it in what is called an Individual Patient Account. Individuals who do not have the Individual Patient Account may receive this information from their primary care physician. It should be noted that in the case of a positive test result, this information is transmitted simultaneously to the Individual Patient Account and the primary care physician.

In the case of COVID-19 patients in home isolation, the primary care physician is required to contact the patient via telephone to monitor the patient's condition. The frequency of these contacts has varied with the emergence of successive waves of the pandemic. Initially, the physician was required to contact the isolated patient a minimum of two times, then once. Patients over the age of 60 years should be handled somewhat differently. The physician must perform a basic physical examination immediately after receiving a positive test result. On the other hand, discussion of the results of the examination can occur through telephone consultation.

The principles of operation of health care institutions aimed at socially responsible reduction of exposure of the public (patients and medical personnel) to the incidence of COVID-19, as stated above, have been changing in Poland with the successive waves of the COVID-19 pandemic. Around mid-2021, the scope of telephone consultation was reduced in favor of conventional consultation, especially for patients over age 60 years and children under age 6 years. This concerns in particular the fact that telephone consultation will never fully replace a conventional doctor's visit.

While pointing out socially responsible activities undertaken for the benefit of patients by non-public health care institutions in Poland, attention should also be drawn to the initiative called the Home Health Care (DOM) program.

For the safety of COVID-19 patients whose health condition allows them to be isolated at home, a Home Health Care program was launched in Poland in November 2021. As part of this program, COVID-19 patients can benefit from a remote monitoring system for patient parameters such as blood saturation and heart rate. Patients receive pulse oximeters as a diagnostic tool and use the PulsoCare system to transmit and monitor data. The program allows for quick detection of patients who, due to deterioration of their health status, with reference to the established results of blood saturation and pulse parameters, should be referred to hospital treatment.

Primary care physicians have a specific role in this program, and if a patient tests positive for COVID-19, they can have the pulse oximeter delivered directly to the patient's home under the designated program. The patient installs the app or the app is installed on the phone by a clinic staff member. The physician then has access to the current measurement results. In a situation of concern, the patient is given a remote consultation, which will take place by phone. If necessary, the doctor refers the patient to the hospital.

The DOM program presented raises the patients' awareness of the need for measurements and shows changes in health status. At the same time, it assists physicians in responding quickly to these changes, especially if they are related to the patient's deteriorating health. It should be emphasized that technological solutions for the increase of social responsibility of health care institutions toward patients are used in the program. However, this program also has some limitations that are related to the patients' ability to take the measurements. This is especially true for people aged 60+, for whom using a phone app is not always an easy task.

Socially responsible actions involving patients of health care institutions also include disinfecting, keeping a safe distance from other people, and using personal protective equipment. This is about the mandatory use of protective masks. The current legislation in this area has evolved over the nearly 2 years of the COVID-19 pandemic. However, it should be noted that until March 28, 2022, it was obligatory in Poland to cover the mouth and nose with a protective mask in all closed public spaces, including stores, places of worship, means of transport, cinemas, theaters, etc. This requirement was removed on March 28, 2022, but not for patients receiving health care services. Thus, in clinics and other places where health services are provided, it is still mandatory to cover the mouth and nose with a mask.

Another manifestation of the socially responsible actions taken by non-public health care institutions during the COVID-19 pandemic was their involvement in vaccination prevention activities. Vaccinations in Poland have been administered since December 2020. Initially, health care workers had access to vaccinations, followed by citizens aged 75+. The next group was teachers. Over time, access to vaccination, around May and June 2021, became fairly unlimited.

The organization of vaccination sites (in addition to other designated locations) at clinic locations appears to be of particular importance in terms of encouraging the public to take advantage of vaccination. It is important to note here the important role of physicians and primary care nursing staff in the promotion of vaccination, especially among those aged 60 years and over and young people. It would seem that interest in vaccination should be huge. In fact, this was the case in the early days when vaccines were relatively scarce. Over time, unfortunately, vaccination in Poland, especially the third dose, began to attract less interest despite a significant increase in the incidence of the disease. It is therefore important to continue preventive measures, especially those closest to the patient, such as activities in non-public health care institutions.

Non-public health care institutions are places that provide medical services, but they are also workplaces for doctors, nurses, and administrative staff. This highlights the second of the dimensions of CSR in non-public health care institutions indicated above, i.e., taking socially responsible actions for the benefit of employees in NZOZs. In particular, these actions are aimed at creating safe and healthy working conditions.

The COVID-19 pandemic has significantly reduced the level of health and safety in NZOZs. With the emergence of COVID-19 cases, several solutions aimed to create the best possible working conditions in the NZOZs have been implemented, especially concerning the use of individual and collective protection of employees.[1] The following measures can be indicated:

- disposable protective clothing, such as medical coveralls and aprons,
- protective masks,
- gloves.

It was also stated that "the minimum requirements for protective clothing recommended by the ECDC for health care personnel during the COVID-19 pandemic while caring for a patient, are respiratory protection equipment such as filtering facepiece respirators (FFP2/FFP3), a waterproof apron, eye or face shield, and gloves" (World Health Organization, 2020; Michalski et al., 2020). It should be noted that before the pandemic, the use of masks by the medical staff of the clinic was not obligatory.

The use of protective clothing and goggles is also accompanied by the basic principle of frequent hygiene and disinfection, especially of hands, but also of surfaces and medical equipment. Disinfectants used include Octenisept, SKINAM SOFT PROTECT, etc.

It should be stressed that disinfectants are compounds that should be used with caution due to their irritating effect. Special attention must be paid to carrying out occupational risk assessments that indicate, among other things, the likelihood of

---

[1]An example of NZOZ where the ways of securing employees are strictly followed is NZOZ Śródmieście in Wrocław and Clinical Research Center in Wrocław, whose representatives are the co-authors of the present paper: P. Napora (doctor) and S. Szromba (nurse)

damage to health and life from the use of these disinfectants in workplaces where these disinfectants are used.

One of the solutions used in these facilities is also germicidal lamps with a flow charter. Such lamps are placed, among others, in waiting rooms and doctor's surgeries. Their advantage is that they can be used while patients and medical staff are in the room.

Creating safe and sanitary working conditions for employees of health care institutions during a pandemic is still a challenge. The need for the actual, constant use of the indicated protective measures and ensuring their availability should be emphasized because only such actions will allow for the limitation of the incidence of diseases among employees and thus enable to maintain the continuity of the NZOZ operations.

3 Conclusion

The concept of social responsibility is receiving increasing attention from organizations and their stakeholders. More and more often, taking into account social responsibility is considered an important aspect of good management of the organization. This issue has taken on increased importance in the last 2 years as the entire world struggles to cope with the consequences of the COVID-19 pandemic. As demonstrated in this chapter, some socially responsible activities are also, or perhaps especially, undertaken by non-public health care centers in Poland. These activities have two main dimensions, i.e., a dimension focused on social responsibility toward patients, and a dimension focused on social responsibility toward employees. The latter is achieved through an appropriate approach to creating safe and hygienic working conditions.

The considerations presented in this chapter certainly do not exhaust the problem of the manifestations of CSR in health care institutions in Poland during the COVID-19 pandemic. Furthermore, some limitations can be indicated, and overcoming them will be the aim of further research. A particularly interesting research problem is the recognition of the problems of estimating the level of occupational risk associated with the creation of safe and healthy working conditions in NZOZs.

References

Cierniak-Emerych, A., & Mazur-Wierzbicka, E. (2022). Safety and health at work as an important interest of employees from the perspective of corporate social responsibility. In *Corporate social responsibility and sustainability* (pp. 131–133). Routledge.

Cierniak-Emerych, A., Mazur-Wierzbicka, E., & Rojek-Nowosielska, M. (2021). Corporate social responsibility in Poland. Economic and political context of development of corporate social responsibility in Poland. In S. O. Idowu (Ed.), *Current global practices of corporate social responsibility. CSR, sustainability, ethics & governance* (pp. 287–310). Springer. https://doi. org/10.1007/978-3-030-68386-3_13

Crozier, M. (1993). *Przedsiębiorstwo na podsłuchu.* PWE.

Durning, A. T., & Ryan, J. C. (1997). *Stuff: The secret lives of everyday things (new report).* Northwest Environment Watch, Sightline Inst.

European Commission. (2001). *Promoting a European framework for corporate social responsibility* (p. 5). http://ec.europa.eu

European Commission, A renewed EU strategy 2011–14 for Corporate Social Responsibility, COMMUNICATION FROM THE COMMISSION TO THE EUROPEAN PARLIAMENT, THE COUNCIL, THE EUROPEAN ECONOMIC AND SOCIAL COMMITTEE AND THE COMMITTEE OF THE REGIONS, Brussels, 25.10.2011 COM(2011) 681 final, p. 6.

Gableta, M. (2012). *Interesy pracowników oraz warunki ich respektowania w przedsiębiorstwach.* Wyd. UE we Wrocławiu.

ISO Advisory Group on Social Responsibility. (2004). *Working report on social responsibility 30 kwietnia 2004* (pp. 28–29).

Konstytucja Rzeczypospolitej Polskiej (The Constitution of the Republic of Poland) z dnia 2 kwietnia 1997 r., Dz.U. Nr 78 poz. 483 z późn. zm.

Koradecka, D. (2000). *Nauka o pracy- bezpieczeństwo, higiena, ergonomia. Zarządzanie bezpieczeństwem i higieną pracy, (Work science – safety, hygiene, ergonomics. Occupational health and safety management)* (Vol. 8, p. 21). CIOP.

Mazur-Wierzbicka, E. (2021). Occupational safety and health from the perspective of the ISO 26000 standard. In K. S. Soliman (Ed.), *Proceedings of the 37th IBIMA* (pp. 9107–9115).

Michalski, A., Bielawska-Drózd, A., Pinkas, J., & Kocik, J. (2020). Środki ochrony indywidualnej personelu medycznego w warunkach pandemii COVID-19. *Wiedza Medyczna*, (Numer Specjalny), 14–23. https://wiedzamedyczna.pl/index.php/wm/article/view/44

NFZ. (2022). *Dla pacjeta.* Accessed March 27, 2022, from https://www.nfz.gov.pl/dla-pacjenta/ informacje-o-swiadczeniach/podstawowa-opieka-zdrowotna/

Pocztowski, A. (1998). *Zarządzanie zasobami ludzkimi. Zarys problematyki i metod. (Human resources management. Outline of issues and methods)* (p. 261). Antykwa. i nast.

PARP. (2015). *Odpowiedzialność się opłaca, czyli CSR w MŚP. Prezentacja dobrych praktyk powstałych w ramach projektu PARP "Zwiększenie konkurencyjności regionów poprzez społeczną odpowiedzialność biznesu (CSR)".* Wyd. PARP. (Responsibility pays off, i.e. CSR in SMEs. Presentation of good practices developed under the PARP project "Increasing the competitiveness of regions through corporate social responsibility (CSR)").

Robbins, R. H. (2006). *Globalne problemy a kultura kapitalizmu, (Global problems and the culture of capitalism).* Pro Publico.

Rojek-Nowosielska, M. (2017). *Społeczna odpowiedzialność przedsiębiorstw. Model, diagnoza, ocena, (Corporate social responsibility. Model, diagnosis, evaluation) Wyd.* Uniwersytet Ekonomiczny we Wrocławiu.

Rozporządzenie Ministra Zdrowia z dnia 12 sierpnia 2020 r. w sprawie standardu organizacyjnego teleporady w ramach podstawowej opieki zdrowotnej (Dz.U. z 2020 r. poz. 1395), (Regulation of the Minister of Health of August 12, 2020 on the organizational standard of teleporting in primary health care).

Rozporządzenie Ministra Zdrowia z dnia 8 października 2020 r. w sprawie standardu organizacyjnego opieki zdrowotnej nad pacjentem podejrzanym o zakażenie lub zakażonym wirusem SARS-CoV-2 (Dz.U. poz. 1749, 1873 i 2043), (Regulation of the Minister of Health of 8 October 2020 on the standard of organizational health care for a patient suspected of being infected or infected with SARS-CoV-2).

Sokołowska, A. (2013). *Społeczna odpowiedzialność małego przedsiębiorstwa Identyfikacja - ocena - kierunki doskonalenia. (Social responsibility of a small enterprise. Identification – evaluation – directions of improvement) Wyd.* UE we Wrocławiu.

Stefańska, M. (2013). *Podstawy teoretyczne i ewolucja pojęcia społeczna odpowiedzialność biznesu (CSR), (Theoretical basis and evolution of the concept of corporate social responsibility (CSR))* (Vol. 288, p. 204). Prace Naukowe Uniwersytetu Ekonomicznego we Wrocławiu.

Ustawa z dnia 27 sierpnia 2004 r. o świadczeniach opieki zdrowotnej finansowanych ze środków publicznych (Dz.U. z 2021 r. poz. 1285, z późn. zm.), (Act of 27 August 2004 on health care services financed from public funds).

World Health Organization. (2020). *Rational use of personal protective equipment for coronavirus disease (COVID-19): Interim guidance, 27 February 2020.* World Health Organization. License: CC BY-NC-SA 3.0 IGO. https://apps.who.int/iris/handle/10665/331215

Zarządzenie Ministra Gospodarki z 9 lipca 2014 r. w sprawie powołania Zespołu do spraw Społecznej Odpowiedzialności Przedsiębiorstw, Dz.U. Ministra Gospodarki z 2014 r. (Ordinance of the Minister of Economy of 9 July 2014 on the establishment of the Team for Corporate Social Responsibility, Journal of Laws of 2014, No. Of the Minister of Economy of 2014).

Anna Cierniak-Emerych is a Professor in the Department of Labour, Capital and Innovation Wroclaw University of Economics and Busines (Poland); head of Department of Labour, Capital and Innovation; Dean for student's affairs, Faculty of Business Wroclaw University of Economics and Busines; director of post-graduate studies "Occupational Safety and Health". The author of more than 180 publications both in Polish and English, among others, monograph entitled "The participation of employees in the management of potential operating company," ed. Wroclaw University of Economics, Wrocław 2012; co-author of the monograph "Labour potential management and the satisfaction of employee interests" ed. Wroclaw University of Economics, Wrocław 2022.

Anna Cierniak-Emerych's research interests relate mainly to the human and his work in the company, particularly such as: Corporate Social Responsibility (CSR),development of tangible and intangible working conditions (health and safety),employee participation, greater flexibility in employment, human resources management, organizational culture, satisfaction of employees, the interests of employees and employers in the enterprise and their respect, contemporary management concepts (e.g. lean management, TQM, etc.).

Ewa Mazur-Wierzbicka is a Professor at the University of Szczecin, Faculty of Economics and Management, Institute of Human Capital Management. She specializes mainly in the field of corporate social responsibility (with emphasis on diversity management, ethical and equality-related actions), sustainable development and human capital management. She also deals with the issues of soft competences. An expert in the field of corporate social responsibility (CSR)—external expert of the Responsible Business Forum. University of Szczecin's plenipotentiary to the Technical Committee no. 305 for Social Responsibility operating at the Polish Committee for Standardization, advisor of the Polish Agency for Enterprise Development in terms of Corporate Social Responsibility. Initiator and Chair of the cycle of Seminars and Conferences titled "Corporate social responsibility—the management and economy perspective" associating both researchers and business practitioners. Expert of the EIGE's (the European Institute for Gender Equality) Experts' Forum (mandate period of 01.12.2018–30.11. 2021 (IV term), 01.12.2021–30.11.2024 (V term)).

Participant of Polish and international research projects. Reviewer of research projects of the National Science Centre. Member of Programme Councils, Scientific Committees, Organizing Committees and a panelist of Polish and international conferences devoted to corporate social responsibility (also addressing equal treatment and diversity management issues) and human capital management. Business consultant. She cooperates with practitioners in the field of management, training companies and institutions, i.a. Polish Entrepreneurs Foundation. Author and co-author of training programmes, numerous scientific studies (including those addressing equal treatment and diversity management issues), analyses, research projects and expert opinions for business.

Piotr Napora, medical doctor, internal medicine and pulmonology specialist, partner of Clinical Research Center in Wroclaw. The author and co-author of more then ten publications both in Polish and English.

Piotr Napora is working primarily as an medical doctor focusing his work on early phases commercial clinical trials. He took an active part in fighting with COVID-19 pandemic both as researcher and medical doctor taking care of infected patients. His research interests currently relate mainly to the human and his work in the medical small companies.

Sylwia Szromba, master of science in nursing, epidemiology specialist, chief nurse in Clinical Research Center in Wroclaw. The co-author of two publications in English and one oral presentation on ERS 2020.

Corporate Social Responsibility: A Solution for Resilience During the COVID-19 Pandemic in Romania

Silvia Puiu

1 Introduction

According to Merriam-Webster dictionary (n.d.), resilience is "an ability to recover from or adjust easily to misfortune or change." The COVID-19 pandemic was one of the most important changes in recent history because of its magnitude. It affected organizations and individuals at many levels: economic, social, psychological, generating more than a health crisis. Organizations had to fight for their survival, learning new ways of developing their activities and finding different solutions to reduce the negative consequences of the pandemic on their results and on their employees. Some were more resilient than others, the situation being different from company to company, from industry to industry, from country to country.

Between 2020 and 2022, there were times between the COVID-19 waves in which companies hoped and anticipated a recovery. There are numerous articles in 2020 and 2021 which implied a recovery after the pandemic, changes in the post-COVID-19 era, coping mechanisms (Karian and Box and the Global Institute for Women's Leadership, 2021; Diedrich et al., 2021; ARUP, 2020). In uncertain times, it is difficult to predict the end. Other international reports, at least those regarding the employment rate, show that we must expect a recovery at the pre-pandemic level only in 2023 (ILO, 2022). Still, even if there will pass some time before a full recovery, there are good signs. Eurostat (2022) shows an increase in productivity in 2021 compared with 2020 almost reaching the level in 2019.

The present paper analyzes the way organizations in Romania used corporate social responsibility (CSR) initiatives to cope with the challenges faced during the pandemic. These initiatives targeted the employees and the community, focusing on ensuring a better and safer environment for their own human resources and on supporting various causes in the society.

S. Puiu (✉)
Faculty of Economics and Business Administration, University of Craiova, Craiova, Romania

© The Author(s), under exclusive license to Springer Nature Switzerland AG 2023
S. O. Idowu et al. (eds.), *Corporate Social Responsibility in the Health Sector*, CSR, Sustainability, Ethics & Governance, https://doi.org/10.1007/978-3-031-23261-9_5

The topic is important because the experience of companies during these difficult times can be a learning lesson for coping and surviving in times of crisis. Organizations can learn from others that were more resilient and thus shape their future strategies in a way that it is more suited to answer to crisis and the uncertainty that comes with it. Also, CSR can represent a solution for the many problems that occurred at so many levels. The CSR projects focused on employees and community can help with recovery after the pandemic and also with creating a better world, a greener and more digitalized one in which organizational activities and processes are oriented to sustainability.

The paper tries to find answers to the question regarding the nature of the CSR initiatives in Romanian organizations used as a way of resilience during and after the pandemic. Some good changes happened, and hopefully, a few of them will remain in place: digitalization, more flexibility and empowerment for the employees, ensured cyber security, a lower carbon footprint.

2 Literature Review and Hypotheses Development

The present paper focuses on four variables: the corporate social responsibility oriented toward organizations employees, the corporate social responsibility oriented toward the community, the pandemic impact on both the organizations and the employees, the changes implemented by the organizations after 2 years in which they faced the pandemic.

2.1 The Corporate Social Responsibility Oriented Toward Organizations Employees

This variable is important for analyzing the resilience capacity of organizations during and after the pandemic. The human resource working within these economic entities were affected in numerous ways but still they struggled and made efforts to continue work in difficult conditions or struggled with reduced income determined by the financial problems of their employers. This variable also known as employee-oriented CSR increases motivation and thus the efficiency and productivity of a company (Suto & Takehara, 2022).

Other authors noticed also some negative consequences of this form of CSR if it is not monitored closely. Thus, Yin et al. (2021) highlight the risk of "unethical pro-organizational behavior" among the employees who might try to be loyal to their organizations even when they do not behave ethically in the community. Mahmud et al. (2021) analyzed the response of companies to COVID-19, stating the important role played by CSR oriented to employees and other stakeholders.

As John et al. (2022) so beautifully put it, "charity begins at home." The authors highlight the importance of CSR practices for keeping the employees motivated especially in difficult times. The CSR oriented toward the communities was also vital during the pandemic but the employees are those who keep the company going forward so the organizations should focus first on them. Mao et al. (2021) made research on employees in the tourism sector in China and concluded that the CSR oriented to their own employees helps the human force contributing to their "self-efficacy, hope, resilience and optimism" and this was noticed during the COVID-19 pandemic. Thus, companies became responsible for the well-being of their employees maybe more than before because everyone was more vulnerable. Aguinis et al. (2020) mention both the positive and negative impact of CSR. Most studies focus on the good side of corporate social responsibility, but the authors emphasize that the impact of CSR projects is more meaningful when the employees are involved in the process and this starts with offering them a background in which they feel safe and appreciated.

2.2 The Corporate Social Responsibility Oriented Toward the Community

Ahmed et al. (2021) argued that CSR during crises such as the pandemic is different that CSR in normal conditions, determining a change in dimensions and the directions that are targeted. The authors highlight that such difficult times might orient the CSR efforts of the companies toward the community because of the many problems encountered during the pandemic.

The community-oriented CSR is also mentioned by García-Sánchez and García-Sánchez (2020) who mention three directions of CSR during the pandemic: one oriented to the community, one to the investors, and one that combines the support offered to the community with the economic interests of the organization. Raimo et al. (2021) presented case studies from Spanish companies which helped the vulnerable communities during the pandemic uniting efforts with NGOs and public institutions.

Appiah (2019) conducted research on the hotel industry in the United States of America and showed that there is a positive relationship between community-oriented CSR and the employees' satisfaction at work. This means that the employees can be an active partner in many CSR projects developed for the community. This would reduce negative outcomes as those mentioned by other authors in the literature (Aguinis et al., 2020; Yin et al., 2021).

Chen and Hang (2021) argue that community-oriented CSR has an important benefit for the organizations because consumers will be more attracted to these businesses and will remember them for the help offered during the challenging period. Their research was conducted in the hotel industry which was one of the most affected sectors because of the restrictions that were imposed. The companies which developed CSR initiatives oriented to helping homeless people or the medical staff might benefit after the pandemic or at least after the relaxation of restrictions.

Gürlek and Kılıç (2021) analyzed the CSR initiatives of the most important hotels in the world during the economic and health crisis in 2020–2021. Thus, it was noticed that only 50% of the initiatives were focused on the community, but

76% were addressed to the employees' needs. Community is important, but the human asset is valuable for the organization's survival in difficult times. Ramya and Baral (2021) conducted research in India about the CSR oriented toward helping the community and noticed a short-term vision in doing this. The authors appreciate the need for a long-term strategy in community-oriented CSR.

2.3   The Pandemic Impact on Both the Organizations and the Employees

This variable takes into account the financial impact of the pandemic on the companies and their employees but also the consequences on the physical and mental health of the human resource. All these dimensions are interconnected because a company cannot function properly if it has employees in bad health. The vice versa is also true because a company affected financially influences the well-being of the employees creating a vicious circle. OECD (n.d.) raises awareness on the risk to go from a job crisis to a social crisis determined by the worsening conditions of the people who lost their job and could not find so easily a new one. Resilience of humans is one of the strengths that should be built and encouraged in an economy that starts to recover after the pandemic.

Deloitte (2020) proposed a set of measures that companies could implement in order to reduce the negative consequences of the pandemic on their employees: greater flexibility granted to the employees, a high emphasis on the safety conditions and hygiene at work, a better communication with the employees to reduce the stress level in uncertain times, offering support for those who might need it. Bartik et al. (2020) highlight that the impact was significant especially for small enterprises which were affected financially and many of them even closed their activities. Companies closing means people who lose their jobs and families with lower incomes.

The resilience of companies and employees during the pandemic increased as time passed and the uncertainty decreased. Also, the vaccine and treatments became available on a larger scale and governments could relax the initial restrictions letting companies return to their activities. Diedrich et al. (2021) mention resilience as being part of the strategic actions of the companies who recovered more rapidly. Resilience as a competitive advantage for some companies is also stated by Reeves et al. (2021).

A report of Eurofound (2021) about the impact of the pandemic on companies in European Union shows that 20% of companies had to give up to a part of their employees and the share is double for those which reduced the number of hours worked per employee. This represents a huge impact on both organizations and their employees. The report also mentions that organization which perceived their employees as an investment on a long term did better than others. Other consequences of the changes generated by the pandemic were related to the increase of teleworking which might bring important benefits but also some disadvantages linked to the poor line between personal and professional lives of the employees working from home.

Cotofan et al. (2021) emphasize the fact the impact was different and in general higher in developing countries compared to the developed ones. The same was noticed at a microeconomic level: the employees with the lowest income were more affected than the employees who had bigger salaries, youngers and women too because of their important share in tourism, restaurants, hotels. ILO (2022) appreciates that the unemployment rate will go down, at the level in 2019, starting with 2023.

2.4   The Changes Implemented by the Organizations After 2 Years in Which They Faced the Pandemic

Even if the COVID-19 pandemic put a significant toll on both the individuals and the organizations, there were also some positive changes that occurred because of the measures implemented in order to cope with these challenging times. Thus, many companies became more friendly with the environment, implemented flexible arrangements of work, took care of cybersecurity and made the transition to digitalization. Of course, there were also many companies who suffered important losses and did not have to power to overcome the difficulties.

United Nations Environment Programme (2021, p. 20) highlights the importance of spending the money for recovery after the pandemic in making a "green transition." ARUP (2020, p. 6) mentions two important positive changes determined by the need to cope: the spread of teleworking and a faster "digital transformation." De Smet et al. (2021) explain the importance of traditional offices for the sense of connection instilled in the employees but show that companies should empower these employees and let them choose the best way of work for them in order to have a higher productivity. Increasing the level of technology leads to digitalization and this offers more possibility of collaborative work.

Bethell (2020) states the importance of incorporating technology and emphasizes the role of choosing closer business partners for both the environmental objective but also for the speed of activities especially in times like the pandemic when everything closes. Karian and Box and the Global Institute for Women's Leadership (2021) analyzed the impact of the COVID-19 pandemic on how 250 companies coped with the changes during the pandemic. They conclude that 78% of the companies in the sample adjusted their strategies to include positive changes such as flexibility in working and a mixed system (remote and from office).

We developed four research hypotheses analyzing the connection between the variables.

Hypothesis 1 (H1): *The corporate social responsibility oriented toward the community has a direct and positive impact on the changes implemented by the organizations after 2 years in which they faced the pandemic.*

Hypothesis 2 (H2): *The corporate social responsibility oriented toward the employees has a direct and positive impact on the changes implemented by the organizations after 2 years in which they faced the pandemic.*

Hypothesis 3 (H3): *The corporate social responsibility oriented toward the employees influences the pandemic impact on both the organizations and the employees.*

Hypothesis 4 (H4): *The pandemic impact on both the organizations and the employees influences the corporate social responsibility oriented toward the community.*

3   Research Methodology

In order to assess the resilience of the organizations in Romania during the COVID-19 pandemic and also at the end of a 2-year period in which there were many restrictions and safety regulations, we launched a questionnaire in February and March 2022. This was addressed to employees in public and private organizations and also entrepreneurs in Romania. The survey was built in Google Forms and sent to 860 employees from different backgrounds. There were 156 valid questionnaires.

The respondents' profile is summarized in Table 1. Most respondents are young (18–35 years old)—71.795%, with a master's degree (79.487%), living in urban areas (87.179%), and working in private companies without a management position (54.487%).

Taking into account the dimension of the sample, we used partial least squares—structural equation modeling with the help of SmartPLS version 3 software (Ringle et al., 2015). For this, we started from four constructs: the corporate social responsibility oriented toward organizations employees (HRCSR), the corporate social responsibility oriented toward the community (COMCSR), PNDI (the pandemic impact on both the organizations and the employees), the changes implemented by the organizations after 2 years in which they faced the pandemic reality (CNG). The model with the constructs and the items for each of the four variables is presented in Fig. 1.

The codification for the four constructs in the model and the items for each of these are presented in Table 2.

The answers were collected through a survey, the questions corresponding to the items in Table 2 being formulated as a Likert scale from total disagreement (1) to total agreement (5).

Table 1 The respondents' profile

Age	%	Status on the labor market	%	Last completed studies	%	Location	%
18–25	45.513	Employee in private sector without management position	54.487	Master	79.487	Urban	87.179
26–35	26.282	Employee in public sector without management position	22.436	Bachelor	19.231	Rural	12.821
36–45	20.513	Employee in private sector with a management position	12.821	PhD	1.282		
46–60	7.692	Employee in public sector with a management position	5.769				
Over 60	–	Entrepreneur	4.487				

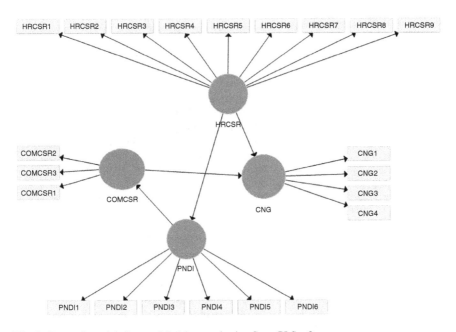

Fig. 1 Research model. *Source*: Model created using SmartPLS v.3

Table 2 Model codification

Constructs (code)	Items	Item codes
CSR directed toward the employees (HRCSR)	The organization implemented teleworking where it was possible.	HRCSR1
	The organization offered the employees the flexibility to work from home, from the office, or in a hybrid arrangement.	HRCSR2
	The organization offered psychological counseling for the employees during the pandemic.	HRCSR3
	The organization offered trainings for remote work to the employees.	HRCSR4
	The organization offered the needed equipment for the employees working remotely.	HRCSR5
	The organization cared about my needs during the pandemic.	HRCSR6
	The organization implemented all the needed safety measures to protect the employees during the pandemic.	HRCSR7
	The organization instructed the employees regarding online security when working from home.	HRCSR8
	The organization respected the personal time of the employees working from home.	HRCSR9
CSR directed toward the community (COMCSR)	The organization donated products, services, or money to support the community during the pandemic.	COMCSR1
	The organization offered the employees the possibility to be volunteers for different social projects in the community.	COMCSR2
	The organizations had partnerships with other organizations in order to increase the support offered to the community during the pandemic.	CMCSR3
The organizational changes at the end of a 2-year pandemic period (CNG)	After 2 years, the organization has a higher level of digitalization	CNG1
	After 2 years, the organization has a more friendly human resources policy.	CNG2
	After 2 years, the organization has a higher level of awareness regarding cyber security.	CNG3
	After 2 years, the organization has a higher level of awareness regarding its impact on the environment.	CNG4
The pandemic impact on the employees and the organization (PNDI)	The pandemic had a negative impact on my income.	PRTN1
	The pandemic had a negative impact on my physical health.	PRTN2
	The pandemic had a negative impact on my mental health.	PRTN3

(continued)

Table 2 (continued)

Constructs (code)	Items	Item codes
	The pandemic had a negative impact on my productivity at work.	PRTN4
	The pandemic had a negative impact on the organizations revenues.	PRTN5
	The pandemic had a negative impact on my social skills and the capacity to work in a team.	PRTN6

Table 3 Outer loadings for the items in the model

	CNG	COMCSR	HRCSR	PNDI
CNG1	0.787			
CNG2	0.820			
CNG3	0.908			
CNG4	0.780			
COMCSR2		0.894		
COMCSR3		0.905		
COMCSR1		0.875		
HRCSR1			0.638	
HRCSR2			0.718	
HRCSR3			0.545	
HRCSR4			0.667	
HRCSR5			0.783	
HRCSR6			0.696	
HRCSR7			0.506	
HRCSR8			0.721	
HRCSR9			0.649	
PNDI1				0.546
PNDI2				0.809
PNDI3				0.669
PNDI4				0.807
PNDI5				0.457
PNDI6				0.627

4 Results

The outer loadings for the items in the model are presented in Table 3. Even if it is better to have all of the items with outer loadings above 0.7, there are many scientific papers (Yana et al., 2015; Chin, 1998; Ghozali, 2014) which highlight the fact that outer loadings between 0.6 and 0.7 are acceptable. Thus, we decided to keep the items in the research model that have outer loadings above 0.6.

For the construct HRCSR, we had to remove from the model two items with outer loadings below 0.6 (HRCSR3 and HRCSR7), and for the construct PNDI, we had to

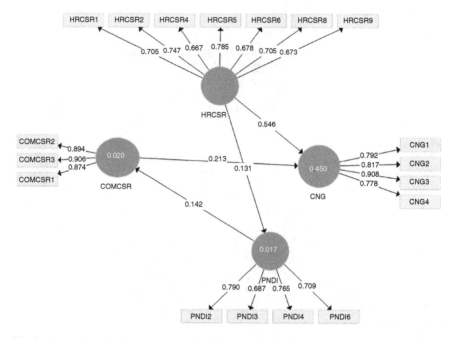

Fig. 2 Research model after removing items with low outer loadings. *Source*: Model created using SmartPLS v.3

remove two items for the same reason (PNDI1 and PNDI5). Thus, the research model changes into the one presented in Fig. 2.

We can notice from Fig. 2 that the strongest impact is from HRCSR to CNG (0.546) followed by COMCSR to CNG (0.213), from PNDI to COMCSR (0.142), and from HRCSR to PNDI (0.131). The great impact of the CSR oriented toward the employees to the changes generated in the organization after 2 years of pandemic restrictions and limitations shows that organizations which care about the employees are also the ones which made positive changes in general (digitalization, green behaviors, friendly HR policies, high cyber security). Also, HRCSR and COMCSR explain 45% of CNG variation.

The convergent validity of the proposed research model is ensured by high outer loadings and VIF lower than 5. As we can notice from both Tables 3 and 4, the values for the outer loadings and VIF show a good collinearity and validity of the model.

The validity and reliability of the four constructs in the model are determined in Table 5. The values for Cronbach's alpha are above 0.7 for PNDI and above 0.83 for CNG, COMCSR, and HRCSR, all being high values which show a high reliability of the model (Taber, 2018).

The AVE values are above 0.5 (good convergent validity) and the composite reliability is above 0.8 for all four constructs which shows a high internal consistency of the model and the constructs (Henseler et al., 2009; Hair Jr. et al., 2021).

Table 4 VIF for the items in the model

Items	VIF
CNG1	1.772
CNG2	1.860
CNG3	2.977
CNG4	1.955
COMCSR2	2.241
COMCSR3	2.514
COMSCR1	2.221
HRCSR1	2.015
HRCSR2	1.969
HRCSR4	1.969
HRCSR5	2.344
HRCSR6	1.489
HRCSR8	1.502
HRCSR9	1.544
PNDI2	1.580
PNDI3	1.520
PNDI4	1.640
PNDI6	1.223

Source: Calculated with smartPLS v.3

Table 5 The validity and reliability for the four constructs in the model

Constructs	Cronbach's alpha	rho_A	Composite reliability	Average variance extracted (AVE)
CNG	0.843	0.852	0.895	0.681
COMCSR	0.871	0.875	0.920	0.794
HRCSR	0.836	0.839	0.876	0.504
PNDI	0.737	0.749	0.827	0.546

Source: Calculated with smartPLS v.3

Table 6 Fornell–Larcker criterion

Constructs of the model	CNG	COMCSR	HRCSR	PNDI
CNG	0.825			
COMCSR	0.465	0.891		
HRCSR	0.644	0.462	0.710	
PNDI	0.031	0.142	0.131	0.739

Source: Calculated with smartPLS v.3

These high values indicate that the constructs and the items are relevant for the model, ensuring consistency and robustness for the results of the research.

Further on, the Fornell-Larcker Criterion was used for determining the discriminant validity of the variables in the model. We can notice from Table 6 that AVE's square roots of each construct are higher than the correlation of the constructs with

Table 7 Bootstrapping test

	Original sample (O)	Sample mean (M)	Standard deviation (STDEV)	T statistics (|O/ STDEV|)	p values	Confidence interval bias corrected
COMCSR →CNG (H1)	0.213	0.210	0.072	2.967	0.003	(0.060, 0.349)
HRCSR →CNG (H2)	0.546	0.552	0.066	8.272	0.000	(0.388, 0.653)
HRCSR →PNDI (H3)	0.131	0.144	0.129	1.019	0.309	(−0.329, 0.266)
PNDI →COMCSR (H4)	0.142	0.169	0.096	1.469	0.143	(−0.224, 0.263)

Source: Calculated with smartPLS v.3

Table 8 Blindfolding test

Constructs	SSO	SSE	Q^2 (=1 − SSE/SSO)
CNG	624.000	440.380	0.294
COMCSR	468.000	463.203	0.010
HRCSR	1092.000	1092.000	
PNDI	624.000	622.195	0.003

Source: Calculated with smartPLS v.3

the other constructs in the model, thus ensuring the discriminant validity of the proposed model.

Afterward, we applied the bootstrapping test. As it can be noticed from Table 7, the t values are above 1.96 and p values are below 0.05 only for the first two correlations: the one from COMCSR to CNG and the one from HRCSR to CNG. Also, taking into account the confidence interval bias corrected, we notice that only the hypotheses H1 and H2 are accepted meanwhile H3 and H4 are rejected.

The blindfolding test was applied to check the predictive relevance of the proposed model. We can notice from Table 8 that Q^2 values are above 0 which reflects the relevance of the variables in the model. CNG has the highest value (0.294), being a variable strongly influenced by HRCSR and COMCSR.

The descriptive statistics for the items retained in the research model is summarized in Table 9.

Asked about the CSR initiatives developed by their organizations during the pandemic, most respondents mentioned CSR projects related to the health system (27.4%), followed by those targeting the employees (23.6%), the environment (19.1%), the disadvantaged community (17.8%), and educational projects (17.8%). Among the respondents, 56.7% mentioned that the organizations had the same amount of CSR projects during the pandemic as before, 19.7% of them that the involvement was higher, 21% said that there was no involvement before and neither during the COVID-19 pandemic, and 2.5% described a lesser involvement between 2020 and 2022.

Table 9 Descriptive statistics of the items

Items	Mean	Standard deviation	Outer loading
CNG1	3.378	1.379	0.792
CNG2	3.224	1.347	0.817
CNG3	3.551	1.331	0.908
CNG4	3.372	1.271	0.778
COMCSR2	2.558	1.538	0.894
COMCSR3	2.622	1.551	0.906
COMCSR1	2.936	1.458	0.874
HRCSR1	3.115	1.752	0.705
HRCSR2	2.846	1.615	0.747
HRCSR4	1.927	1.371	0.667
HRCSR5	2.338	1.604	0.785
HRCSR6	2.744	1.463	0.678
HRCSR8	3.500	1.555	0.705
HRCSR9	3.283	1.499	0.673
PNDI2	2.833	1.353	0.790
PNDI3	3.135	1.359	0.687
PNDI4	2.237	1.229	0.765
PNDI6	2.055	1.212	0.709

Source: Calculated with smartPLS v.3 and JASP Team (2021) v. 0.16

5 Discussions

Only the first two hypotheses (H1 and H2) were validated, both showing a direct and positive impact from the community-oriented CSR and employee-oriented CSR to the changes implemented by organization between 2020 and 2022.

H1: The corporate social responsibility oriented toward the community has a direct and positive impact on the changes implemented by the organizations after 2 years in which they faced the pandemic. The hypothesis was accepted ($p = 0.003$, $t = 2.967$) as it can be seen in Table 7. This means that the organizations which implemented CSR projects for the community were also more willing to implement other positive changes internally (become greener, more friendly with the employees, more secure, and more digital). This result is in accordance with other similar studies which show that the health crisis was seen by some companies as an opportunity to develop "more authentic CSR" oriented toward the community and the environment (He & Harris, 2020). Ikram et al. (2020) also connected the various dimensions of CSR during the pandemic. The authors show that the crisis challenged companies to address multiple domains related to community, but also to the environment, to employees, or to digitalization. These were in many cases motivated by an urge to cope with the challenges brought by the pandemic and the inherent uncertainty.

H2: The corporate social responsibility oriented toward the employees has a direct and positive impact on the changes implemented by the organizations after 2 years in which they faced the pandemic. The hypothesis was accepted ($p = 0.000$,

$t = 8.272$) which means that companies focused on employees tend to be proactive and become greener and more digital. The health crisis was a driver for many organizations. This result is in accordance with other studies which analyzed the CSR behavior of companies during the COVID-19 pandemic (Khattak & Yousaf, 2021; Glavas, 2016). In many cases, the CSR initiative focused on the employees includes other components referring to digitalization and adopting behaviors that are friendlier with the environment. As Glavas (2016) states, CSR can empower the employees to adopt positive changes at work, thus contributing to their motivation and satisfaction.

H3: The corporate social responsibility oriented toward the employees influences the pandemic impact on both the organizations and the employees. The hypothesis was not supported ($p = 0.309$, $t = 1.019$) which means that there was not found any connection between the employee-oriented CSR and the pandemic impact. A validation would have meant that organizations focused on employees would have the capacity to diminish the negative impact of the pandemic. At least for the case of Romanian organizations, this relationship was not identified. The negative impact included items referring to the financial loss, the mental and physical health problems, deficiencies regarding online security.

H4: The pandemic impact on both the organizations and the employees influences the corporate social responsibility oriented toward the community. The hypothesis was not supported ($p = 0.143$, $t = 1.469$) which means that there was not identified a relationship between the negative impact of the pandemic and the community-oriented CSR strategy in Romanian organizations.

These results are helpful for organizations that can understand the way their CSR actions influence other relevant areas such as sustainability, digitalization, and cyber security. Most of the organizations in the sample were small and medium enterprises (31.8%) which might explain the results of hypotheses H3 and H4. SMEs are having fewer financial resources for the CSR initiatives and were more affected during the pandemic (Bartik et al., 2020; Engidaw, 2022).

6   Conclusions

The paper analyzes the capacity of Romanian organizations to cope and survive to a crisis like the COVID-19 pandemic. Corporate social responsibility can be a tool for raising the resilience of the organizations, their employees, and the community. The study was conducted taking into account the period 2020–2022 with the hope that 2022 represents the end of the pandemic and the start for recovery and a post-pandemic era. As we have shown, during these last 2 years, many articles were written speaking about the end and that proved to be wrong.

The results show a direct and positive relationship between the employee-oriented and community-oriented CSR and other positive changes that happened at the organizational level (digitalization, green policies, friendlier human resources policies, and increased online security). The practical implications of the research refer to the fact that managers in various organizations start to understand the role played by CSR in becoming more resilient. An organization that takes care of its employees will create a climate of trust where the employees might feel more motivated and satisfied at work even in difficult times.

There should be a partnership between the employer and the employees. Also, an organization which invests in the community taking care of it will receive benefits in the future when people in the community could return the help by choosing to buy products and services from the organization that helped them. A CSR strategy must be developed with the future in mind especially in challenging times like the COVID-19 pandemic when everyone was affected in numerous ways.

Future research directions can include an extension of the present research to specific industries in Romania because there were important differences in the way the organizations were affected. The novelty of the present research consists in the variables that were chosen. There are no other studies in Romania regarding the way CSR influences the positive changes that were triggered by the restrictions during the pandemic. Even if, at first, everyone had to adjust and change was difficult, after 2 years, people and organizations adjusted and they remained with some positive behaviors. Working from home incentivized companies to digitalize faster, to increase their online security, and to create friendlier human resources policies. At the same time, this also led to a reduction in the carbon footprint of the organizations.

References

Aguinis, H., Villamor, I., & Gabriel, K. P. (2020). Understanding employee responses to COVID-19: A behavioral corporate social responsibility perspective. *Management Research, 18*(4), 421–438. https://doi.org/10.1108/MRJIAM-06-2020-1053

Ahmed, J. U., Islam, Q. T., Ahmed, A., Faroque, A. R., & Uddin, M. J. (2021). Corporate social responsibility in the wake of COVID-19: Multiple cases of social responsibility as an organizational value. *Society and Business Review, 16*(4), 496–516. https://doi.org/10.1108/SBR-09-2020-0113

Appiah, J. K. (2019). Community-based corporate social responsibility activities and employee job satisfaction in the US hotel industry: An explanatory study. *Journal of Hospitality and Tourism Management, 38*, 140–148. https://doi.org/10.1016/j.jhtm.2019.01.002

ARUP. (2020). *Future of offices: In a post-pandemic world.* Accessed April 6, 2022, from https://www.arup.com/perspectives/publications/research/section/future-of-offices-in-a-post-pandemic-world

Bartik, A. W., Bertrand, M., Cullen, Z., Glaeser, E. L., Luca, M., & Stanton, C. (2020). The impact of COVID-19 on small business outcomes and expectations. *Proceedings of the National Academy of Sciences, 117*(30), 17656–17666. https://doi.org/10.1073/pnas.2006991117

Bethell, M. (2020). *Five lessons learned from COVID-19 recovery.* Accessed April 6, 2022, from https://www2.deloitte.com/global/en/blog/responsible-business-blog/2020/five-lessons-learned-from-covid-19-recovery.html

Chen, Z., & Hang, H. (2021). Corporate social responsibility in times of need: Community support during the COVID-19 pandemics. *Tourism Management, 87*, 104364. https://doi.org/10.1016/j.tourman.2021.104364

Chin, W. W. (1998). The partial least squares approach for structural equation modeling. In G. A. Marcoulides (Ed.), *Modern methods for business research* (pp. 295–336). Lawrence Erlbaum Associates Publishers.

Cotofan, M., de Neve, J. E., Golin, M., Kaats, M., & Ward, G. (2021). *Work and well-being during COVID-19: Impact, inequalities, resilience, and the future of work.* World happiness report. Accessed April 4, 2022, from https://worldhappiness.report/ed/2021/work-and-well-being-during-covid-19-impact-inequalities-resilience-and-the-future-of-work/

De Smet, A., Mysore, M., Reich, A., & Sternfels, B. (2021). *Return as a muscle: How lessons from COVID-19 can shape a robust operating model for hybrid and beyond.* Accessed April 6, 2022, from https://www.mckinsey.com/business-functions/people-and-organizational-performance/our-insights/return-as-a-muscle-how-lessons-from-covid-19-can-shape-a-robust-operating-model-for-hybrid-and-beyond

Deloitte. (2020). *How to reduce the pandemic impact on employees: A guide for company leaders.* Accessed April 2, 2022, from https://www2.deloitte.com/ua/en/pages/human-capital/articles/impact-of-covid-19.html

Diedrich, D., Northcote, N., Roder, T., & Sauer-Sidor, K. (2021). *Strategic resilience during the COVID-19 crisis.* McKinsey & Company. Accessed April 3, 2022, from https://www.mckinsey.com/business-functions/strategy-and-corporate-finance/our-insights/strategic-resilience-during-the-covid-19-crisis

Engidaw, A. E. (2022). Small businesses and their challenges during COVID-19 pandemic in developing countries: In the case of Ethiopia. *Journal of Innovation and Entrepreneurship, 11*(1), 1–14.

Eurofound. (2021). *Business not as usual: How EU companies adapted to the COVID-19 pandemic.* Publications Office of the European Union.

Eurostat. (2022). *Key indicators of labour productivity show recovery.* Accessed April 3, 2022, from https://ec.europa.eu/eurostat/web/products-eurostat-news/-/ddn-20220314-1

García-Sánchez, I. M., & García-Sánchez, A. (2020). Corporate social responsibility during COVID-19 pandemic. *Journal of Open Innovation: Technology, Market, and Complexity, 6*(4), 1–21. https://doi.org/10.3390/joitmc6040126

Ghozali, I. (2014). *Structural equation modeling, metode alternatif dengan partial least square (PLS)* (4th ed.). Badan Penerbit Universitas Diponegoro.

Glavas, A. (2016). Corporate social responsibility and employee engagement: Enabling employees to employ more of their whole selves at work. *Frontiers in Psychology, 796.* https://doi.org/10.3389/fpsyg.2016.00796

Gürlek, M., & Kılıç, I. (2021). A true friend becomes apparent on a rainy day: Corporate social responsibility practices of top hotels during the COVID-19 pandemic. *Current Issues in Tourism, 24*(7), 905–918. https://doi.org/10.1080/13683500.2021.1883557

Hair, J. F., Jr., Hult, G. T. M., Ringle, C. M., Sarstedt, M., Danks, N. P., & Ray, S. (2021). *Partial least squares structural equation modeling (PLS-SEM) using R: A workbook.* Springer.

He, H., & Harris, L. (2020). The impact of Covid-19 pandemic on corporate social responsibility and marketing philosophy. *Journal of Business Research, 116,* 176–182.

Henseler, J., Ringle, C. M., & Sinkovics, R. R. (2009). The use of partial least squares path modeling in international marketing. In *New challenges to international marketing.* Emerald Group Publishing.

Ikram, M., Zhang, Q., Sroufe, R., & Ferasso, M. (2020). The social dimensions of corporate sustainability: An integrative framework including COVID-19 insights. *Sustainability, 12*(20), 8747. https://doi.org/10.3390/su12208747

ILO. (2022). *ILO downgrades labour market recovery forecast for 2022.* Accessed April 5, 2022, from https://www.ilo.org/global/about-the-ilo/newsroom/news/WCMS_834117/lang%2D%2Den/index.htm

JASP Team. (2021). *JASP, version 0.16, computer software.*

John, A., Shahzadi, G., Khan, K. I., Chaudhry, S., & Bhatti, S. (2022). Charity begins at home: Understanding the role of corporate social responsibility and human resource practices on employees' attitudes during COVID-19 in the hospitality sector. *Frontiers in Psychology, 13,* 828524–828524.

Karian and Box and the Global Institute for Women's Leadership. (2021). *New ways of working: The lasting impact and influence of the pandemic.* Accessed April 7, 2022, from https://www.kcl.ac.uk/giwl/assets/New-ways-of-working.pdf

Khattak, A., & Yousaf, Z. (2021). Digital social responsibility towards corporate social responsibility and strategic performance of Hi-Tech SMEs: Customer engagement as a mediator. *Sustainability, 14*(1), 131. https://doi.org/10.3390/su14010131

Mahmud, A., Ding, D., & Hasan, M. M. (2021). Corporate social responsibility: Business responses to Coronavirus (COVID-19) pandemic. *SAGE Open, 11*(1). https://doi.org/10.1177/2158244020988710

Mao, Y., He, J., Morrison, A. M., & Andres Coca-Stefaniak, J. (2021). Effects of tourism CSR on employee psychological capital in the COVID-19 crisis: From the perspective of conservation of resources theory. *Current Issues in Tourism, 24*(19), 2716–2734. https://doi.org/10.1080/13683500.2020.1770706

Merriam Webster dictionary. (n.d.). *Definition of resilience.* Accessed April 1, 2022, from https://www.merriam-webster.com/dictionary/resilience

OECD. (n.d.). *The impact of COVID-19 on employment and jobs.* Accessed April 2, 2022, from https://www.oecd.org/employment/covid-19.htm

Raimo, N., Rella, A., Vitolla, F., Sánchez-Vicente, M. I., & García-Sánchez, I. M. (2021). Corporate social responsibility in the COVID-19 pandemic period: A traditional way to address new social issues. *Sustainability, 13*(12), 6561. https://doi.org/10.3390/su13126561

Ramya, S. M., & Baral, R. (2021). CSR during COVID-19: Exploring select organizations' intents and activities. *Corporate Governance: The International Journal of Business in Society, 21*(6), 1028–1042.

Reeves, M., Shmul, Y., & Martinez, D. Z. (2021). *How resilient businesses created advantage in adversity during COVID-19.* Accessed April 4, 2022, from https://www.bcg.com/publications/2021/how-resilient-companies-created-advantages-in-adversity-during-covid

Ringle, C. M., Wende, S., & Becker, J. M. (2015). *SmartPLS 3.* SmartPLS GmbH. http://www.smartpls.com

Suto, M., & Takehara, H. (2022). Employee-oriented corporate social responsibility, innovation, and firm value. *Corporate Social Responsibility and Environmental Management*, 1–14. https://doi.org/10.1002/csr.2232

Taber, K. S. (2018). The use of Cronbach's alpha when developing and reporting research instruments in science education. *Research in Science Education, 48*, 1273–1296.

United Nations Environment Programme. (2021). *Adapt to survive: Business transformation in a time of uncertainty.* UNEP.

Yana, A. G. A., Rusdhi, H. A., & Wibowo, M. A. (2015). Analysis of factors affecting design changes in construction project with partial least square (PLS). *Procedia Engineering, 125*, 40–45.

Yin, C., Zhang, Y., & Lu, L. (2021). Employee-oriented CSR and unethical pro-organizational behavior: The role of perceived insider status and ethical climate rules. *Sustainability, 13*(12), 6613. https://doi.org/10.3390/su13126613

Silvia Puiu is an Associate Professor PhD Habil. at the Department of Management, Marketing and Business Administration within the Faculty of Economics and Business Administration, University of Craiova, Romania. She has a PhD in Management and teaches Management, Marketing, Ethics Management in Business, Creative Writing in Marketing. In 2014, she conducted research on *Ethics Management in Higher Education System of Romania.* In 2015, Silvia Puiu graduated postdoctoral studies on *Ethics Management in the Public Sector of Romania.* During last years, Silvia Puiu published more than 60 articles in national and international journals or in the proceedings of international conferences. Her research covers topics from corporate social responsibility, ethics management, sustainability, marketing and management. She is also the founder of an NGO in Romania—Building Hopes—which organizes workshops for non-formal education of youngsters.

Responsiveness, Strategy and Health as Diplomacy: The Unlikely Case of Serbia

Milan Todorovic

1 Beyond Introduction: Conceptual Foundations

Three long winters and another one coming. The spectre of viral resurgence looms again.

At the time of completing this text in July 2022, another 2520 cases of covid were identified in Serbia (Politika, 2022[1]). As compared with the approximate 400 per day only a week before, this seems like the evidence of The Virus's resurgence in the summer of 2022; the informally labelled, current *Centaurus* strain being the likely culprit, and with the likelihood of new variants emerging as the disease mutates into the future.

Why "The Virus"? Why such capitalisation? This paper is about discourse just as much as it is about practice.[2] The global discursive, cultural and policy changes already brought about by the COVID-19 outbreak are greater in scope, scale and reach than those that came in the aftermath of many more deadly diseases of the past, including cholera, *Variola Vera*[3] (smallpox), the bubonic plague and the most recent disastrous agent, the Spanish Flu. While the plague radically altered the demographics and very history of Europe from the fourteenth century, the Plague's cultural impact is still felt in visual and textual narratives spreading across centuries

---

[1] The figure cited on 16 July 2022 above has already risen to over 3000 two days later, indicating new concerns at the time. However, the pandemic seems to be over, at least for now and at least with this agent (Reuters, 2022a, b; WHO, 2022c)

[2] The author shall aim to invoke the works of Michel Foucault to formulate some likely insights in this respect.

[3] This particular infectious agent is a matter of special significance in Serbia, historically, factually and culturally.

M. Todorovic (✉)
London Metropolitan University, London, UK
e-mail: m.todorovic@londonmet.ac.uk

© The Author(s), under exclusive license to Springer Nature Switzerland AG 2023
S. O. Idowu et al. (eds.), *Corporate Social Responsibility in the Health Sector*, CSR, Sustainability, Ethics & Governance, https://doi.org/10.1007/978-3-031-23261-9_6

and geographies, the impact of covid is most evidently an economic and cultural one. Compared to the relative death rate recorded, the economic and cultural effects of covid are highly significant.

Serbia was one of the first countries in Europe to act decisively. Indeed, Serbia declared a national state of emergency effective 15 March 2020, and its borders were shut that day, very nearly hermetically (US Embassy in Serbia, 2020). The most unyielding of lockdowns was introduced, preventing intercity travel without special permission from the police. Foreigners non-domiciled in the land were barred from entry and its own citizens found to be entering and in breach of the 14-day entry quarantine were to be fined and/or imprisoned for up to 3 years (exit.al, 2020; Šantić and Antić, 2020). There was a special curfew for those aged 65 or over (Blic.rs, 2020a; eVršac, 2020; Informer.rs, 2020; Kurir.rs, 2020; RTV, 2020). With slight easing in late April 2020, it lasted 7 weeks and the elderly were allowed out only at night and very early in the morning, once a week, for up to 500 m outside of their residential address, with select shops especially open for the senior citizens exposed to such restrictions, unprecedented in peacetime. All this on the day with only 55 cases and no fatalities reported. (Swift response in the wake of disturbing reports from abroad, and as recent data substantiates, modelled on the 1972 smallpox outbreak when the virus was brought into the country for the first time in 42 years. This major threat was resolutely suppressed and beaten by the most uncompromising of actions (e.g. Мијалковић, 2022, among various others[4]).)

As will be seen, the consensus among authors and journalists in Serbia is that the traumatic experience in early 1972, with the smallpox import affecting Serbia had been decisive in the radical even overbearing moves to protect the populace during the early days of COVID-19. However, there were many pragmatic issues. Online retail, which is still in relative infancy and remains comparatively pricey even in the most developed of countries, was hardly developed in Serbia at the time, complicating the lives of those subjected to the harshest of restrictions (Антељ & Албуновић, 2020; B92.net, 2021a; Blic, 2020b; RTS, 2020). The visual distresses of those early days of the pandemic were particularly pronounced in the form of covid hospitals being set up in sports venues, and on the premises of Belgrade City Fair,[5] otherwise a major event, trade show and exhibition space (Alo, 2020; paragraf. rs, 2020); the same occurred with the Novi Sad City Fair, as an impromptu covid clinic[6] (Savanović, 2020).

Those, latter aspects of battling the COVID-19 pandemic quite frequently occurred across the world, and in some cases, over a rather significantly prolonged period, by comparison.

---

[4] As will be seen below, this was a significant event in the country's twentieth Century history and will be treated with due attention.

[5] See Radio Slobodna Evropa (2021) concerning its closure as an impromptu covid hospital.

[6] China supplied Serbia with much-needed ventilators at the onset of the pandemic in the country, when no other help of such scale or significance was forthcoming (see Wang, 2021).

1.1 Where Does Serbia Then Differ and Stand out?

There are several aspects to the claim of such uniqueness explored in this text.

1.1.1 Historical Backdrop: *Variola Vera*—Major Disease Staved Off— Ingrained in Collective Memory

"Variola" is the name of the virus that causes smallpox (WHO, 2022b). The historical and cultural significance of its Latin species name will be explored in this section of the text. A disease of extreme mortality, it was considered near-eradicated in Europe and indeed in Yugoslavia until the 1972 outbreak (mortality, ranging in this instance from approx. 9 to 35% and at 20% on aggregate as reported[7]). The man identified as Patient Zero Ibrahim Hoti (Давидов-Кесап, 2022c; Ferhadbegović, 2020; Миладиновић/BBC, 2020; Telegraf.rs, 2020) returned from pilgrimage in February 1972. Давидов-Кесап (2022c) states that it is presumed he had contracted it in a Baghdad street market on 6 February as the mean incubation period is 11.3 days (Litvinjenko et al., 1973: 3). According to the Yugoslav paper submitted on behalf of the WHO in 1973, the disease had started to spread from 15 to 16 February of that year after his return to his native village of Dranjane in Kosovo, near the town of Djakovica/Gjakove (Litvinjenko et al., 1973: 1) and the final infection was detected on 11 April 1972. Litvinjenko et al. (1973: 2) further assert that Patient Zero may also likely have been infected in Basra on 3 February, as both cities were reported to have had a measurable number of cases of smallpox.

As Patient Zero was vaccinated, it is believed he had developed fewer identifiable symptoms of the smallpox disease and had been spreading the virus undetected for a while until the cause of his and other cases were identified on 14 March 1972. As a non-typical case, he had developed no lesions on his skin a month after the initial symptoms of fatigue but serological evidence pointed at him being the one who imported the virus into the country (Litvinjenko et al., 1973: 2). Reportedly, opinions are divided as to the effectiveness of inoculation he had received in Skopje as anecdotal narratives hint that he may have rubbed off the site of the injection with medicinal alcohol immediately after receiving the jab (Telegraf.rs, 2020). While this cannot be verified, both Ferhadbegović (2020) and Litvinjenko et al. (1973) state the mortality ranged considerably, dependent on whether or not a patient was vacci-nated, with Ferhadbegović stating that for the unvaccinated the mortality rate was as high as 35% *"and 9 per cent in those with the scars of old vaccinations."* The comparative severity of smallpox and COVID-19 is starkly different. Despite the enormous, comprehensive effort involving the entire country in Yugoslavia, there were 175 confirmed cases of the disease and 35 deaths or 20% in total (Давидо-в-Кесап, 2022b, c; Ferhadbegović, 2020; Пејовић, 2022; Litvinjenko et al., 1973; BBC, 2018, however, states that 40 fatalities had occurred in total). It is less known

---

[7] Such a broad range will be explained below.

that a Yugoslav patient had also arrived in Hannover, which led to a local quarantine involving 645 people and the inoculation of another 70,000 (Ferhadbegović, 2020; also, as implicit in a graph—Fig. 4—presented in Litvinjenko et al., 1973: 12; Пејовић, 2022). However, in the end the same patient was the only one registered in Germany as positive (Ferhadbegović, 2020; Пејовић, 2022).

In Yugoslavia in the spring of 1972, the deadly epidemic had spread across several towns and cities (Ilic & Ilic, 2022; Litvinjenko et al., 1973; Trifunović, 2017) before eventually being contained through nationwide concerted effort. This, however, could not have been achieved without vastly effective measures involving the strictest state of emergency. Startlingly, over the 55 days of crisis, the health service had vaccinated 18 million people out of the then Yugoslav population of 20 million. The government had asked the WHO for a supply of another 2 million more effective vaccines half-way through the process (Henderson, 2008: 918); the entire city of Belgrade, then counting 1.2 million inhabitants was inoculated in 10 days (Давидов-Кесар, 2022c); the effectiveness of the campaign to contain the virus and protect the populace is just as significant, noting that it took 4 and 5 weeks respectively to identify the cause in two distinct contexts (Давидов-Кесар, 2022a, c; Пејовић, 2022). D A Henderson, then WHO's Chief of Global Smallpox Eradication Programme, stated that Yugoslavian medical staff and his policy-lead counterparts in the country:

> ...took very heroic steps to stop the spread. They stopped cars along the road to vaccinate people; they went from village to village, vaccinating almost the entire country—18 million out of 20 million people. The secretary of health was deeply concerned as to whether they were succeeding so he asked me to come and review what was going on. I did, and I told him they had done a really fine job and I agreed to say so on the radio to reassure the country.
> Henderson (2008: 918)

In addition to the initial, main outbreak in Kosovo, Belgrade as the capital was the second most affected part of the country. The virus had failed to spread to the west, with one case also reported in Montenegro. According to historic data (Пејовић, 2022), the authorities of Slovenia, then a republic of federal Yugoslavia, would not allow entry to anyone from other parts of the country unless they possessed "the yellow booklet," i.e. the document proving vaccination—rather reminiscent of the far more recent "covid passports," vaccination certificates, COVID-19 track & trace and testing apps, etc.

The reported harrowing experience affecting Belgradians is reflected in the fact that two popular leisure spots "Čarapić's Elm" and the romantically named "1000 Roses" ("Чарапићев брест," "1000 ружа") were among the four sites of the quarantine, the other two being respective major clinics for infectious, tropical diseases and skin/STI infections (BBC, 2018; Новости онлајн, 2022). In the feature series published in March and April 2022, to coincide with the semi-centenary of the smallpox outbreak, Belgrade's respected "Politika" daily had interviewed some of the key surviving protagonists among the health care professionals who led the fight against the smallpox virus, as health authorities in Serbia were the ones who bore the brunt: Dr. Ana Gligić, Prof. Dr. Radmilo Petrović, Prof. Dr. Zoran Radovanović (Давидов-Кесар, 2022a–c). Dr. Gligić provides an exhaustive description on how the *Variola* virus was thoroughly destroyed, sealed and buried in the grounds of the Torlak Virology Institute where she worked during the 1972 smallpox epidemic, adding that a live virus of the Spanish Flu was isolated in Greenland from a deceased sufferer buried in permafrost and thus urging exceptional caution with such agents. The WHO, she claims, called five times to check that the procedure had been followed through in Belgrade (Давидов-Кесар, 2022a).

Belgrade's leading "broadsheet" also revisited the fact that *reading between the lines* was required at the time, not least as the leadership of Yugoslavia had aimed to pre-empt any sense of panic before effective measures were in place (Пејовић, 2022), which would threaten to add to the sense of uncertainty and concern. Пејовић (2022) also asserts that immediately upon learning of the smallpox diagnosis, the two lead specialists and department heads from the Belgrade Infectious Diseases Clinic were sent to Djakovica; Dr. Miomir Kecmanović; and Dr. Vojislav Šuvaković, both with extensive experience of treating smallpox in India on behalf of the WHO were there from the outset. BBC (2018) emphasises that Serbia had implemented its first smallpox vaccination drive in 1839 (at the time still a semi-autonomous principality under the auspices of the Ottoman Empire) and 14 years before Britain did so. Moreover, in an interview wittily entitled "How Variola defeated Corona," historian Dr. Radina Vučetić reflects on the fact that Prince Miloš Obrenović, who ruled Serbia at the time had implemented an ambitious modernisation programme, contained a plague outbreak and passed laws that no-one could enter into marriage, attain a scholarship or get accepted into a grammar school without being inoculated, having seen three of his own children succumb to smallpox. Known as an autocrat, he elicited the assistance of civil servants and the Church to promote vaccination every 3 months (Мијалковић, 2022[8]).

Smallpox was declared eradicated by the WHO by 1980 (WHO, 2022b); the Yugoslav outbreak was the last of its kind in Europe and the most significant one in 20–25 years (Henderson, 2008: 918). The broad variety of sources consulted here agree that containing this outbreak was possible due to the resolute, organised and highly disciplined response involving health, the armed forces, police and civil service, engaging and educating the population.

This was the factual side of things, relating to the 55 days that led to the mass vaccination of 18 million [then] Yugoslavians.

The cultural, discursive impact of these crucial events is equally curious.

On the subject of *Variola Vera*, the named agent of the disease, acclaimed Serbian film director Goran Marković, had produced the eponymous movie that affected scores of viewers across Serbia and quite possibly across former Yugoslavia (IMDb, n.d./2022; Lazic, 2012; prlekija-on.net, 2012; and, for instance, r/TrueFilm, 2021[9]). The feature film *Variola Vera* straddles genres; the author himself states that the

---

[8] [Vaccination practice of the time was as 'grafting' in Serbian, an apt allegory for the method in use then.]

[9] A US film enthusiast who discovered *Variola Vera* during the COVID-19 pandemic reflects on it in this reference.

primary aim was to present a satire that would use the epidemic as a metaphor for society and that he had been long inspired by Albert Camus's *The Plague*, aiming to create his own filmic rendition of the theme (Lazic, 2012). The result is a curious mix of genres: satire, horror, disaster movie, social drama and chiller. Released at the tenth anniversary of the outbreak, it had an impact well-described as a "cult" following. Mandić (2019: 495) takes issue with some aspects of the liberal poetic licence which suggested that there was corruption, soap-like intrigue, dishonesty and negligence among the staff who treated the outbreak at one of Belgrade's quarantined clinics. Indeed, such is the example of courageous nurse Dušica Spasić, after whom an award for outstanding nurses in the land was named following her passing when fitting the outbreak (Давидов-Кесар, 2022c).

Regardless of one's stance on such interpretations of sombre facts within the arts, the film has left a deep mark on media audiences within Serbia and has been a frequent referential point in pop culture and slang, for example relating to one-liners, characters, citations, etc. The significance of this may be somewhat related to contemporary matters discussed in this paper. Fifty years is a relatively short period of time for a society to absorb and reflect on such experience and cultural content tends to reinforce, reinterpret and [re]direct such reflexions.

1.1.2   Discipline, Libertarian Responses, Regulation and Popular Scepticism

In 2020, Serbia's first, exceptionally strict initial lockdown, curfew and border closure policy proved a success. As was the case in the 1972 smallpox outbreak, this elicited the involvement of the border force, police, the military, in addition to health care sectors normally assigned the task. While Ritchie et al. (2022), WHO (2022a) and others provide very useful infographics and/or dashboards to monitor the spread and incidence of COVID-19 infections, recoveries and fatalities, it is rather ironic that most easily accessible collated data relating to the timeline and key events seems provided by Wikipedia (!). Irrespective of its improved functionality, this source is not one that can be considered sufficiently dependable for a matter of this gravity. (Perhaps this is something that can be of use to researchers and policymakers alike as a call for enhanced simplicity and readability of data pre- sentations in accessible formats.) Following 7.5 weeks of measures among the strictest in Europe, an easing of those was announced on 6 May 2020 (srbija.gov. rs, 2020; Đokić & Čvorović, 2021). While some highly critical texts have appeared to question the motivation behind the easing of measures (e.g. EWB, 2020; Ioniță, 2021), the WHO (2022a) dashboard can be triangulated to such treatises. There are indications that June 2020 saw a rise in the numbers of cases though further comparisons do not appear to be definitively causal. Moreover, the coverage of protests in Serbia in early July 2020 may be indicative of a discursive practice seeking to identify the causes within perspectives known to accustomed external observers of the region. Once more, it is worth engaging in a reflexive, contextual comparison, especially now that sufficient time has elapsed to prevent the metaphorical *"Heisenbergian"* blurring that occurs in dealing with contemporary events[10] (Katz & Csordas. 2003; Savić, 2001; Todorovic, 2022). The features of the protests of July 2020 in Belgrade may be contrasted with other reported instances worldwide[11] so as to arrive at a comparative perspective rooted in adequate contexts (e.g. BBC, 2020a, b; CNN, 2022; Convery, 2021; DW.com, 2021b; EWB, 2020; Ferhadbegović, 2020; Ioniţă, 2021; Jones, 2020; Özdüzen et al., 2021; Reuters, 2020). The July 2020 short lockdown was cancelled and yet, the pattern of lockdowns, mandatory tests and mandatory safety measures continued to be implemented over the following months in ways that appear to comply with the overall strategies elsewhere, though arguably in a less weighty manner than in some instances. However, as stated in The Heritage Foundation Economic Freedom Index (2022), the country ranks as 111th in terms of the stringency of its response to the covid pandemic. This contrasts with its 59th global rank in terms of overall economic freedom and progress since 2017. Similar indicators are worth looking at in the context of sections to follow, which deal with the significant drive to procure, supply, distribute and donate COVID-19 vaccines.

Two years on, an AP article neatly sums up all the internal contradictions of varied inescapable suppositions about the July 2020 protest (Stojanovic, 2020) and its aftermath being the reversal of an earlier decision to impose a new lockdown at that time. Conversely, and in relation to the perceived libertarianism—or scepticism of the Serbian public—in an almost prophetic article by the BBC (2018) dealing with the circumvention of measles vaccination, citing Dr. Radovanović, a key protagonist of the 1972 fight against *Variola,*[12] that "[people] even perceive doctors as *the authorities,*[13] which have let them down many times before." The term "libertarian" in its multiple meanings plays a significant role in Serbian language. The word *"slobodarski,"* as the literal translation would have it, would score fairly high in content analysis of media texts in the language. In this respect, once more as contrasted with other media texts relating to other lockdown protests worldwide, the nonconformist response of parts of the populace can be explored by a dispassionate reader: e.g. BBC (2020a, b), CNN (2022), Convery (2021), DW.com (2021a, b), EWB (2020), Ferhadbegović (2020), Ioniţă (2021), Jones (2020),

---

[10] A note on methodology: This text is decisively based on the systematic collection of a wide array of secondary sources, comprising, academic treatises in select key areas; documentary and archival data applicable to the context; journalistic accounts and documents of relevance. The author has consciously steered clear of data collection by means of interviews with key protagonists. This, to some extent thanks to methodological precautions concerning contemporary events (Savić, 2001), and once more, as invoking the level of decision-making involved here would steer the text closer to convoluted areas which may not actually be of use here. Elements of immersive reflexive ethnography (Alvesson & Sköldberg, 2017), not intended at the time of uncontrived data collection, may be of use to merely verify or contest some of the accounts of summer 2020.

[11] See also Vériter et al. (2022) for further context.

[12] The connotations of the Latin term run deep in Serbia as explained, hence the use here as opposed to a more commonplace term 'smallpox'.

[13] Author's emphasis.

Özdüzen et al. (2021), Reuters (2020), once more raising the age-old question about the balance between individual liberties and the common good invoked so many times since early 2020.

In terms of the breadth of discourse, there are a few more observations to be made. Never has this been truer than with the public role of celebrities as role models. And with the benefit of hindsight, it appears that some luminaries can get away with more than others (Allen, 2020; Kuenssberg, 2022; Pietras, 2020). Arguably, however, it is the degree of publicity and the reach of someone's position as a positive role model of *health, as in sports,* that played a role in the following case which in turn had no less than a diplomatic effect. The terming of *libertarianism* and its multiple interpretations including individual choice is a feature that does seem to play a part here. Media narratives surrounding one of the world's leading tennis players (Clarey & Peltier, 2020a, b; Mitchell, 2020; Rathborn, 2020; Rogulj, 2020) and his erroneous initiative to hold an international tennis tournament in Zadar, Croatia, in June 2020 seems to play a major role in subsequent outcomes that continue to affect his standing in the public perception around the world. Despite his profuse apologies, Serbian tennis player Novak Djoković was caught within a controversy that evaded the likes of musicians et al. which seems to continue and is indicative of the matters of freedom of movement and such like, in the emerging *new normal* and foreseeable oscillations between fresh policy measures.[14]

Argued as a matter of principle, public health and adherence to standards, regulation and equal rights for all under the law, the continued narrative produced significantly passionate overtones in January 2022 concerning the player's involvement—or lack thereof—in the Australian Open Grand Slam tournament (Agarwal et al., 2022). Articulated along the discursive proclamations of *sovereignty, border control, equal rights, and duties under rule of law, [in]considerate behaviour,* etc., and risking an international diplomatic spat, was the enforced quarantine detention in a hotel normally designated to the least privileged of humans: refugees and asylum seekers. This being followed by the unvaccinated[15] [post-recovery] athlete's subsequent deportation. Analysts, commentators and critics from the full diversity and breadth of discursive spectrums claimed that the positive aftermath of the very public row involving border agencies, state and federal ministers and the judiciary had pointed at the flaws that far surpass the fortunes of an otherwise, rather fortunate individual (e.g. Abramsky, 2022; BBC, 2022; Byers, 2022; DS, 2022; Freckelton, 2022; Higgins, 2022; Miller, 2022; O'Sullivan, 2022).

Having looked at the facts of the matter, journalists and others (e.g. Horton, 2022) could not find a definite legal flaw in what the player had done, though as Abramsky (2022) points out, sundry find him *obnoxious*, which does remain a matter of personal taste, not policy, or law. The [perceived?] faux pas of the

---

[14] *The New Normal* indeed appears to be a discursive development likely to shape the future as a *precedent.*

[15] A diverse view in principle, explored in scientific ethics terms by Pugh et al. (2022).

player's very public statements seem to be the crux of what is *interpretive not factual* and perhaps, a discursive matter bordering pointed public relations bluster[16] (astutely identified by Attree, 2022 in advance of the event)—gone wrong. Discursive context is key, and it is therefore worth re-examining contextual constellations where appropriate.[17]

The contexts of diplomacy and vaccination, indeed logically lead to *vaccine diplomacy*, next.

1.2   Substantive or Instrumental[18] Vaccination Strategy: Pragmatism, Diplomacy or Altruism?

Relative to its size, economic status and rank, world position, impact and overall prosperity (Legatum Institute, 2021; Moody's Analytics, 2022; The Heritage Foundation, 2022; The World Bank, 2022), Serbia achieved a great deal more than many far wealthier nations during the first few months of the global vaccination drive, than was to be plausibly expected.

An illuminating quote from article by The Economist (2021):

Serbia may not have had such glowing press coverage since the first world war

The said article also features representations on autocracy and pointed references as to the significant origin of its vaccination programme being beyond the established pharmaceutical offerings in the west.[19] Related global press coverage includes other insightful commentary, some of it somewhat satirical:

Belgrade's purchase of Chinese, Russian and western jabs also recalls non-aligned[20] movement's golden days
(Hopkins, 2021)

This, together with the reporter's musings suggesting that the *"autocratic"* leadership, had used its own people as a testing ground by offering non-western vaccines before they were approved by the EU. Irrespective of related commentary

---

[16] For broader policy contexts, see Taflaga (2022).

[17] Reports such as those by AP (2022) point at the public outcry in Serbia and high tensions emerging.

[18] Pertinent notions of instrumentality and substantive motives in strategy, based on Weber's Rationality concept, are explored in Bakir and Todorovic (2010); also, Todorovic (2016).

[19] A rhetorical device used in *othering* [not to be mistaken for its artistic, creative or representational form], *Orientalism*, had been explored by various critical scholars (e.g. Bakić-Hayden, 1995 did so in the context of [former] Yugoslavia). See also Marciacq (2019).

[20] Todorovic (2022) revisits the notion of Serbia's standing as one of the successors to Yugoslavia, the Non-Aligned Movement's co-founder state, which does appear to play a role in modern Serbian diplomacy.

hinting at varied ulterior motives (the two texts are not an exception[21]), the result was one which meant that inoculation—*notably offered but not mandated*[22]—had in its initial phase reached more people in the small Balkan country than many others in its far wealthier EU neighbourhood (Delauney, 2021; Juncos, 2021; Karčić, 2021; Öztürk, 2021a; Petrov, 2021; Vuksanovic, 2021b). This, moreover, to the extent to which Serbian vaccination programme was almost immediately offered to the citizens of neighbouring countries (ABC News, 2021; Schlappig, 2021). Among the first to visit Belgrade and accept the immunisation offer was the prominent Croatian journalist and author, Vedrana Rudan (Kerbler, 2021; B92.net, 2021b). Habitually outspoken on any subject of interest, Rudan animatedly proclaimed her wonder with the Serbian *"buffet menu laden with vaccines"* (B92.net, 2021b).

Indeed, Euronews (2021c) confirms that, at the time of Serbia's early immunisation drive, it had as many as four vaccines to offer its citizens and those willing to venture in. Emmanouilidou (2021) further goes on to suggest that such breadth of variety has political connotations, *a ballot* as she frames the term, claiming that this outcome which enabled choice and secured access to diverse vaccines needs to be subjected to critical scrutiny on *"politicization."* In an interview for Euronews, Prime Minister Ana Brnabić reasoned against contemplations of such nature as she had stated that her government put health before politics, racing to secure the immunisation of its citizens:

> We were one of the first five countries to sign a contract with Pfizer/BioNtech. And we also signed for Sputnik V also for the Chinese vaccine Sinopharm. We were the second country in Europe to get the first dispatches of the Pfizer vaccine immediately after the UK.
> (Euronews, 2021b)

It is, nonetheless, noteworthy that an impact on EU's soft power was detected in the region (Juncos, 2021; Safi & Pantovic, 2021) and that a drive to create a legacy of reconciliation in the wake of a common threat was identified. While Kovacic (2021) focuses on the goodwill effect in the West Balkans region, and Reuters (2021a) points out that the country donated thousands of vaccines to "former enemy," there were regional responses that pointed out, somewhat in line with Hopkins's (2021) critique, that the AstraZeneca vaccine batch donated had not been approved by the EU at the time (Öztürk, 2021a). As stated in a number of reports, the donations of tens of thousands of vials to Bosnia (Öztürk, 2021a; Reuters, 2021a, b; srbija.gov.rs, 2021), Montenegro (Tirana Times, 2021) and North Macedonia (Radio Free Europe, 2021; zdravlje.gov.rs, 2021) have been followed by a number of similar initiatives far surpassing the region of Southeast Europe.

Not contrasting the notion of Non-Aligned Movement logic, Serbia's vaccine donations across the world included those sent to Angola (mfa.gov.rs, 2021a, b), Iran (Öztürk, 2021b), Lebanon (OLJ, 2021), Namibia (mfa.gov.rs, 2021a, b), Uganda (mfa.gov.rs, 2021a, b, c) and, rather notably, an EU member state, the Czech

---

[21] See also: Karčić (2021).

[22] Another note of libertarian approaches not too dissimilar to those adopted so far in the United Kingdom.

Republic (Johnston, 2021; republicworld.com, 2021; Telegraf.rs, 2021). Given recent history and the country's less than enviable economic standing, the question of motives beyond altruism remains plausible. Some analyses seem to point at convoluted diplomatic ruses allegedly aimed at the tactics of its own tensions with Serbia's [former] province (Vuksanovic, 2021b), however, even a cursory examination of the varied recipients of such aid invalidates such supposition. Namely, Czech Republic, Montenegro and North Macedonia had all established diplomatic relations with Kosovo years ago. While it is not implausible that an altruistic motive genuinely exists (e.g. Savkovic, 2022), when combined with a low uptake of vaccines past the initial stage of adoption among Serbia's populace now stalling at 67% wholly vaccinated[23] (Boytchev, 2021; Our World in Data, 2022), a diplomatic effort would be conclusive. Such efforts would not appear to be of limited aim, scale and scope as suggested by Vuksanovic (2021b) and authors of like inclination. The notion of goodwill; the *soft power* seemingly reserved for those who may afford it; the concept of value beyond immediate gain, all seem plausible. Not least the legacy of a post-colonial club of equals that is the Non-Aligned Movement (NAM) where Serbia is now a respected observer, hosting the 50th Anniversary Summit[24] of the *NAM* (Dragojlo, 2021).

Such unexpected results as those that lead to an improved position on the world stage appear to inform and part-formulate some rather critical commentary (DW. com, 2021a) while Conley and Saric (2021) take a more balanced view in their comprehensive analysis of these contexts in the West Balkan region. Kovacic (2021) cites Serbian publisher Robert Coban: *"this is Serbia's best soft power move in the past 50 years,"* and reiteration by the PM Brnabić, that members of her government are *"not interested in politics, we are interested in saving lives."* In the text with an elegantly formulated title, revealing the contexts of soft power, health strategies and goodwill that balances complex constellations of interest, Bechev (2021) refers to the Serbian President Vučić *"stealing the vaccine diplomacy show."*

In several treatises and commentaries which indicate that Serbia had attained a great deal of unexpected *soft power,* Dinic (2021), Filipovic (2022), Subotić (2021) and Tzifakis and Prelec (2021) connect the shift from *mask diplomacy*[25] *to vaccine diplomacy* (Vuksanovic, 2021a, b; Verma, 2020; Wang, 2021) in the Balkans [and beyond], with the broader and increasingly more challenging matters involving major players on the world stage.[26]

---

[23] Another non-mandating vaccination drive, that of the UK now stands at 74.8% fully inoculated persons (Our World in Data, 2022).

[24] Singh and Chattu (2021) address global vaccine equity which was a core theme of the 50th NAM Summit.

[25] The lack of PPE equipment at the start of the crisis now feels largely forgotten.

[26] *[NOTE:]* This text consciously eludes current affairs aspects of increasingly intensifying international relations where the notion of remaining non-aligned becomes uncertain. The author's decades long adherence to nonviolence demand so, and while interdependent concepts may eventually lead to the need for a broader, and more contested analysis, one of the main

As the acme of such efforts, defined as *immunological autonomy* (an *independence* would be contingent on own, self-developed intellectual property based on proprietary patents), Serbia initially secured the licensing rights to two vaccines, Sputnik V and Sinovac. The former is being produced by the Torlak Virology Institute, instrumental in the 1972 battle against smallpox (torlakinstitut. com/en, 2022; Euractiv.rs, 2021; Ralev, 2021; TANJUG, 2021) and the latter by teleSUR (2021) Vuksanovic and Vladisavljev (2021).

The discipline of health diplomacy, and its most recent tributary, that of vaccine diplomacy are very new additions to the larger contexts of social sciences, critical thought, CSR and sustainability as they stand at the intersection of those more established concepts. As stated in a somewhat paradoxical manner, this text is decisively *not about politics* as such. However, the notions of policy, regulation, communications; goodwill, control, discourse and others as seemingly disparate notions all converge in a challenging time like the one, we occupy. Discourse and narrative play an often decisive part in formulating what something *is, and such ontology determines the trajectories and outcomes of practice*. Therefore, the opportunities awarded by crises are no exception.

Sometimes it is necessary to quote rather than paraphrase an original source as it is so well formulated that interpretations may not do it justice. One such example is Jorgen Samso (2021), reporting from Serbia in 2021 for the American Public Broadcasting Service:

> By March, Serbia hadn't received a single Western vaccine through the European Union or the International COVAX programme.[27] Its tens of thousands of Pfizer and AstraZeneca vaccines were secured independently from manufacturers. Russia sent hundreds of thousands of vaccines, but nowhere near what China could offer.[28]
>
> . . .
>
> Because of the Chinese vaccine deliveries, by mid-March, Serbia had vaccinated more adults than any of the other 27 countries in the European Union, becoming continental Europe's best vaccinator. The country raced ahead with inoculations at double the rate of Spain and Germany, internationally trailing only behind Israel, the U.K. and the U.S.

The later slowdown in the uptake of vaccination opportunities may be attributed to a number of factors; as explored above, the libertarian approach (of which there may be many forms) is one possible root[29] of this. Discursive notions explored above, including those distrusting authority embodied in the hazmat suit and arguably ingrained through pop culture representations of *The Plague/Variola Vera* are not dissimilar to the Foucauldian (1975) notion of *panopticism* and *body/discipline*

---

reverberations here is that seemingly desperate matters may hold key to later concord, however implausible that may seem in the eye of the storm.

[27] Among others, UNICEF (2021) did deliver, albeit with delay as recorded and evidenced, as did COVAX.

[28] See Wang (2021).

[29] That libertarian and controlling may not be polar opposites is explored in the text by Bourgeron (2022).

further asserted by the *track/trace* logic of the *new normal*. Experience of oppression as an inoculator from docility is counterbalanced by what is equally perceived as an absence of responsibility that influential individuals rather than groups tend to be identified with in public discourse. To complicate matters further, the term itself is not homogeneous (Barthes & Duisit, 1975) and while much innovation had occurred in technology of devices and content management tools since 1975 and the two treatises' (Barthes and Foucault) rise to prominence, the ontological features of the two persist.[30] Coercion and mandatory behaviour modification led to reactionary responses recorded, while a proactive and educational approaches do not seem to have been inferior in effect. There is no doubt that *ugly emanations of rejection of reason* appeared across the globe, and in the case of Serbia that involved offensive chants against a prominent epidemiologist, a co-author of a study on Influenza (Radovanović & Kon, 2010) and much credited for the rational and comprehensive strategy of containment and control of the spread, much in the same way as his predecessors and counterparts in the alarming outbreak of 1972.[31] Yet despite all such incongruity in the face of the unprecedented,[32] and without aiming to unrewardingly try to dissect the unknowable motivations of major decision makers at a time of contemporaneous developments, the meaning of which may only be revealed once the narratives, facts, layers of meaning and delicate action are disentangled after the event, there are measurable outcomes to be had which may be left to the historians of posterity.

That there were in existence some foundations of *the new, the novel and the definitive* qualitative change may be seen from the growing literature that had been— and should be consulted to benefit related research and novel treatises on, say, a digitised health policy of near-future incremental development (or disruptive for that matter, as the two converge in crises).

In the specific context of the Serbian response [an ongoing matter indeed as we live the history thus described and interacted with by means of minute observation], there are texts that would be of considerable use to the concrete developments at hand. These include: Barovic and Cardenas (2021) on the concrete case at hand explored here; Rokvić (2016) addressing principal matters; Maglajlija (2021) providing a contemporary critical angle; Penev et al. (2014) on commercial diplomacy in Serbia; Petrović-Ristić (2016) on the same subject, closely aligned to the current debate; Prvulović (2015), also providing theorisation in national/historical contexts preceding the crisis but formulating the logic of *soft power* [without "power"?]

Closer to the challenges and finding interim solutions but on a much grander scale than can ever be conceived to be done by a small country with fewer inhabitants than

---

[30] Andersen et al. (2021) providing the conceptual, contextual and factual underpinning for comparative analysis engaging Denmark, Serbia and Sweden as key cases for an exploration of *surveillance as a force for good*.

[31] And, as challenges are utilised as opportunities, where science acts as standard-bearer, Kolaković (2021).

[32] Without any conscious recourse to Zuboff's concepts in this text.

central boroughs of London, the response of which is inevitably tangled within these emerging trends: Rudolf (2021), Rofii (2020), Giusti and Tafuro Ambrosetti (2022), Kazharski and Makarychev (2021), Kurecic and Haluga (2021), Lee (2021), Indraswari and Lestari (2022), Suzuki and Yang (2022), Bachulska (2020), Brown and Ladwig (2020), Chattu and Chami (2020), Chattu et al. (2021), Fayyaz and Siddiqui (2021). Of value in this are also the works of Al Bayaa (2020), providing an interim future perspective with a conceptual extrapolation; Godinho et al. (2022) analysing the tools that may assist implementation of such emerging trends.

Last not least, texts and treatises that lay the foundations of contemporary questions of health diplomacy that rapidly came to the fore as the result of crisis: Feldbaum and Michaud (2010), Kickbusch and Kökény (2013), Kickbusch and Liu (2022), Kickbusch et al. (2021), Kickbusch et al. (2007), Lee and Smith (2011), Michaud and Kates (2013). And as a logical conclusion, an array of texts on soft power, the power of culture to engage, educate, influence, mobilise and permeate, some of which include: Ang et al. (2015), Nisbett (2016) and Pajtinka (2014): Cultural diplomacy (e.g. Zamorano, 2016), forming a conceptual foundation, context-dependent and emphasis-sensitive—in this specific case veiled beyond the more concrete action but influencing it and being modified and articulated by it[33] as it unfolds.

1.2.1 A *Summary*: The Specificity and *Uniqueness* of Serbia's Response to the COVID-19 Pandemic

First, as stated above, it was the speed—and notable strictness—of the state reaction to the pandemic. This may, to some extent be explained by the enduring memory of the last European outbreak of smallpox in 1972 (e.g. Давидов-Кесар, 2022а–с; Мијалковић, 2022; Миладиновић, 2022; Новости онлајн, 2022; Пејовић, 2022). It is quite opportune for the purposes of this text that this year (2022) Serbia marks the 50th anniversary of the terrifying outbreak of a most contagious and deadly virus which it has successfully beaten back in '72.

Indeed, contemporary Serbian journalists, policymakers and medical professionals have not failed to draw parallels and point out at the effective utilisation of such extensive, valuable and costly experience of a major virology and immunology challenge that Serbian health care professionals had faced decades ago earning invaluable understanding of the matter. The smallpox event is indelibly carved into the national psyche, such memory enduring to some degree thanks to a feature film rendition framed within the borderline horror aesthetic,[34] which earned itself cult status in Serbia and across former Yugoslavia (e.g. prlekija-on.net, 2012).

---

[33] Articulation as a valuable concept surpassing its original formulations (Grossberg & Hall, 1986).

[34] The hazmat suit played a powerful symbolical role in that visual narrative. The subject of poetic licence was debated in recent treatises (Mandić, 2019; Стојановић, 2003; the latter being a precocious essay on media coverage then written by a high-school student).

Secondly, the comparatively temperate response to the Serbians' rapid and severe public rejection of a second curfew during the summer 2020. The reading of which may indeed contrast the scrutiny of global media organisations reporting on the said events at the time of their occurrence[35] (e.g. Reuters, 2020). An aspect of such uniqueness is the imposition of rules both exacting regulatory control and their subsequent, sudden and significant easing, somewhat at the dismay of experts both established and emerging, during the early stages of the global pandemic.

Third: The enormous scale, scope and reach of Serbia's early vaccination effort, relative to its size, global position, economic facility and perceived status, Serbia's effort to secure, enable, supply, share and donate the vaccines in the immediate term as inoculation became a realistic prospect. Also, and extending the invitation across the region and beyond its borders to foreigners wishing to get vaccinated at a time when its powerful neighbouring trading bloc, the EU, struggled to procure and distribute vaccines to its citizens (e.g. B92, 2021b; EURACTIV with AFP, 2021; Euronews, 2021a; The Economist, 2021).

Fourth, the unexpectedly effective libertarian drive of its populace in the wake of the pandemic, comparable to only the most advanced democracies of the west, as the extensive reporting on the protests held in early July 2020 can be further studied in comparative terms (e.g. France24, 2021). Thus, the international publicity effects of such libertarianism, latterly oft perceived as irresponsible even illegitimately inconsiderate by those whose subsequent mandating strategies appeared far more stringent, and the impact on the tiniest of handful of global celebrities that the small country may claim (e.g. Higgins, 2022; Kacer et al., 2022).

Fifth, the unusual drive towards enhancing the nation state's access to such intellectual property rights—albeit by licensing in from abroad—as well as its own production opportunities coming from the knowledge transfer from the antiviral vaccines. Those, to be produced and delivered to the populace, interested foreigners and donated to international partners. This appears now, *subsequently but (arguably) not consequently*, in July 2022 not without controversy due to the interim [likely long-term?] intensification of international affairs. And lastly, more as a footnote but not to be overlooked. A cultural impact, registered and identified in national treatises (Šuvaković, 2020; Vukomanović, 2020), relating to the notion of social distancing as well as broader societal effects such impact being articulated, quite uniquely, in a form of popular art which enjoyed international exposure quite recently. An unexpected and unlikely cultural outcome that had spread from the country outwards, though largely undetected and understated beyond its niche: That of a critical stance embodied in a song performed at one of the most inclusive of global media events, oft-dismissed as superficial, capricious and self-absorbed, the Eurovision Song Contest. International audiences seem to have recognised the value of those lyrics, both in front of their laptops and those who could afford the ticket and the education

---

[35] Worthy of mention are some suggestions of perceived repression (e.g. The Economist, 2021) which will be given due attention but shall not be discussed at length, as this text is not oriented towards the political sciences.

to read the cryptic lyrics about health care, equality of opportunity and mental health in the wake of the pandemic.

"In Corpore Sano," a post-modernist art-pop composition reportedly selected by accident was hailed as *Serbia's hymn to health* (Dotto, 2022). A YouTube *"reaction genre"* clip of an American couple viewing the performance reveals some of the *"viral"* aspects of the song (TK Top Tunes, 2022). In this instance, the meaning of the term is indeed related to the unpredictable spread of publicity, of content and of information across the Internet.

1.2.2 An Aftermath, the Epilogue, or Merely *"Act One"*: Emerging Cultures of Reflection and Insight

The shared experience that divides and unites the world at once. A song that discusses mental health emerging in the wake of the pandemic, films that do not construe horror as entertainment, the art that reveals deep anxieties about disease and frailty of life may all act as a counterweight to the extreme self-preservation logic of pseudo-evolutionary survival. At a time when education and meaning are just as effective as *benevolent coercion* and may perhaps act as coherent and consistent supplementary materials to formulate the need for individual responsibility, language is no longer an insurmountable barrier where digital tools enable translation simultaneously and the underpinning of Latin, the language of medicine, much philosophy and historical wisdom may contribute to a shared experience of humanity. Even in a miniscule way: *Variola Vera in Corpore Sano*—1972–2022, a footnote in history. Irrespective of whether or not there is much future history to be had, we do leave a trace.

References

ABC News. (2021, March 27). Foreigners flock to Serbia to get coronavirus vaccine shots. [online]. *ABC News/The Associated Press*. Accessed April 14, 2022, from https://abcnews.go.com/Health/wireStory/foreigners-flock-serbia-coronavirus-vaccine-shots-76723620.

Abramsky, S. (2022, January 15). Djokovic is obnoxious, but his story highlights cruelty of Australian detention. [online]. *Truthout*. Accessed April 14, 2022, from https://truthout.org/articles/djokovic-is-obnoxious-but-his-story-highlights-cruelty-of-australian-detention/?utm_source=headtopics&utm_medium=news&utm_campaign=2022-01-15

Agarwal, A., Anderson, M., & Zhu, J. (2022, January 24). *Week under spotlight# 23*. [online]. Accessed February 21, 2022, from https://infectiousdiseasespotlight.org/informational/week-under-spotlight-23

Al Bayaa, A. (2020). *Global health diplomacy and the security of nations beyond COVID-19*. E-International Relations.

Allen, S. N. (2020, December 14). Celebrity rule breaking is not new, but hosting lockdown parties is different to throwing a TV out the window. [online]. *The Independent*. Accessed December 27, 2020, from https://www.independent.co.uk/voices/celebrities-lockdown-rules-rita-ora-b1773441.html

Alo [daily]. (2020, March 24). Sajam pretvoren u bolnicu za lečenje obolelih od korona virusa. [online]. *Alo.rs*. Accessed February 21, 2022, from https://www.alo.rs/vesti/drustvo/hala-sajam-korona-virus-bolnica/298292/vest

Alvesson, M., & Sköldberg, K. (2017). *Reflexive methodology: New vistas for qualitative research*. Sage.

Andersen, P. T., Loncarevic, N., Damgaard, M. B., Jacobsen, M. W., Bassioni-Stamenic, F., & Eklund Karlsson, L. (2021). Public health, surveillance policies and actions to prevent community spread of COVID-19 in Denmark, Serbia and Sweden. *Scandinavian Journal of Public Health*, 14034948211056215.

Ang, I., Isar, Y. R., & Mar, P. (2015). Cultural diplomacy: Beyond the national interest? *International Journal of Cultural Policy, 21*(4), 365–381.

AP. (2022, January 7). Djokovic's fans in Serbia protest his detention in Australia. [online]. *AP*. Accessed April 23, 2022, from https://apnews.com/article/immigration-coronavirus-pandemic-novak-djokovic-sports-health-223e1935947ff2c1f099d96aca32cf10

Attree, R. (2022, January 5). Australia's Djokovic double fault – How to create a public relations disaster [blog]. *Roja Marketing*. Accessed April 23, 2022, from https://rojamarketing.com/australias-djokovic-double-fault-how-to-create-a-public-relations-disaster/

B92.net. (2021a, September 22). Internet trgovina - naša nova navika. *Beograd*: B92.net. https://www.b92.net/tehnopolis/pr/internet-trgovina-nasa-nova-navika-1926431

B92.net. (2021b, February 1). Vedrana Rudan se vakcinisala u Beogradu: "Što je ovo? U Srbiji švedski stol krcat cjepiva?". *Beograd*: B92.net. Accessed April 14, 2022, from https://www.b92.net/zivot/vesti.php?yyyy=2021&mm=02&dd=01&nav_id=1804219

Bachulska, A. (2020). *Exploring China's diplomatic offensive in Central and Eastern Europe amidst its 'mask diplomacy'*. WORLD ORDER POST COVID-19, 63.

Bakić-Hayden, M. (1995). Nesting orientalisms: The case of former Yugoslavia. *Slavic Review, 54*(4), 917–931.

Bakir, A., & Todorovic, M. (2010, September 5). A hermeneutic reading into "what strategy is": Ambiguous means-end relationship. *The Qualitative Report, 15*, 1037–1057.

Barovic, A., & Cardenas, N. C. (2021). COVID-19 diplomacy: Analysis of Serbia COVID-19 vaccine strategy in the western Balkans. *Journal of Public Health*.

Barthes, R., & Duisit, L. (1975). An introduction to the structural analysis of narrative. *New Literary History, 6*(2), 237–272.

BBC. (2018, March 25). Variola vera: strašna rođaka malih boginja. [online]. *BBC News*. Accessed April 4, 2022, from https://www.bbc.com/serbian/lat/srbija-43512417

BBC. (2020a, May 19). Coronavirus: Chile protesters clash with police over lockdown. [online]. *BBC News*. Accessed May 19, 2022, from https://www.bbc.co.uk/news/world-latin-america-52717402

BBC. (2020b, April 19) Coronavirus: US protests against and for lockdown restrictions. [online]. *BBC News*. Accessed May 19, 2022, from https://www.bbc.co.uk/news/av/world-us-canada-52344540

BBC. (2022, February 16). Djokovic saga highlights Australia asylum seekers held for record 689 days. [online]. *BBC News*. Accessed May 19, 2022, from https://www.bbc.co.uk/news/world-australia-60398029

Bechev, D (2021, June 30). How Aleksandar Vučić stole the vaccine-diplomacy show. [online]. *Atlantic Council*. Accessed April 14, 2022, from https://www.atlanticcouncil.org/blogs/new-atlanticist/how-aleksandar-vucic-stole-the-vaccine-diplomacy-show/

Blic. (2020a, March 19). 97 ZARAŽENIH U SRBIJI Policija legitimiše starije građane, letovi obustavljeni, "pljušte" kazne za KRŠENJE IZOLACIJE. [online]. *Blic.rs*. Accessed May 18, 2022, from https://www.blic.rs/vesti/drustvo/97-zarazenih-u-srbiji-policija-legitimise-starije-gradane-letovi-obustavljeni-pljuste/hxp55w4

Blic. (2020b, November 17). Šta kupovati u DOMAĆIM, a šta u STRANIM ONLINE ŠOPOVIMA? Nije svejedno i nekoliko je stvari važno znati. [online]. *Blic.rs*. Accessed May 18, 2022, from https://www.blic.rs/biznis/internet-trgovina/rljy9gr

Bourgeron, T. (2022). 'Let the virus spread'. A doctrine of pandemic management for the libertarian-authoritarian capital accumulation regime. *Organization, 29*(3), 401–413.

Boytchev, H. (2021). Covid-19: Why the Balkans' vaccine rollout lags behind most of Europe. *BMJ, 375*.

Brown, T. M., & Ladwig, S. (2020). COVID-19, China, the World Health Organization, and the limits of international health diplomacy. *American Journal of Public Health, 110*(8), 1149–1151.

Byers, M. (2022). Lessons to be learnt from the Djokovic debacle. *TheANZSLA Commentator, 112*, 31–33.

Chattu, V. K., & Chami, G. (2020). Global health diplomacy amid the COVID-19 pandemic: A strategic opportunity for improving health, peace, and well-being in the CARICOM region—A systematic review. *Social Sciences, 9*(5), 88.

Chattu, V. K., Pooransingh, S., & Allahverdipour, H. (2021). Global health diplomacy at the intersection of trade and health in the COVID-19 era. *Health Promotion Perspective, 11*(1), 1.

Clarey, C., & Peltier, E. (2020a, June 23). Novak Djokovic tests positive for the Coronavirus. [online]. *The New York Times*. Accessed May 18, 2022, from https://www.nytimes.com/2020/0 6/23/sports/tennis/novak-djokovic-coronavirus.html

Clarey, C., & Peltier, E. (2020b). Djokovic says he's sorry about all those infections. *The New York Times*, B9-L.

CNN [with Reuters]. (2022, January 2). Dutch police disperse thousands protesting against lockdown measures. [online]. *CNN*. Accessed May 19, 2022, from https://edition.cnn. com/2022/01/02/europe/netherlands-amsterdam-lockdown-protestors-intl/index.html

Conley, H. A., & Saric, D. (2021, March 24). *Serbia's vaccine influence in the Balkans*. [online]. Center for Strategic and International Studies. Accessed May 19, 2022, from https://www.csis. org/analysis/serbias-vaccine-influence-balkans

Convery, S. (2021, November 20). Australia Covid protests: Threats against 'traitorous' politicians as thousands rally in capital cities | Australian politics. [online]. *The Guardian*. Accessed May 18, 2022, from https://www.theguardian.com/australia-news/2021/nov/20/australia-covid-pro tests-threats-against-traitorous-politicians-as-thousands-rally-in-capital-cities

Delauney, G. (2021, February 10). Covid: How Serbia soared ahead in vaccination campaign. [online]. *BBC News*. Accessed March 19, 2021, from https://www.bbc.com/news/world-europe-55931864

Dinic, L. (2021, April 8). *Serbia mimics China's 'vaccine diplomacy' in the Balkans*. [online]. chinausfocus.com. Accessed May 19, 2022, from https://www.chinausfocus.com/society-culture/serbia-mimics-chinas-vaccine-diplomacy-in-the-balkans

Đokić, I., & Čvorović, D. (2021). Criminal legal challenges in Republic of Serbia during COVID-19 pandemic. *Crimen (Beograd), 12*(3), 259–276.

Dotto, N. (2022, May 5) "In corpore sano", Serbia's hymn to health for Eurosong in Turin/Serbia/ Areas/Homepage. [online]. *Osservatorio Balcani e Caucaso Transeuropa*. Accessed May 19, 2022, from https://www.balcanicaucaso.org/eng/Areas/Serbia/In-corpore-sano-Serbia-s-hymn-to-health-for-Eurosong-in-Turin-217419

Dragojlo, S. (2021, October 11). Belgrade non-aligned summit condemns West's 'vaccination nationalism'. [online]. *Balkan Insight/BIRN*. Accessed April 14, 2022, from https:// balkaninsight.com/2021/10/11/belgrade-non-aligned-summit-condemns-wests-vaccination-nationalism/

DS, C. (2022). Novak Djokovic's temporary detention highlights cruelty of mandatory detention. *Green Left Weekly*, (1329), 6.

DW.com. (2021a, February 13). *Geopolitics and vaccines in the Balkans – Balkans: Are geopolitics getting in the way of COVID-19 vaccines?* DW.com. Accessed April 14, 2022, from https:// www.dw.com/en/balkans-are-geopolitics-getting-in-the-way-of-covid-19-vaccines/a-56542620

DW.com. (2021b, August 1). *Berlin anti-lockdown protesters clash with police*. DW.com. Accessed April 14, 2022, from https://www.dw.com/en/coronavirus-berlin-anti-lockdown-protesters-clash-with-police/a-58722453

Emmanouilidou, L. (2021). Serbia lets people choose their COVID-19 vaccine. Some call it a 'political ballot.' [online]. *The World from PRX*. Accessed April 14, 2022, from https://theworld. org/stories/2021-03-03/serbia-lets-people-choose-their-covid-19-vaccine-some-call-it-political-ballot

EURACTIV with AFP. (2021, March 26). *Foreign vaccine-seekers flock to Serbia for COVID-19 shots*. [online]. EURACTIV.com. Accessed May 18, 2022, from https://www.euractiv.com/ section/enlargement/news/foreign-vaccine-seekers-flock-to-serbia-for-covid-19-shots/

Euractiv.rs. (2021, May 20). *Belgrade's Torlak Institute's Sputnik vaccine approved in Russia*. [online]. Euractiv.rs. Accessed April 14, 2022, from https://www.euractiv.com/section/politics/ short_news/belgrades-torlak-institutes-sputnik-vaccine-approved-in-russia/

Euronews. (2021a, March 28). *Serbia innoculates neighbours as other Balkan countries receive doses of China's vaccine*. [online]. Euronews.com. Accessed May 29, 2021, from https://www. euronews.com/2021/03/28/serbia-innoculates-neighbours-as-other-balkan-countries-receive-doses-of-china-s-vaccine

Euronews. (2021b, February 2). *Serbian PM: Vaccine success down to prioritising health over politics*. [online]. Euronews.com. Accessed March 19, 2021, from https://www.euronews.com/ video/2021/02/02/serbia-s-pm-says-vaccine-success-down-to-prioritising-healthcare-over-politics

Euronews. (2021c, February 24). *Which vaccine should I choose? Serbia gives citizens choice of four coronavirus jabs*. [online]. Euronews.com. Accessed March 19, 2021, from https://www. euronews.com/2021/02/24/which-vaccine-should-i-choose-serbia-gives-citizens-choice-of-four-coronavirus-jabs

eVršac. (2020). *NOVE MERE: Karantin za Prvomajske praznike, olakšanja za starije od 65*.

EWB. (2020, July 8). Violent protests in Serbia as Vučić announces another lockdown after weeks of alleged pre-election cover-ups. [online]. *European Western Balkans*. Accessed July 8, 2022, from https://europeanwesternbalkans.com/2020/07/08/violent-protests-in-serbia-as-vucic-announces-another-lockdown-after-weeks-of-alleged-pre-election-cover-ups/

Exit.al. (2020, March 16). *Serbia declares state of emergency, closes borders*. [online]. Exit.al. Accessed March 14, 2021, from https://exit.al/en/2020/03/16/serbia-declares-state-of-emergency-closes-borders/

Fayyaz, S., & Siddiqui, T. (2021). China's vaccine diplomacy amid COVID-19 pandemic: A case study of South Asia. *South Asian Studies (1026-678X), 36*(2).

Feldbaum, H., & Michaud, J. (2010). Health diplomacy and the enduring relevance of foreign policy interests. *PLoS Medicine, 7*(4), e1000226.

Ferhadbegović, S. (2020, July 22). Past and present health crises: How Yugoslavia managed the smallpox epidemic of 1972. [online]. *Cultures of History Forum*. Accessed April 28, 2022, from https://www.cultures-of-history.uni-jena.de/focus/kleio-in-pandemia/past-and-present-health-crises-how-yugoslavia-managed-the-smallpox-epidemic-of-1972/

Filipovic, A. (2022). Vaccine diplomacy during the COVID-19 pandemic on the example of the Republic of Serbia. *SENTENTIA. European Journal of Humanities and Social Sciences*, (1), 1–16.

Foucault, M. (1975). *Discipline and punish*. A. Sheridan, Tr., Paris, FR, Gallimard.

France24. (2021, May 5). Serbia to offer cash to those who get Covid vaccine. [online]. *France 24*. Accessed April 14, 2022, from https://www.france24.com/en/live-news/20210505-serbia-to-offer-cash-to-those-who-get-covid-vaccine

Freckelton, I. (2022). Pandemics, polycentricity and public perceptions: Lessons from the Djokovic Saga. *Journal of Law and Medicine, 29*(1), 9–22.

Giusti, S., & Tafuro Ambrosetti, E. (2022). Making the best out of a crisis: Russia's health diplomacy during COVID-19. *Social Sciences, 11*(2), 53.

Godinho, M. A., Martins, H., Al-Shorbaji, N., Quintana, Y., & Liaw, S. T. (2022). "Digital health diplomacy" in global digital health? A call for critique and discourse. *Journal of the American Medical Informatics Association, 29*(5), 1019–1024.

Grossberg, L., & Hall, S. (1986). On postmodernism and articulation: An interview with Stuart Hall. *Journal of Communication Inquiry, 10*(2), 45–60.

Henderson, D. A. (2008). Smallpox: Dispelling the myths. *Bulletin of the World Health Organization, 86*(12).

Higgins, A. (2022). Novak Djokovic through Australia's pandemic looking glass: Denied natural justice, faulted by open justice and failed by a legal system unable to stop the arbitrary use of state power. *Civil Justice Quarterly, 42.*

Hopkins, V. (2021, April 13). *Bounty of Serbian vaccine diplomacy shames the EU.* [online]. FT. com. Accessed April 14, 2022, from https://www.ft.com/content/81fc28aa-04a9-4108-a69b-80 dc93a9e985

Horton, J. (2022, January 12). Djokovic Covid timeline: Did he break rules after testing positive? [online]. *BBC News.* Accessed May 19, 2022, from https://www.bbc.co.uk/news/59939122

Ilic, I., & Ilic, M. (2022). Historical review: Towards the 50th anniversary of the last major smallpox outbreak (Yugoslavia, 1972). *Travel Medicine and Infectious Disease, 48,* 102327.

IMDb. (n.d./2022). *Variola vera (1982).* Accessed May 19, 2022., from https://www.imdb.com/ title/tt0083275/

Indraswari, F. V., & Lestari, L. E. (2022, January). The impact of Covid-19 pandemic towards belt road initiative of China in Europe region: Development studies approach. In *Universitas Lampung international conference on social sciences (ULICoSS 2021)* (pp. 447–456). Atlantis Press.

Informer. (2020, April 3). *VAŽNA VEST ZA STARIJE GRAĐANE: Krizni štab je promenio dan za kupovinu namirnica, evo zašto.* [online]. Informer.rs. Accessed April 14, 2022, from https:// informer.rs/zastitnik/vesti/505907/vazna-vest-starije-gradjane-krizni-stab-promenio-dan-kupovinu-namirnica-evo-zasto

Ioniţă, D. (2021, February 9). COVID-19, elections and protests: A testing year for EU-Serbia relations. [online]. *EuropeNow.* Accessed April 8, 2022, from https://www.europenowjournal. org/2021/02/08/covid-19-elections-and-protests-a-testing-year-for-eu-serbia-relations/

Johnston, R. (2021, May 20). *Coronavirus update, May 20, 2021: Serbia ready to donate 100,000 vaccines to the Czech Republic – Prague, Czech Republic.* [online]. expats.cz. Accessed April 14, 2022, from https://www.expats.cz/czech-news/article/coronavirus-update-may-20-2021

Jones, S. (2020, November 20). Spain's PM calls for calm after violent anti-lockdown protests. [online]. *The Guardian.* Accessed April 14, 2022, from https://www.theguardian.com/ world/2020/nov/01/spain-pm-calls-for-calm-after-violent-anti-lockdown-protests

Juncos, A. E. (2021, July 8). Vaccine geopolitics and the EU's ailing credibility in the Western Balkans. [online]. *Carnegie Europe – Carnegie Endowment for International Peace.* Accessed April 14, 2022, from https://carnegieeurope.eu/2021/07/08/vaccine-geopolitics-and-eu-s-ailing-credibility-in-western-balkans-pub-84900

Kacer, H., Marcan, K. G., & Marcan, D. (2022). Red card for Novak Djokovic, (at least) Yellow for the State of Australia. *Zb. Radova, 59,* 25.

Karčić, H. (2021, April 4). How Serbia's coronavirus vaccines are resetting Balkan animosities. [online]. *The National Interest.* Accessed May 19, 2022, from https://nationalinterest.org/blog/ coronavirus/how-serbia%E2%80%99s-coronavirus-vaccines-are-resetting-balkan-animosi ties-181961

Katz, J., & Csordas, T. J. (2003). Phenomenological ethnography in sociology and anthropology. *Ethnography, 4*(3), 275–288.

Kazharski, A., & Makarychev, A. (2021). Russia's vaccine diplomacy in Central Europe: Between a political campaign and a business project. *Mezinárodní vztahy, 56*(4), 131–146.

Kerbler, J. (2021, February 7). INTERVJU Vedrana Rudan: Srbija je svetu održala lekciju iz patriotizma! [online]. *Novosti.* Accessed April 14, 2022, from https://www.novosti.rs/kultura/ vesti/962373/intervju-vedrana-rudan-srbija-svetu-odrzala-lekciju-patriotizma

Kickbusch, I., & Kökény, M. (2013). Global health diplomacy: Five years on. *Bulletin of the World Health Organization, 91,* 159–159.

Kickbusch, I., & Liu, A. (2022). Global health diplomacy—Reconstructing power and governance. *The Lancet*.

Kickbusch, I., Silberschmidt, G., & Buss, P. (2007). Global health diplomacy: The need for new perspectives, strategic approaches and skills in global health. *Bulletin of the World Health Organization, 85*, 230–232.

Kickbusch, I., Nikogosian, H., Kazatchkine, M., & Kökény, M. (2021). *A guide to global health diplomacy* (No. BOOK). Graduate Institute of International and Development Studies, Global Health Centre.

Kolaković, A. (2021). Scientists and cultural diplomacy of Serbia. *Kultura, 173*, 175–197.

Kovacic, J. (2021, March 5). Serbia donates vaccines and spreads goodwill in the region. [online]. *New Europe*. Accessed April 14, 2022, from https://www.neweurope.eu/article/serbia-donates-vaccines-and-spreads-goodwill-in-the-region/

Kuenssberg, L. (2022, May 24). Partygate: Insiders tell of packed no 10 lockdown parties. [online]. *BBC Panorama*. Accessed May 29, 2022, from https://www.bbc.co.uk/news/uk-politics-61 566410

Kurecic, P., & Haluga, V. (2021). Health diplomacy as a soft power tool of the PR China during the covid 19 pandemic. *Economic and Social Development: Book of Proceedings*, 237–243.

Kurir. (2020, April 3). *NOVA PRAVILA ZA STARIJE OD 65: Evo da li će im biti zabranjeno kretanje i kada se ukine policijski čas*. [online]. Kurir.rs. Accessed April 23, 2022, from https://www.kurir.rs/vesti/drustvo/3458097/nova-pravila-za-starije-od-65-evo-da-li-ce-im-biti-zabranjeno-kretanje-i-kada-se-ukine-policijski-cas

Lazic, J. (2012, Mart 14). Kako je sniman film Variola vera. [online]. *Vreme*. Accessed April 23, 2022, from https://www.vreme.com/mozaik/kako-je-sniman-film-variola-vera/

Lee, S. T. (2021). Vaccine diplomacy: Nation branding and China's COVID-19 soft power play. *Place Branding and Public Diplomacy*, 1–15.

Lee, K., & Smith, R. (2011). *What is 'global health diplomacy'? A conceptual review*.

Legatum Institute. (2021). *Serbia country profile*. [online]. THE LEGATUM PROSPERITY INDEX. Accessed May 29, 2022, from https://www.prosperity.com/globe/serbia

Litvinjenko, S., Arsic, B., & Borjanovic, S. (1973). *Epidemiologic aspects of smallpox in Yugoslavia in 1972 (no. WHO/SE/73.57)*. World Health Organization.

Maglajlija, V. (2021, November 3). Serbia and China: 'Steel Friendship' in the EU's backyard [online]. *Queen Mary University of London, Centre for European Research*. Accessed April 14, 2022, from http://www.cer.qmul.ac.uk/europemattersblog/items/serbia-and-china-steel-friendship-in-the-eus-backyard.html#

Mandić, M. (2019). Između stvarnog i zamišljenog: infektivna oboljenja u kinematografiji na primeru filma Variola Vera. *Етноантрополошки проблеми, 14*(2), 487–505.

Marciacq, F. (2019). Serbia: Looking east, going west? In *The Western Balkans in the world* (pp. 61–82). Routledge.

mfa.gov.rs. (2021a, August 21). *Serbia also donated Covid-19 vaccines to Namibia*. Accessed April 14, 2022, from https://msp.gov.rs/en/press-service/statements/serbia-also-donated-covid-19-vaccines-namibia

mfa.gov.rs. (2021b, August 21). *Serbia donated 50,000 doses of Covid-19 vaccine to Angola*. Accessed April 14, 2022, from https://mfa.gov.rs/en/press-service/statements/serbia-donated-50-000-doses-covid-19-vaccine-angola

mfa.gov.rs. (2021c, October 12). *Serbia donates 40,000 coronavirus vaccines to Uganda*. Accessed April 14, 2022, from https://mfa.gov.rs/en/press-service/statements/serbia-donates-40-000-coronavirus-vaccines-uganda

Michaud, J., & Kates, J. (2013). Global health diplomacy: Advancing foreign policy and global health interests. *Global Health: Science and Practice, 1*(1), 24–28.

Miller, M. E. (2022). Novak Djokovic loses visa challenge, says he will 'cooperate' and leave Australia. *The Washington Post*, NA-NA.

Mitchell, K. (2020, June 23). Novak Djokovic tests positive for Covid-19 amid Adria Tour fiasco. [online]. *The Guardian*. Accessed January 22, 2022, from https://www.theguardian.com/sport/2020/jun/23/novak-djokovic-tests-positive-for-covid-19-amid-adria-tour-fallout

Moody's Analytics. (2022). *Serbia – Economic indicators*. [online]. Moody's. Accessed March 14, 2022, from https://www.economy.com/serbia/indicators

Nisbett, M. (2016). Who holds the power in soft power? *Arts and International Affairs, 1*(1), 1–24.

O'Sullivan, M. (2022). Game, set but no match-irrationality, visa cancellations and Novak Djokovic: A commentary on 'Djokovic v Minister for Immigration, Citizenship, Migrant Services and Multicultural Affairs' (2022). *Australian Journal of Administrative Law, 29*(1), 13–20.

OLJ. (2021, August 17). The Serbian government has donated 20,000 Sputnik V COVID-19 vaccines to Lebanon, the country's Foreign Affairs Ministry has announced. It is the first batch of a total 40,000 vaccines the country plans to donate to Lebanon, the ministry said. [online]. *L'Orient Today*. Accessed April 14, 2022, from https://today.lorientlejour.com/article/1271981/the-serbian-government-has-donated-20000-sputnik-v-covid-19-vaccines-to-lebanon-the-countrys-foreign-affairs-ministry-has-announced-it-is-the-first-ba.html

Our World in Data. (2022, July 20). *Coronavirus (COVID-19) Vaccinations*. [online]. Accessed July 20, 2022, from https://ourworldindata.org/covid-vaccinations?country=OWID_WRL

Özdüzen, Ö., Ianosev, B., & Ozgul, B. A. (2021, September 14). *Freedom or self-interest?: Motivations, ideology and visual symbols uniting anti-lockdown protesters in the UK*. [online]. The Political Studies Association (PSA). Accessed April 14, 2022, from https://www.psa.ac.uk/psa/news/freedom-or-self-interest-motivations-ideology-and-visual-symbols-uniting-anti-lockdown

Öztürk, T. (2021a, March 2). Serbia donates COVID-19 vaccines to Bosnia: Bosnian Foreign Minister Bisera Turkovic criticizes Serbia's donation of AstraZeneca vaccines for not having EU approval. [online]. *Anadolu Ajansı*. Accessed April 14, 2022, from https://www.aa.com.tr/en/health/serbia-donates-covid-19-vaccines-to-bosnia/2162384

Öztürk, T. (2021b, August 6). Serbia to donate 50,000 COVID-19 vaccine doses to Iran. [online]. *Anadolu Ajansı*. Accessed April 14, 2022, from https://www.aa.com.tr/en/latest-on-coronavirus-outbreak/serbia-to-donate-50-000-covid-19-vaccine-doses-to-iran/2326632

Pajtinka, E. (2014). Cultural diplomacy in theory and practice of contemporary international relations. *Politické vedy, 17*(4), 95–108.

paragraf.rs. (2020, March 24). *Odluka o otvaranju privremenog objekta za smeštaj i lečenje Sajam - virus korona*. paragraf.rs. Accessed May 14, 2022, from https://www.paragraf.rs/propisi/odluka-o-otvaranju-privremenog-objekta-za-smestaj-i-lecenje-sajam.html

Penev, S., Udovič, B., & Đukić, M. (2014). Commercial diplomacy in Serbia: Characteristics and areas for improvement. *Economic Themes, 52*(3), 263–280.

Petrov, A. (2021, March 17). Serbia's vaccine diplomacy in China's shadow | European Union. [online]. *Al Jazeera*. Accessed April 14, 2022, from https://theconversation.com/small-countries-and-covid-19-vaccination-the-example-of-serbia-157159

Petrović-Ristić, D. (2016). Modern economic diplomacy as instrument for achieving economic objectives of the Republic of Serbia. *Megatrend Revija, 13*(3), 131–154.

Pietras, E. (2020) CHUMPS & CHAMPS from Adele to Madonna – The coronavirus lockdown celebrity winners and losers. [online]. *The Sun*. Accessed April 8, 2021, from https://www.thesun.co.uk/tvandshowbiz/11806471/covid-19-celebrity-winners-losers/

Politika [daily]. (2022, July 16). У последња 24 сата 2.520 позитивних на корона вирус, двоје преминуло. [online]. *Politika*. Accessed July 16, 2022, from https://www.politika.rs/scc/clanak/512162/U-poslednja-24-sata-2-520-pozitivnih-na-korona-virus-dvoje-preminulo

prlekija-on.net. (2012, July 20). *Goran Marković in projekcija filma Že videno: Pogovor z letošnjim častnim gostom Grossmannovega festivala Goranom Markovićem*. [online]. prlekija-on.net. Accessed July 16, 2022, from https://www.prlekija-on.net/lokalno/4409/goran-markovic-in-projekcija-filma-ze-videno.html

Prvulović, V. (2015). The beginning of economic diplomacy in Serbia. *The Review of International Affairs, 66*(1158–1159), 2015120.

Pugh, J., Savulescu, J., Brown, R. C., & Wilkinson, D. (2022). The unnaturalistic fallacy: COVID-19 vaccine mandates should not discriminate against natural immunity. *Journal of Medical Ethics, 48*(6), 371–377.

r/TrueFilm. (2021, November 29). *Variola Vera (1982) – A stark depiction of a virus outbreak, which hits very strongly today.* [blog]. Accessed November 29, 2021, from https://www.reddit.com/r/TrueFilm/comments/r56p4f/variola_vera_1982_a_stark_depiction_of_a_virus/

Radio Free Europe. (2021, February 14). Serbia donates thousands of doses of COVID-19 vaccine to North Macedonia. [online]. *Radio Free Europe.* Accessed April 14, 2022, from https://www.rferl.org/a/serbia-donates-covid-19-vaccine-macedonia/31102339.html

Radio Slobodna Evropa. (2021, June 4). Zatvara se privremena COVID bolnica na Beogradskom sajmu. [online]. *Radio slobodna Evropa.* Accessed April 14, 2022, from https://www.slobodnaevropa.org/a/30653153.html

Radovanović, Z., & Kon, P. (2010). *Grip.* Arhipelag.

Ralev, R. (2021, June 7). Serbia starts production of Russia's Sputnik V vaccine – govt. [online]. *SeeNews.* Accessed April 14, 2022, from https://seenews.com/news/serbia-starts-production-of-russias-sputnik-v-vaccine-govt-743662

Rathborn, J. (2020, June 23). Novak Djokovic 'extremely sorry' to those infected at Adria Tour after tennis star tests positive for coronavirus. [online]. *The Independent.* Accessed January 22, 2022, from https://www.independent.co.uk/sport/tennis/novak-djokovic-coronavirus-positive-test-adria-tour-apology-a9581086.html

republicworld.com. (2021, May 31). *Serbia donates 100,000 vaccines to Czech Republic: Serbia donated 100,000 doses of Pfizer vaccines to Czech Republic on Monday.* Accessed April 14, 2022, from https://www.republicworld.com/world-news/europe/serbia-donates-100000-vaccines-to-czech-republic.html

Reuters. (2020, July 7). Demonstrators storm Serbian parliament in protest over lockdown. [online]. *Reuters.* Accessed April 14, 2022, from https://www.reuters.com/article/us-health-coronavirus-serbia-protests-idUSKBN24835R

Reuters. (2021a, March 2). Former enemy Serbia donates COVID-19 vaccines to Bosnia's Muslims, Croats. [online]. *Reuters.* Accessed April 14, 2022, from https://www.reuters.com/article/us-health-coronavirus-serbia-bosnia-idUSKBN2AU1HR

Reuters. (2021b, March 2). Serbia donates COVID-19 vaccines to Bosnia's Bosniak-Croat Federation. [online]. *Reuters.* Accessed April 14, 2022, from https://www.reuters.com/business/healthcare-pharmaceuticals/serbia-donates-covid-19-vaccines-bosnias-bosniak-croat-federation-2021-03-02/

Reuters. (2022a, July 15). COVID-19 TRACKER: Serbia. [online]. Accessed July 20, 2022, from https://graphics.reuters.com/world-coronavirus-tracker-and-maps/countries-and-territories/serbia/

Reuters. (2022b). *COVID-19 TRACKER: United Kingdom.* [online]. Accessed July 20, 2022, from https://graphics.reuters.com/world-coronavirus-tracker-and-maps/countries-and-territories/united-kingdom/

Ritchie, H. et al. (2022). *Serbia: Coronavirus pandemic country profile – Our world in data.* [online]. Our World in Data. [n.d.]. Accessed July 14, 2022, from https://ourworldindata.org/coronavirus/country/serbia

Rofii, M. S. (2020). UNDERSTANDING TURKISH AND RUSSIAN HEALTH DIPLOMACY DURING THE COVID-19 PANDEMIC. *Ilkogretim Online, 19*(4), 3371–3375.

Rogulj, D. (2020, May 23). Adria Tour: Zadar to Host New Tennis Tournament Organized by Novak Djokovic. [online]. *Total Croatia News.* Accessed May 29, 2020, from https://www.total-croatia-news.com/sport/43769-zadar

Rokvić, V. I. (2016). Sekuritizacija zdravlja: da li je javno zdravlje pitanje nacionalne bezbednosti u Republici Srbiji? *Међународни проблеми, 68*(2–3), 225–241.

RTS. (2020, July 29). Onlajn kupovina u doba korone. [online]. *RTS*. Accessed August 2, 2020, from https://www.rts.rs/page/radio/sr/story/25/beograd-202/4032483/onlajn-kupovina-u-doba-korone.html

RTV. (2020). *Uputstva za starije građane 65+. mart 2020*. [online]. Accessed April 14, 2022, from https://rtvcityub.rs/video-arhiva/uputstva-za-starije-gradjane-65-mart-2020/

Rudolf, M. (2021). *China's health diplomacy during Covid-19: The Belt and Road Initiative (BRI) in action*.

Safi, M., & Pantovic, M. (2021, February 19). Vaccine diplomacy: West falling behind in race for influence | Coronavirus. [online]. *The Guardian*. Accessed April 14, 2022, from https://www.theguardian.com/world/2021/feb/19/coronavirus-vaccine-diplomacy-west-falling-behind-russia-china-race-influence

Samso, J. (2021, May 8). Serbia's winning fight against COVID-19 raises questions about 'vaccine diplomacy'. [online]. *PBS*. Accessed April 14, 2022, from https://www.pbs.org/newshour/show/serbias-winning-fight-against-covid-19-raises-questions-about-vaccine-diplomacy

Šantić, D., & Antić, M. (2020). Serbia in the time of COVID-19: Between "corona diplomacy", tough measures and migration management. *Eurasian Geography and Economics, 61*(4–5), 546–558.

Savanović, A. (2020, December 4). Srbija: Zbog rasta broja obolelih od koronavirusa Novosadski sajam ponovo postaje COVID bolnica. [online]. *Anadolu Ajansı*. https://www.aa.com.tr/ba/balkan/srbija-zbog-rasta-broja-obolelih-od-koronavirusa-novosadski-sajam-ponovo-postaje-covid-bolnica/2065917

Savić, M. (2001). Event and narrative (on judging the character of contemporary events). In I. Spasić & M. Subotić (Eds.), *Revolution and order* (pp. 11–20). Institute for Philosophy and Social Theory.

Savkovic, M. (2022). Serbia: Shared loyalties amidst the pandemic. In *European solidarity in action and the future of Europe* (pp. 157–160). Springer.

Schlappig, B. (2021, June 1). Serbia vaccinating some visitors for free. [online]. *One Mile at a Time*. Accessed April 14, 2022, from https://onemileatatime.com/serbia-vaccinating-visitors/

Singh, B., & Chattu, V. K. (2021). Prioritizing 'equity' in COVID-19 vaccine distribution through Global Health diplomacy. *Health Promotion Perspective, 11*(3), 281.

srbija.gov.rs. (2020, May 6). *Serbia lifts state of emergency*. srbija.gov.rs. Accessed April 14, 2022, from https://www.srbija.gov.rs/vest/en/155727/serbia-lifts-state-of-emergency.php

srbija.gov.rs. (2021, March 2). *Serbia donates 5,000 AstraZeneca vaccines to Bosnia and Herzegovina*. srbija.gov.rs. Accessed April 14, 2022, from https://www.srbija.gov.rs/vest/en/168699/serbia-donates-5000-astrazeneca-vaccines-to-bosnia-and-herzegovina.php

Stojanovic, D. (2020, July 13). AP explains: Why Serbs are protesting against virus lockdown. *ABC News with AP*. Accessed August 2, 2020, from https://abcnews.go.com/Health/wireStory/ap-explains-serbs-protesting-virus-lockdown-71755092

Subotić, S. (2021). From mask diplomacy to vaccine diplomacy: The rise of Sino-Serbian relations during the COVID-19 pandemic. In AIES (Ed.), *China's engagement in Central and Eastern European countries, AIES study 2021* (pp. 78–90).

Šuvaković, U. V. (2020). On the methodological issue of uncritical adoption of concepts using the example of the concept of "social distance" during the COVID-19 pandemic. *Социолошки преглед, 54*(3), 445–470.

Suzuki, M., & Yang, S. (2022). Political economy of vaccine diplomacy: Explaining varying strategies of China, India, and Russia's COVID-19 vaccine diplomacy. *Review of International Political Economy*, 1–26.

Taflaga, M. (2022). Australia's COVID politics. *Political Insight, 13*(1), 7–9.

TANJUG. (2021, September 1). Sputnik V vaccine doses from Torlak approved for use. [online]. *Beograd: TANJUG*. Accessed April 14, 2022, from http://www.tanjug.rs/full-view_en.aspx?izb=679673

Telegraf. (2020, April 15). Variola vera u Jugoslaviji: Zato što je Ibrahim Hoti naseo na laž o vakcini, zato je umrlo 40 ljudi. [online]. *Belgrade: Telegraf*. Accessed April 14, 2022, from https://www.telegraf.rs/zanimljivosti/zabavnik/3177893-variola-vera-u-jugoslaviji-zato-sto-je-ibrahim-hoti-naseo-na-laz-o-vakcini-zato-je-umrlo-40-ljudi

Telegraf.rs. (2021, May 31). Serbia donates 100,000 Pfizer vaccines to Czech Republic. Brnabic: 2 reasons for vaccine factory. [online]. *Belgrade: Telegraf.rs*. Accessed April 14, 2022, from https://www.telegraf.rs/english/3345176-serbia-donates-100000-pfizer-vaccines-to-czech-republic-brnabic-2-reasons-for-vaccine-factory

teleSUR. (2021, September 10). China & Serbia to build first COVID-19 vaccine plant in Europe. [online]. *teleSUR*. Accessed April 14, 2022, from https://www.telesurtv.net/news/China%2D% 2DSerbia-to-Build-First-COVID-19-Vaccine-Plant-in-Europe-20210910-0020.html

The Economist. (2021, April 3). Serbia is outpacing nearly every country in the EU at vaccination: Poor, autocratic and happy to take vaccines from Russia and China. [online]. *The Economist*. Accessed April 14, 2022, from https://www.economist.com/europe/2021/04/03/serbia-is-outpacing-nearly-every-country-in-the-eu-at-vaccination

The Heritage Foundation. (2022). *2022 index of economic freedom. [online]*. The Heritage Foundation. Accessed July 19, 2022, from https://www.heritage.org/index/country/serbia

The World Bank. (2022) *Serbia country profile*. [online]. Accessed April 23, 2022, from https:// databank.worldbank.org/views/reports/reportwidget.aspx?Report_Name=CountryProfile&Id= b450fd57&tbar=y&dd=y&inf=n&zm=n&country=SRB.

Tirana Times. (2021, February 17). Montenegro to receive Sputnik V vaccines from Serbia. [online]. *Tirana Times*. Accessed April 14, 2022, from https://www.tiranatimes.com/?p=1492 85

TK Top Tunes. (2022, April 21). *Konstrakta - In Corpore Sano - Serbia - Eurovision 2022 I AMERICAN COUPLE REACTION*. [vlog]. Accessed April 26, 2022, from https://www. youtube.com/watch?v=PPcVgCIDWYM

Todorovic, M. (2016). *Rethinking strategy for creative industry: Innovation and interaction*. Routledge.

Todorovic, M. (2022). Sustainable food production in Serbia, an exploration of discourse/practice in early 2020s. In S. Idowu & R. Schmidpeter (Eds.), *Handbook of sustainability in the food industry*. Springer.

Torlak Virology Institute Homepage. (2022). Accessed July 18, 2022, from http://www. torlakinstitut.com/en

Trifunović, V. (2017). Temporality and discontinuity as aspects of smallpox outbreak in Yugoslavia. *Гласник Етнографског института САНУ, 65*(1), 127–145.

Tzifakis, N., & Prelec, T. (2021). *From mask to vaccine diplomacy: Geopolitical competition in the Western Balkans*. ISPI.

UNICEF. (2021, May 12). *New delivery of COVID-19 vaccines procured through the COVAX Facility arrives in Serbia*. [online]. UNICEF.org. Accessed May 29, 2021, from https://www. unicef.org/serbia/en/press-releases/new-delivery-covid-19-vaccines-procured-through-covax-facility-arrives-serbia

US Embassy in Serbia. (2020). *State of emergency declared in Serbia: Widespread travel restrictions in effect*. Accessed May 19, 2022, from https://rs.usembassy.gov/state-of-emergency-declared-in-serbia-widespread-travel-restrictions-in-effect/

Vériter, S. L., Bjolab, C., & Koops, J. A. (2022). Tackling COVID-19 disinformation: Internal and external challenges for the European Union. *The Hague Journal of Diplomacy, 15*(2020), 569–582.

Verma, R. (2020). China's 'mask diplomacy' to change the COVID-19 narrative in Europe. *Asia Europe Journal, 18*(2), 205–209.

Vukomanović, D. (2020). New modes of acculturation and democratic institutional change during COVID-19 crisis. In I. Bondarevskaya & B. Todosijević (Eds.), *Proceedings of the VIII international scientific and practical seminar*. Institute of Social Sciences.

Vuksanovic, V. (2021a, April 16). In the Balkans, Serbia has its own vaccine diplomacy. *CEPA*. Accessed April 14, 2022, from https://cepa.org/in-the-balkans-serbia-has-its-own-vaccine-diplomacy/

Vuksanovic, V. (2021b, September 10) *Kosovo: The goal of Serbia's global 'vaccine diplomacy'*. euobserver.com. Accessed April 14, 2022, from https://euobserver.com/opinion/152849

Vuksanovic, V., & Vladisavljev, S. (2021, September 13). It's smooth sailing for Sinopharm in Serbia. [online]. *China observers*. Accessed April 14, 2022, from https://chinaobservers.eu/its-smooth-sailing-for-sinopharm-in-serbia/

Wang, P. (2021). China's vaccine diplomacy during Covid-19 pandemic: When it worked and when it did not work? In *2021 3rd international conference on literature, art and human development (ICLAHD 2021)* (pp. 150–159). Atlantis Press.

WHO. (2022a). *Serbia: WHO Coronavirus disease (COVID-19) dashboard with vaccination data*. Accessed July 20, 2022, from https://covid19.who.int/region/euro/country/rs

WHO. (2022b). *Smallpox*. [online]. who.int. Accessed April 14, 2022, from https://www.who.int/healthtopics/smallpox#tab=tab_1

WHO. (2022c). *WHO Coronavirus (COVID-19) Dashboard*. Available at: https://covid19.who.int. Accessed 28 December 2022.

Zamorano, M. M. (2016). Reframing cultural diplomacy: The instrumentalization of culture under the soft power theory. *Culture Unbound, 8*(2), 165–186.

zdravlje.gov.rs. (2021, February 24). *Serbia donates additional doses of vaccine to North Macedonia*. [online]. Ministry of Health, Republic of Serbia. Accessed April 14, 2022, from https://www.zdravlje.gov.rs/vest/en/376/serbia-donates-additional-doses-of-vaccine-to-north-macedonia-.php

Антељ, Ј., & Албуновић, И. (2020, August 8). Како избећи ризике онлајн куповине. [online]. *Politika*. Accessed April 23, 2022, from https://www.politika.rs/scc/clanak/459692/Kako-izbeci-rizike-onlajn-kupovine

Давидов-Кесар, Д. (2022a, March 13). Вирус закопан у "Торлаку". [online]. *Politika*. Accessed April 23, 2022, from https://www.politika.rs/scc/clanak/501894/Virus-zakopan-u-Torlaku

Давидов-Кесар, Д (2022b, March 13) Комплетан Београд вакцинисан за десет дана. [online]. *Politika*. Accessed April 23, 2022, from https://www.politika.rs/scc/clanak/501893/Kompletan-Beograd-vakcinisan-za-deset-dana

Давидов-Кесар, Д. (2022c, March 13). Нико није одбио вакцину. [online]. *Politika*. Accessed April 23, 2022, from https://www.politika.rs/scc/clanak/501891/Niko-nije-odbio-vakcinu

Мијалковић, А. (2022, April 8). Како је вариола поразила корону. [online]. *Politika*. Accessed April 23, 2022, from https://www.politika.rs/articles/details/504342

Миладиновић, А/ВВС. (2020). Вариола вера: Шта смо научили од велике епидемије. [online]. *BBC News in Serbian*. 14 March 2022. Available at: https://www.bbc.com/serbian/cyr/balkan-51875842. Accessed 23 April 2022.

Новости онлајн. (2022, March 13). "И ПОСЛЕ ПОЛА ВЕКА ПАМТИМ АГОНИЈУ УМИРАЊА" Проф. др Милан Шашић у исповести за "Новости" о драматичној борби са вирусом великих богиња. [online]. *Novosti*. Accessed April 23, 2022, from https://www.novosti.rs/c/drustvo/vesti/1095583/variola-vera-milan-sasic-ispovest

Пејовић, Д. (2022, March 13). Терапија истином у доба вариоле. [online]. *Politika*. Accessed April 23, 2022, from https://www.politika.rs/scc/clanak/501896/Terapija-istinom-u-doba-variole

Стојановић, В. (2003). Епидемија вариоле у Србији 1972. године и понашање медија. *Гимназија "Светозар Марковић", Нови Сад*.

Milan Todorovic PhD MA BA (Hons.) HNC rather audaciously asserts to be a polymath. Evidentially: with a background in natural sciences and a qualification in Laser Physics he still utilises in methodology and theorisation alike (2016). An active Fine [and Visual] Art painter, practitioner, educated and classically trained in ancient, traditional and modern techniques; who has a proven track record in music and media enterprise, having spearheaded major change in alternative/'underground' popular culture scenes of Belgrade, in music, video, live events, media PR, before embarking on an academic career which is now his main domain. Having anticipated the major changes that the internet would bring about in 1999 (MA). Meticulously mapping out the multidisciplinary scenes he helped define (2004, PhD) and progressing on to a thorough investigation and advancement of strategic management of the Creative Industries (2016) through the prism of interaction, innovation and disruption. Milan's engagement with sustainability is of a fundamental and principal nature, as he had addressed sustainability, business ethics and the future of humanity through his art, creative work across media, forms and expressive dimensions, and finds profound realisation in exploring the keys, ontological intricacies and juxtaposition/convergence of challenges and solutions that involve transformative thinking and altruistic motivation of free-thought.

Corporate Social Responsibility and Profitability in Spanish Private Health Care During the COVID-19 Period

María del Carmen Valls Martínez, Rafael Soriano Román, Mayra Soledad Grasso, and Pedro Antonio Martín-Cervantes

1 Introduction

The modern social welfare system is located in Europe and rests on four pillars: health care, social security, education, and social services. According to a number of factors such as the financing source (employer and employee contributions or taxes), the provider of health services (state or private sector), and the degree of social coverage, four distinct models can be identified in Europe (Valls Martínez et al., 2021a).

Almost all hospitals are publicly owned in the Nordic system, which offers the highest coverage. In the *continental system*, which is modeled on Bismarck's social security system, the state acts as a mere regulator of services, which are provided by private companies. The *Anglo-Saxon system*, founded by Sir William Beveridge in the United Kingdom in the mid-twentieth century, covers the population's health needs through public hospitals. Still, private health care is also highly developed and operates in a complementary way to the public system. Finally, the *Mediterranean healthcare model*, where Spain is located, is in an intermediate situation between the continental and Anglo-Saxon models.

The Spanish National Health System, created in 1908, provides health coverage to the entire population through public hospitals and health centers. People must go to their family doctor, located in small health centers, for daily care and everyday health problems. The family doctor will be the one who, after studying and assessing the patient, will refer the patient to the specialist or the emergency department of the corresponding territorial hospital. However, in a health emergency, the patient can also go directly to the hospital.

M. d. C. Valls Martínez (✉) · R. Soriano Román · M. S. Grasso · P. A. Martín-Cervantes
Mediterranean Research Center on Economics and Sustainable Development (CIMEDES),
University of Almería, Almería, Spain
e-mail: mcvalls@ual.es; sr2466@icaalmeria.com; pmc552@ual.es

© The Author(s), under exclusive license to Springer Nature Switzerland AG 2023
S. O. Idowu et al. (eds.), *Corporate Social Responsibility in the Health Sector*, CSR,
Sustainability, Ethics & Governance, https://doi.org/10.1007/978-3-031-23261-9_7

Initially, health care depended directly on the nation's central government. However, with the Spanish Constitution of 1978, which is the country's supreme law, the state of the autonomous regions was created. Thus, Spain is divided territorially and administratively into 17 regions with their own autonomous governments. Health care competencies, among others, are currently transferred to autonomous communities. For this reason, public management is very different depending on the political sign of the autonomous government. This fact, together with the differences in per capita income in each geographical area, means that the development and spirit of private health care differ greatly throughout the country.

Consequently, Spain does not have a single, common health care system for the whole country, but rather 17 health care systems with different levels of spending, investment, management, and, in short, health care services for citizens.

Therefore, the quality of health care is uneven across the country. For example, the number of doctors, hospital beds, operating rooms, computerized magnetic resonance equipment, etc., per inhabitant varies greatly from one region to another. In some autonomous communities, patients suffer from long waiting lists to be examined by a specialist or to undergo surgery. In contrast, in others, the attention is very rapid. In short, the quality of health care varies widely throughout Spain (Valls Martínez et al., 2021b; Valls Martínez & Ramírez-Orellana, 2019).

The Spanish national health system is one of the most developed in the world and has international prestige. However, private health care in Spain plays a major role since the number of public employees is significant, and most of them choose private health care as an alternative to public health care.

While employees are compulsorily covered by the public social security system, civil servants, members of the judiciary, and the military have the possibility of choosing private health care instead of public, an option preferred by a large majority. Therefore, during the COVID-19 period, a significant fraction of the population was assisted in private hospitals, which are performing a fundamental role in the fight against the pandemic.

In recent decades, there has been a growing demand for companies to take responsibility for the adverse social and environmental effects caused by their activities. As a result, they are no longer only accountable for their economic performance but also their corporate social responsibility.

At the end of the twentieth century, as a result of a series of international financial, environmental, and social scandals, codes of ethics and conduct began to emerge as a way of controlling such risks, requiring companies to behave sustainably. However, there was previously a latent interest in corporate social responsibility (Aparicio & Valdés, 2009), the origin of which is placed by most authors with the publication of the book *Social Responsibilities of the Businessman* (Bowen, 1953).

There is no obligation to report on corporate social responsibility activities in Spain, nor is there a reporting model that companies must follow when disclosing this type of information. At the international level, the best-known attempt at harmonization is carried out by the Global Reporting Initiative. However, few Spanish hospitals follow these standards. Therefore, this study will consider reports issued and published by hospitals in any format.

Hospitals are companies with unique characteristics that make them more prone to corporate social responsibility, which takes on a particular dimension. As in other companies, the external image is important to achieve legitimacy in the eyes of society (Jia, 2019; O'donovan, 2002; Scherer & Palazzo, 2007). But hospitals must pay special attention to patients and health care professionals, who must be involved in hospital policies and are ultimately responsible for quality health care (Medina-Aguerrebere, 2012; Meneu & Ortún, 2011). However, despite this involvement of hospitals in social responsibility, these types of companies joined this disclosure practice later than other sectors such as banking, energy, etc. It was not until the last decade that Spanish hospitals began to issue corporate social responsibility reports (Valls Martínez, 2019).

The board of directors is the company's highest decision-making and control body. For this reason, corporate governance studies have proliferated in recent years, focusing on characteristics such as the number of annual meetings, the size of the board, the percentage of independent directors, the percentage of executive members, chairman-CEO duality, etc. (Adams & Ferreira, 2009; Ben-Amar et al., 2017; Gallego-Álvarez et al., 2010; Rodríguez Fernández et al., 2013; Sial et al., 2018; Velte, 2017). Above all, the percentage of female directors has been the subject of research.

How gender diversity contributes to a company's profitability or corporate social responsibility activities is a hot topic of interest. The results are not homogeneous, so research on the subject can shed light on the desirability of gender-parity boards (Post & Byron, 2015; Rao & Tilt, 2016), especially considering that the private health care sector has been scarcely studied.

The COVID-19 pandemic has altered the world in every way, affecting the survival of individuals and businesses. The year 2020 was the hardest year of the pandemic. Contagions were numerous in all areas of the planet, some health care systems collapsed, and all hospital capacity seemed scarce.

As mentioned above, the private health care system in Spain is a substitute for the public system for a large number of public employees and complementary for the rest of the population who can afford private insurance. Therefore, during the pandemic, a good number of cases were treated in private hospitals, disrupting their normal functioning. It is logical to think that the profitability of these hospitals, the fundamental objective of their existence, was altered. The question is in what sense. Did the hospitalizations caused by COVID-19 make it possible to increase the profitability of this hospital service, or, on the contrary, did they cause a loss in corporate profits?

This study has two objectives. First, to analyze the profitability of Spanish private hospitals in relation to implementing and disseminating corporate social responsibility measures, gender diversity on boards of directors, and the COVID-19 pandemic. Second, to identify the distribution of the COVID-19 pandemic in the Spanish territory, differentiating by regions (autonomous communities) according to public health expenditure, GDP per capita, and the number of inhabitants, trying to find a relationship between wealth and risk of contracting the disease. For this purpose, a previous 3-year period, from 2017 to 2019, will be studied and compared with the data corresponding to 2020.

The remainder of this paper is organized as follows. Section 2 reviews the main literature on the subject. Section 3 describes the methodology applied for each objective. Section 4 shows the results obtained. Section 5 discusses them. Finally, Sect. 5 draws the conclusions of the study.

2 Literature Review

The relationship between corporate social responsibility and profitability is not clearly defined in practice (Garcia-Castro et al., 2010). Studies mostly find a positive relationship (Allouche & Laroche, 2005; Orlitzky et al., 2003), but sometimes the findings show a negative relationship or even no relationship (Hou, 2019; Lin et al., 2018; Nollet et al., 2016).

The main theories that explain the relationship between corporate social responsibility and corporate profitability are agency theory and resource management theory, although stakeholder theory and legitimacy theory are noteworthy (Fernández-Gago et al., 2018). Following most of the previous research, and with an integrating character, we will adopt a multi-theoretical criterion (Nicholson & Kiel, 2007).

According to agency theory (Fama & Jensen, 1983; Fama & French, 2000; Jensen & Meckling, 1976), the costs required to implement and communicate corporate social responsibility policies would reduce the company's short-term profits but would have a positive effect in the long term. Thus, managers, who are more interested in immediate results, might reject such sustainability practices against the interests of shareholders. On the other hand, disclosing information on corporate social responsibility would reduce information asymmetries between managers and owners of the company, leading to an increase in performance.

Resource management theory holds that the firm gains a competitive advantage from the proper management of accumulated resources (Sirmon et al., 2007). In this way, adequate management of corporate social responsibility activities can increase the appreciation received from the different stakeholders, increasing confidence in the company and strengthening its competitive position, which would lead to an increase in the result (Jia, 2019).

The stakeholder theory (Freeman, 1984) considers that the company is responsible not only to its shareholders but also to its customers, suppliers, employees, lenders, the State, and society. Thus, good long-term relationships with these stakeholders will require the application of corporate social responsibility practices, which will translate into stable relationships and, ultimately, continued benefits for the company.

In relation to the company, legitimacy is the perception from the outside that the company's actions are appropriate in accordance with the prevailing system of values and beliefs. (Suchman, 1995). Based on the theory of legitimacy, the company cannot survive on its own if it is not connected to the society that gives it the legitimacy to exist. Corporate social responsibility activities help the company achieve legitimacy in society's eyes by gaining approval for its actions. In this way, its survival and profits can be sustained (Fernández-Gago et al., 2018; Reverte, 2009).

In recent years, a large number of European countries have enacted gender laws that, on a mandatory or voluntary basis, determine a certain quota of women on boards of directors, with the aim of achieving gender diversity at the highest corporate management level. After Norway, Spain was the second country in Europe to enact a gender law in 2007, establishing parity between men and women.

The establishment of gender quotas is a controversial issue, with reasons both for and against. The literature establishes that heterogeneous groups bring a greater variety of ideas and alternatives based on their members' different skills and experiences, which translates into more effective and innovative solutions to complex problems. Indeed, studies show that women are more cautious in taking risks and making less drastic decisions than men (Eagly et al., 2003; Francoeur et al., 2019; Nielsen & Huse, 2010).

Women also consider themselves more sensitive to social and environmental problems, showing a more empathetic character and acting more participatory and democratic. This greater sensitivity also leads them to a better understanding of stakeholder demands and a better knowledge of the market (Miller & Triana, 2009). It has even been shown that women show more ethical behavior and less earnings-management practices (Osma & Noguer, 2007).

In addition, investors are increasingly considering corporate social responsibility and social justice practices among their investment selection criteria, with the result that companies with gender diversity are more highly valued (Velte, 2017).

On the other hand, homogeneous groups have more fluid communication and fewer conflicts when making decisions, which leads to faster decisions that allow the company to adapt quickly to new market conditions, saving unnecessary costs (Earley & Mosakowski, 2000).

However, the literature is not conclusive on the relationship between gender diversity on the board of directors and firm profitability (Erhardt et al., 2003; Francoeur et al., 2008; Valls Martínez & Cruz Rambaud, 2019). Most studies find a positive relationship between the two variables (He & Huang, 2011; Shrader et al., 1997). However, some research has found a negative or no relationship (Haslam et al., 2010; Valls Martínez, 2019).

Previous research has based the relationship between the percentage of women on boards of directors and financial performance on a wide range of theories. We will highlight those mentioned above.

According to agency theory, the board of directors plays an essential role in monitoring management decisions to avoid actions contrary to shareholders' interests. Board control will be more effective the greater the members' skills, knowledge, and experience (Adams & Ferreira, 2009; Hillman & Dalziel, 2003). Therefore, diverse boards will reduce agency costs, leading to higher profitability.

According to the resource dependence theory, more diverse boards will have more excellent relationships with suppliers, investors, and lenders, providing more favorable linkages for obtaining resources from outside the company, which will result in higher economic benefits (Stiles, 2001).

Based on the stakeholder theory, incorporating women on boards of directors, with their different skills and experience, will give the company a better disposition to meet the demands of all those stakeholders, favoring the company's profits (Kaufman & Englander, 2011; Webb, 2004).

Gender diversity is increasingly seen in advanced societies as a sign of compliance with women's rights. Therefore, according to the legitimacy theory, companies with more gender-diverse boards of directors will be better regarded by customers, investors, and even the state, which will favor the company's sustainability in the market and, ultimately, its profitability (Lückerath-Rovers, 2013).

3 Methodology

The sample used in this study includes Spanish private hospitals with a revenue of more than 10 million euros and a number of employees of more than 50. The period of analysis covers the years 2017–2020. The final sample comprises a total of 269 observations.

The independent variable was the Return on Assets ratio, a variable frequently used in economic studies. The dependent variables were three: (1) COVID-19 period, that is, a dummy variable that took the value 1 if the year corresponded to 2020 and 0 otherwise; (2) corporate social responsibility, another dummy variable that took the value 1 if the hospital reports on its corporate social responsibility policies and 0 otherwise; (3) gender diversity on the hospital's board of directors, measured by the percentage of women on the board.

Five variables related to the hospital's business characteristics were used as control variables: size, measured through the number of employees, asset turnover, indebtedness, age of the company, and legal form. In addition, two macroeconomic variables relating to the autonomous community in which the hospital is located were used: public health expenditure per inhabitant and gross domestic product per capita.

Table 1 shows a description of all the variables used in the research.

Hospital data were collected from the AMADEUS database of the Bureau Van Dijk, a database that compiles economic and financial information on European companies. On the other hand, macroeconomic data were obtained from the Spanish National Institute of Statistics (INE, 2022).

After performing a descriptive analysis of the variables, their bivariate correlations were analyzed using Pearson's correlation coefficient to ensure no subsequent multicollinearity problems. Next, ordinary least squares regression was performed, considering that this was the best way to identify the relationships of the dependent variable with the different regressors. To determine the goodness of fit of the model,

Table 1 Variable description

Abbreviation	Variable	Definition
ROA	Return on assets	Earnings before interest and taxes divided by total assets
EMP	Size	Logarithm of the number of employees
TUR	Asset turnover	Operating income divided by total asset
IND	Indebtedness	Liabilities divided by total assets
AGE	Age	Age of the hospital in years
EXP	Public expenditure	Logarithm of the public health expenditure per inhabitant in the autonomous community where the hospital is located
GDP	Gross domestic product	Gross domestic product per capita in the autonomous community where the hospital is located
COV	COVID-19	Dummy variable, which takes the value 1 in 2020 and 0 otherwise
LEG	Legal form	Dummy variable, which takes the value 1 if private limited companies (hospitals) and 0 in public limited companies (hospitals)
CSR	Corporate social responsibility	Dummy variable, which takes the value 1 if the hospital informs about corporate social responsibility policies and 0, otherwise
BGD	Board gender diversity	Percentage of women on board of directors

Source: Own elaboration

in addition to the R^2 fit coefficient and the F-statistic, the residuals were analyzed, verifying that they verified the conditions of normality and homoscedasticity (Breusch & Pagan, 1979; Jarque & Bera, 1980).

In addition, a cluster analysis was performed on the different autonomous communities of the country to analyze the different behavior in the expansion of the COVID-19 pandemic according to public health expenditure, GDP per capita, and the number of inhabitants of the territory.

Cluster analysis is a set of multivariate statistical techniques that group a set of cases or individuals into clusters. The objective is to make the data in each cluster as similar as possible and as different as possible in relation to the other groups. As the data of the different variables do not have the same unit of measurement, the statistical process requires prior standardization. Among the different hierarchical grouping methods, Ward's method, which sums the squares of the deviations between each individual and the mean of its cluster, was used in this study to avoid loss of information (Bock, 2008; Ward, 1963).

Finally, employing a factor analysis, the four variables were reduced to only two dimensions and then the clusters obtained were represented to provide a visual interpretation of the findings (Granato et al., 2018; Kline, 1993).

4   Results

Table 2 shows the descriptive statistics of the sample. It can be seen that the average profitability of the hospitals in the period considered is 4.52%, ranging from a minimum of −11.73% to a maximum of 24.65%. Regarding the number of employees, the smallest hospital has 77 workers, while there are 44 observations with a number of employees over 1000. These figures show the difference in size among Spanish hospitals. On average, the level of indebtedness is 52.64% of total resources, although the range of this variable is high, from 5.37 to 99.18%.

The average age of the hospitals is 30.53 years. In general, they are relatively new centers. More than 60% are private limited companies regarding the legal form, while slightly less than 40% are public limited companies. The average number of women on the board of directors is 21.17%, but there are 43 observations where the female presence is null, which is a noteworthy fact.

Concerning the implementation and disclosure of corporate social responsibility policies, 71.75% of Spanish hospitals fulfilled their social and environmental commitment, compared to 28.25% that did not show such solidarity behavior. The average profitability of hospitals that did disclose their corporate social responsibility activities was 5.02%. Those who did not disclose their activities had average profitability of 3.25%, a significantly lower figure. The test of means showed that this difference of 1.77% yielded a p-value of 0.0504, so it can be considered significant, although with a low significance level.

Public health expenditure per inhabitant ranged from 1199.24 to 1918.90 euros, which shows the great differences between the different autonomous communities in Spain in terms of health policies. Similarly, GDP per capita indicates the unequal wealth between regions, ranging from 17,448 to 36,049 euros.

Table 3 shows the Pearson correlations between the study variables. All variables except GDP per capita showed a significant correlation with profitability. In particular, asset turnover and disclosure of corporate social responsibility policies were

Table 2 Descriptive statistics of the continuous variables

Variable	Mean	SD	Minimum	Maximum
ROA	4.52412	6.69511	−11.72604	24.6529
EMP	5.96000	0.74774	4.34380	8.00470
TUR	1.31038	0.77716	0.33626	4.36441
IND	52.63843	24.02508	5.37252	99.17978
AGE	30.53147	20.97856	1.29315	86.54521
EXP	7.31093	0.10926	7.08944	7.55951
GDP	10.15304	0.22366	9.76698	10.49263
COV	0.25279	0.43542	0	1
LEG	1.60595	0.48956	0	1
CSR	0.71747	0.45107	0	1
BGD	0.21166	0.16199	0	0.6

Number of observations: 269
Source: Own elaboration

Table 3 Pearson correlations of the continuous variables

	ROA	EMP	TUR	IND	AGE
EMP	-0.0293 (0.6326)				
TUR	0.1283** (0.0355)	-0.1879*** 0.0020			
IND	-0.1465** (0.0162)	0.1263** (0.0384)	0.0225 (0.7136)		
AGE	-0.1261** (0.0387)	-0.1271** (0.0373)	-0.1313** (0.0313)	-0.2258*** (0.0002)	
EXP	-0.1592*** (0.0089)	-0.1642*** (0.0070)	0.0225 (0.7134)	-0.0745 (0.2234)	0.2184*** (0.0003)
GDP	0.0001 (0.9985)	0.1035* (0.0902)	0.1044* (0.0875)	0.1836** (0.0025)	-0.0419 (0.4937)
COV	-0.1849*** (0.0023)	0.0434 (0.4789)	0.0560 (0.3607)	0.0560 (0.3607)	0.0485 (0.4284)
LEG	-0.1396* (0.0220)	0.1224** (0.0449)	-0.0929 (0.1284)	-0.0929 (0.1284)	0.3510*** (0.0000)
CSR	0.1194* (0.0504)	0.1823*** (0.0476)	0.0208 (0.7337)	0.0208 (0.7337)	0.0052 (0.9328)
BGD	-0.2717*** (0.0000)	0.1229** (0.0440)	-0.0407 (0.5061)	0.0892 (0.1447)	0.1108* (0.0695)

(continued)

Table 3 (continued)

	EXP	GDP	COV	LEG	RSC
EMP					
TUR					
IND					
AGE					
EXP					
GDP	−0.0347 (0.5714)				
COV	0.2143*** (0.0004)	−0.1618*** (0.0078)			
LEG	−0.0367 (0.5489)	0.2513*** (0.0000)	−0.0036 (0.9534)		
CSR	−0.1739*** (0.0042)	−0.0683 (0.2643)	−0.0150 (0.8069)	0.0685 (0.2631)	
BGD	0.1181* (0.0531)	0.0658 (0.2819)	0.0142 (0.8170)	0.0521 (0.3951)	−0.0654 (0.2849)

Number of observations: 269

Source: Own elaboration

***, **, and * indicate less than 1% significance level, less than 5%, and less than 10%, respectively

Table 4 Regression analysis

Variable	Coefficient	p-value	95% confidence interval		VIF
Intercept	0.2248725	0.993	−51.38909	51.83883	
ROA (1lag)	0.7479801***	0.000	0.6504242	0.845536	1.22
COV	−2.019761***	0.002	−3.315748	0.7237744	1.12
CSR	0.4986301	0.466	−0.8476933	1.844953	1.09
BGD	−3.273693*	0.088	−7.037005	0.4896193	1.11
EMP	−0.4009101	0.353	−1.250685	0.4488646	1.22
TUR	0.6431893	0.122	−0.1726809	1.45906	1.14
IND	0.0211526	0.118	−0.0054096	0.0477148	1.22
AGE	0.0020243	0.904	−0.0309158	0.0349645	1.40
EXP	−2.599006	0.366	−8.255018	3.057005	1.24
GDP	2.031580	0.156	−0.778893	4.842052	1.22
LEG	0.0382416	0.956	−1.343101	1.419584	1.38
Adjusted R^2	0.6157				
F-statistic	29.25***	0.000			

Number of observations: 195
Source: Own elaboration
***, **, and * indicate less than 1% significance level, less than 5%, and less than 10%, respectively

positively correlated with ROA. In contrast, the level of indebtedness, the company's age, the legal form of the private limited company in the face of the public limited company, gender diversity on the board of directors, the existence of COVID-19, and public health spending showed a negative correlation with ROA.

None of the regressors showed a high correlation with any other, so there should be no subsequent multicollinearity problems in the linear regression model. This fact led us not to eliminate any of the proposed variables, keeping all of them in the multivariate analysis.

Other remarkable relationships are that the largest hospitals are the most indebted. Likewise, the newest centers are larger, and regions with lower per capita health care spending tend to have larger hospitals.

Disclosure of corporate social responsibility policies is higher in larger hospitals and regions with lower per capita health care spending. Perhaps the latter is due to the fact that it is in these regions that the largest hospitals are located. Therefore, we could directly relate corporate social responsibility to the size of the hospital.

Gender diversity on the board of directors is higher in hospitals with the highest number of employees, i.e., the largest hospitals. A particular positive and significant relationship is also found with age so that older hospitals have more women on their boards.

Table 4 shows the regression analysis, in which the dependent variable with a lag has been used as a regressor in order to address possible endogeneity problems. The results show how COVID-19 had a negative influence on profitability, with the highest significance level. Indeed, the average ROA was 5.015% in 2017, 5.762% in 2018, and 4.950% in 2019, while the average ROA in 2020, the year corresponding to COVID-19, fell to 2.399%.

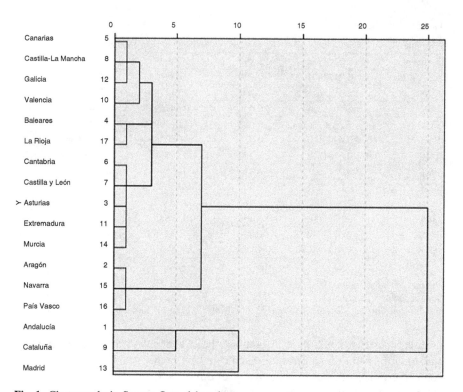

Fig. 1 Cluster analysis. Source: Own elaboration

On the other hand, gender diversity also showed a negative relationship with ROA but with a low significance level.

The model managed to explain 61.57% of the variance of the ROA and was shown to be a good model, as reflected in the F-statistic. Additionally, the residuals were analyzed to confirm the goodness of fit, obtaining through the Jarque–Bera test the certainty that the residuals were normal and, through the Breush–Pagan test, the certainty that they were homoscedastic.

Figure 1 shows the cluster analysis results of the different autonomous communities in Spain based on the following variables: public health expenditure per inhabitant, GDP per capita, number of inhabitants, and number of COVID-19 cases per 100,000 inhabitants.

Selecting level 5, we find four distinct clusters. The first cluster includes the autonomous communities of Andalucía and Cataluña. The second comprises the regions of Aragon, Navarra, and País Vasco. The third comprises the autonomous communities of the Canarias, Castilla-La Mancha, Galicia, Valencia, the Baleares, La Rioja, Cantabria, Castilla y León, Asturias, Extremadura, and Murcia. Finally, the fourth cluster is integrated only by the region of Madrid.

Table 5 shows the mean values of each variable for each of the grouping clusters. It can be observed that the autonomous community of Madrid, which is an isolated

Table 5 Average of variables by cluster

Cluster	Public expenditure per inhabitant	Gross domestic product per inhabitant	COVID-19 in 2020	Inhabitants
1	1421.38	24262.37	443.22	7996.76
2	1726.96	30235.83	236.01	1382.90
3	1608.93	22468.68	216.18	1816.18
4	1344.95	34318.00	1957.94	6633.57

Source: Own elaboration

Table 6 Factor matrix

Variable	Factor 1 (risk)	Factor 2 (wealth)
COVID-19	0.856	0.348
Public expenditure	−0.791	0.493
Gross domestic product	0.420	0.866
Inhabitants	0.849	−0.320
Total variance explained	86.789%	

Source: Own elaboration

case, has the lowest public health expenditure per capita, but has the highest GDP per capita, the largest number of inhabitants, and the most COVID-19 cases. Clusters 2 and 3 had the fewest COVID-19 cases in 2020 and the highest public health care expenditure in the 2017–2020 period, but it is true that they are also the least populated areas.

Table 6 shows the data corresponding to the factor matrix. Factor 1 presents higher values for COVID-19 and the number of inhabitants; therefore, we can call this factor *Risk*. Factor 2, on the other hand, has higher values for public health expenditure and GDP per capita, so we will call it *Wealth*.

With the component analysis, we explained 86.789% of the variance, which is a very good approximation. In addition, Bartlett's test of sphericity yielded a *p*-value of 0.001, clearly lower than 0.05, which indicates that the application of the factor analysis is correct. Likewise, the Kaiser–Meyer–Olkin test gives a *p*-value that is also correct.

Figure 2 shows the graphical representation of the Spanish autonomous communities according to risk and wealth factors, differentiating the four clusters with different colors. Cluster 2, grouping Aragón, Navarra, and País Vasco, is the richest and has a low level of risk. Cluster 3, which includes the largest number of regions, has a moderate–low risk and a medium level of wealth. Cluster 1 presents medium–high risk and includes the communities of Andalucía, with the lowest level of wealth, and Cataluña, with medium–high wealth. Finally, cluster 4, comprising only the region of Madrid, is an atypical case since it has a high wealth and a very high risk.

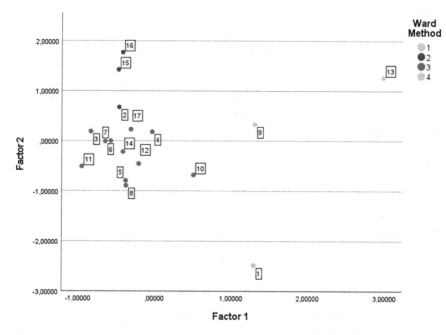

Fig. 2 Scatter plot of the four resulting clusters. Source: Own elaboration

5 Discussion

This paper analyzed the performance of Spanish private hospitals in relation to the application and dissemination of corporate social responsibility practices, gender diversity on boards of directors, and the influence of the COVID-19 pandemic. In addition, considering the existing differences between the diverse regions of the Spanish territory in terms of public health management, income level, and population, a cluster analysis was performed, including the cases of COVID-19 infection during the year 2020. Finally, by means of component analysis, the clusters obtained were represented according to two factors: the risk of contracting COVID-19 and the level of wealth of the geographic region.

COVID-19 cases were those occurring in 2020. The 3 years before the pandemic were also considered for the rest of the variables. Therefore, the total period 2017–2020 was analyzed.

With respect to corporate social responsibility, it can be seen that Spanish hospitals have been incorporating these practices only recently, but to a large extent. In comparison with a previous study on this sector published in 2019 and referring to the year 2015 (Valls Martínez, 2019), it is found how while in 2015, only 12.5% of hospitals carried out corporate social responsibility practices and conducted divulgation about them, in the period 2017–2020 the figure reaches 71.75%.

The profitability of socially responsible hospitals exceeds those that do not have this social and environmental concern by 1.77%. This fact could lead us to think that corporate social responsibility practices bring greater benefits to companies, in accordance with the theories argued in the literature review section. However, although this has been the case in many studies carried out in this sense in samples corresponding to other countries, sectors, and/or time periods, our study does not corroborate a causal relationship. There may be an external cause that influences both variables simultaneously and in the same direction. It is also possible that the early incorporation of hospitals into the implementation of corporate social responsibility policies has not given sufficient time for the effects of having settled in society in such a way that patients choose the particular hospital or are willing to pay more for the services received.

It is true that health care services are special and, at least in certain senses, not comparable to other types of services. What the patient values most, without a doubt, is the prestige of the doctor, the trust generated by her/his professional career, as well as the technical means (facilities and diagnosis equipment) of the hospital. In this sense, large hospitals would be more highly valued by patients, although they would have more expensive resources than small hospitals, which would result in a reduction in their profits. Indeed, large hospitals are also the ones with the most corporate social responsibility policies.

In relation to gender diversity on the board of directors, Spanish private hospitals have 21.17% women, which is in line with the rest of the country's sectors. Effectively, the percentage of women in Spanish listed companies stood at 20% in 2017 and 22% in 2020. However, while the relationship is negative in the hospital sector, although with low significance, in the country's entire sectors, the relationship was found to be positive and highly significant (Valls Martínez & Cruz Rambaud, 2019).

In the sample analyzed in this research, it is noteworthy that 43 of the 269 observations showed the absolute absence of women on the board of directors. Considering that in 2005 Spain enacted its Equality Law, which advocates a balance between men and women, the reality shows that the inequality between the two sexes, in favor of men, is real, although it is not legal. Unlike the Nordic European countries, where the concept of equality is rooted in society, Spanish society has a strong culture in favor of men for positions of responsibility.

The COVID-19 pandemic, which disrupted all aspects of society, had a major influence on the profitability of private hospitals. A priori, one might think that if hospitals had to take in a vast number of COVID-19 patients, their turnover must have increased and, consequently, their financial results.

However, there are two important aspects to consider that would negatively influence this regard. The first is the increased care needed by COVID-19 patients, a disease that in many cases requires admission to intensive care units, with the surveillance and care costs that this entails. In addition, extra sanitary material, with masks and special protective equipment to be used. The second is the suppression of consultations and treatments for other illnesses, including surgical interventions. In 2020, at the height of the pandemic, a new and unknown disease that caused the death of thousands of people worldwide, hospitals ceased outpatient consultations in all specialities, suppressed scheduled surgical interventions, and only attended to hospital emergencies. This fact resulted in a reduction in billing and, consequently, a decline in the hospital's income statement.

The cluster analysis shows a positive relationship between COVID-19 cases and the number of inhabitants, a relationship that seems logical and expected. In addition, it shows how those communities that invest more in health were able to deal better with the pandemic and better control its spread. A particular case is the Community of Madrid, which had the highest COVID-19 rates. Still, it is also true that it is the community with not only the largest population in absolute value but also the highest population density. Madrid has a low public investment. However, it has a high GDP per capita and a good number of private hospitals. These are large hospitals available to the local population, although it is true that people from all over the country also go to them for the most severe cases of ailments and illnesses. Madrid has the best medical specialists in the country and the most advanced treatments. Perhaps because of this, advanced public health care and high GDP, public spending is less influential than in other areas of the country where there are few or even no private hospitals.

The main limitation of this work lies in the absence of an assessment of corporate social responsibility practices by an international rating agency, which has been considered as a dummy variable based on the disclosure or non-disclosure made by the hospitals. Therefore, a future line of research is to carry out an assessment of corporate responsibility practices based on the three dimensions of corporate responsibility: environmental, social, and corporate governance. In this way, more precise results could be obtained on the sustainability of this sector of companies.

6 Conclusion

Corporate social responsibility has only recently been incorporated into the private hospital sector, although most hospitals apply socially and environmentally responsible policies today. Those socially responsible hospitals show higher profitability, but no causal relationship has been established. Gender diversity negatively influences the profitability of the private hospital sector, although it can be considered non-significant. The COVID-19 pandemic significantly affected the profitability of hospitals, causing a sharp drop. The spread of the COVID-19 pandemic was mainly influenced by the population density of the territories but also by public health investment, showing a greater propensity to control the pandemic in those regions that allocate more funds to health care.

References

Adams, R. B., & Ferreira, D. (2009). Women in the boardroom and their impact on governance and performance. *Journal of Financial Economics, 94*(2), 291–309. https://doi.org/10.1016/j. jfineco.2008.10.007

Allouche, J., & Laroche, P. (2005). A meta-analytical investigation of the relationship between corporate social and financial performance. *Revue de Gestion Des Ressources Humaines, 57*(July), 18–41.

Aparicio, J., & Valdés, B. (2009). Sobre el concepto de responsabilidad social de las empresas. Un análisis comparado. *Cuadernos de Relaciones Laborales, 27*(1), 53–75.

Ben-Amar, W., Chang, M., & McIlkenny, P. (2017). Board gender diversity and corporate response to sustainability initiatives: Evidence from the carbon disclosure project. *Journal of Business Ethics, 142*(2), 369–383. https://doi.org/10.1007/s10551-015-2759-1

Bock, H. (2008). Origins and extensions of the k-means algorithm in cluster analysis. *Electronic Journal for History of Probability and Statistics, 4*(2), 1–18.

Bowen, H. R. (1953). *Social responsibilities of the businessman.* Harper.

Breusch, T. S., & Pagan, A. R. (1979). A simple test for heteroscedasticity and random coefficient variation. *Econometrica, 47*(5), 1287–1294. https://doi.org/10.2307/1911963

Eagly, A. H., Johannesen-Schmidt, M. C., & Van Engen, M. L. (2003). Transformational, transactional, and laissez-faire leadership styles: A meta-analysis comparing women and men. *Psychological Bulletin, 129*(4), 569–591. https://doi.org/10.1037/0033-2909.129.4.569

Earley, P. C., & Mosakowski, E. (2000). Creating hybrid team cultures: An empirical test of transnational team functioning. *Academy of Management Journal, 43*(1), 26–49. https://doi. org/10.2307/1556384

Erhardt, N. L., Werbel, J. D., & Shrader, C. B. (2003). Board of director diversity and firm financial performance. *Corporate Governance: An International Review, 11*(2), 102–111. https://doi.org/ 10.1111/1467-8683.00011

Fama, E. F., & French, K. R. (2000). Forecasting profitability and earnings. *Journal of Business, 73*(2), 161–175. https://doi.org/10.1086/209638

Fama, E. F., & Jensen, M. C. (1983). Separation of ownership and control. *Journal of Law and Economics, 26*(2), 301–325.

Fernández-Gago, R., Cabeza-García, L., & Nieto, M. (2018). Independent directors' background and CSR disclosure. *Corporate Social Responsibility and Environmental Management, 25*(5), 991–1001. https://doi.org/10.1002/csr.1515

Francoeur, C., Labelle, R., & Sinclair-Desgagné, B. (2008). Gender diversity in corporate governance and top management. *Journal of Business Ethics, 81*(1), 83–95. https://doi.org/10.1007/ s10551-007-9482-5

Francoeur, C., Labelle, R., Balti, S., & El Bouzaidi, S. (2019). To what extent do gender diverse boards enhance corporate social performance? *Journal of Business Ethics, 155*, 343–357. https://doi.org/10.1007/s10551-017-3529-z

Freeman, R. E. (1984). *Strategic management: A stakeholder approach.* Pitman.

Gallego-Álvarez, I., García-Sánchez, I. M., & Rodríguez-Dominguez, L. (2010). The influence of gender diversity on corporate performance. *Revista de Contabilidad-Spanish Accounting Review, 13*(1), 53–88. https://doi.org/10.1016/S1138-4891(10)70012-1

Garcia-Castro, R., Ariño, M. A., & Canela, M. A. (2010). Does social performance really lead to financial performance? Accounting for endogeneity. *Journal of Business Ethics, 92*(1), 107–126. https://doi.org/10.1007/s10551-009-0143-8

Granato, D., Santos, J. S., Escher, G. B., Ferreira, B. L., & Maggio, R. M. (2018). Use of principal component analysis (PCA) and hierarchical cluster analysis (HCA) for multivariate association between bioactive compounds and functional properties in foods: A critical perspective. *Trends in Food Science and Technology, 72*(October 2017), 83–90. https://doi.org/10.1016/j.tifs.2017. 12.006

Haslam, S. A., Ryan, M. K., Kulich, C., Trojanowski, G., & Atkins, C. (2010). Investing with prejudice: The relationship between women's presence on company boards and objective and subjective measures of company performance. *British Journal of Management, 21*(2), 484–497. https://doi.org/10.1111/j.1467-8551.2009.00670.x

He, J., & Huang, Z. (2011). Board informal hierarchy and firm financial performance: Exploring a tacit structural guiding boardroom interaction. *Academy of Management Journal, 54*(6), 1119–1139. https://doi.org/10.5465/amj.2009.0824

Hillman, A. J., & Dalziel, T. (2003). Board of directors and firm performance: Integrating agency and resource dependence perspective. *Strategic Management Journal, 28*(3), 383–396.

Hou, T. C. T. (2019). The relationship between corporate social responsibility and sustainable financial performance: Firm-level evidence from Taiwan. *Corporate Social Responsibility and Environmental Management, 26*(1), 19–28. https://doi.org/10.1002/csr.1647

INE. (2022). *Instituto Nacional de Estadística.* https://www.ine.es/

Jarque, C. M., & Bera, A. K. (1980). Efficient tests for normality, homoscedasticity and serial independence of regression residuals. *Economics Letters, 6*(3), 225–229. https://doi.org/10.1016/0165-1765(80)90024-5

Jensen, M. C., & Meckling, W. H. (1976). Theory of the firm: Managerial behavior, agency costs, and ownership structure. *Journal of Financial Economics, 3*(4), 305–360. https://doi.org/10.1016/0304-405X(76)90026-X

Jia, X. (2019). Corporate social responsibility activities and firm performance: The moderating role of strategic emphasis and industry competition. *Corporate Social Responsibility and Environmental Management, 27*(1), 65–73. https://doi.org/10.1002/csr.1774

Kaufman, A., & Englander, E. (2011). Behavioral economics, federalism, and the triumph of stakeholder theory. *Journal of Business Ethics, 102*(3), 421–438. https://doi.org/10.1007/s10551-011-0822-0

Kline, P. (1993). *An easy guide to factor analysis.* Routledge.

Lin, W. L., Ho, J. A., & Sambasivan, M. (2018). Impact of corporate political activity on the relationship between corporate social responsibility and financial performance: A dynamic panel data approach. *Sustainability, 11*(1), 1–22. https://doi.org/10.3390/su11010060

Lückerath-Rovers, M. (2013). Women on boards and firm performance. *Journal of Management and Governance, 17*(2), 491–509. https://doi.org/10.1007/s10997-011-9186-1

Medina-Aguerrebere, P. (2012). La responsabilidad social corporativa en hospitales: un nuevo desafío para la comunicación institucional. *Revista Española de Comunicación En Salud, 3*(1), 77–87.

Meneu, R., & Ortún, V. (2011). Transparencia y buen gobierno en sanidad. También para salir de la crisis. *Gaceta Sanitaria, 25*(4), 333–338.

Miller, T., & del Carmen Triana, M. (2009). Demographic diversity in the boardroom: Mediators of the board diversity–firm performance relationship. *Journal of Management Studies, 46*(5), 755–786. https://doi.org/10.1111/j.1467-6486.2009.00839.x

Nicholson, G. J., & Kiel, G. C. (2007). Can directors impact performance? A case-based test of three theories of corporate governance. *Corporate Governance: An International Review, 15*(4), 585–608. https://doi.org/10.1111/j.1467-8683.2007.00590.x

Nielsen, S., & Huse, M. (2010). The contribution of women on boards of directors: Going beyond the surface. *Corporate Governance: An International Review, 18*(2), 136–148. https://doi.org/10.1111/j.1467-8683.2010.00784.x

Nollet, J., Filis, G., & Mitrokostas, E. (2016). Corporate social responsibility and financial performance: A non-linear and disaggregated approach. *Economic Modelling, 52*(Part B), 400–407. https://doi.org/10.1016/j.econmod.2015.09.019

O'donovan, G. (2002). Environmental disclosures in the annual report: Extending the applicability and predictive power of legitimacy theory. *Accounting, Auditing & Accountability Journal, 15*(3), 344–371. https://doi.org/10.1108/09513570210435870

Orlitzky, M., Schmidt, F. L., & Rynes, S. L. (2003). Corporate social and financial performance: A meta-analysis. *Organization Studies, 24*(3). https://doi.org/10.1177/0170840603024003910

Osma, B. G., & Noguer, B. G. D. A. (2007). The effect of the board composition and its monitoring committees on earnings management: Evidence from Spain. *Corporate Governance: An International Review, 15*(6), 1413–1428. https://doi.org/10.1111/j.1467-8683.2007.00654.x

Post, C., & Byron, K. (2015). Women on boards and firm financial performance: A meta-analysis. *Academy of Management Journal, 58*(5), 1546–1571. https://doi.org/10.5465/amj.2013.0319

Rao, K., & Tilt, C. (2016). Board composition and corporate social responsibility: The role of diversity, gender, strategy and decision making. *Journal of Business Ethics, 138*(2), 327–347. https://doi.org/10.1007/s10551-015-2613-5

Reverte, C. (2009). Determinants of corporate social responsibility disclosure ratings by Spanish listed firms. *Journal of Business Ethics, 88*(2), 351–366. https://doi.org/10.1007/s10551-008-9968-9

Rodríguez Fernández, M., Fernández Alonso, S., & Rodríguez Rodríguez, J. (2013). Estructura del consejo de administración y rendimiento de la empresa española cotizada. *Revista Europea de Dirección y Economía de La Empresa, 22*(3), 155–168. https://doi.org/10.1016/j.redee.2012.10.002

Scherer, A. G., & Palazzo, G. (2007). Towards a political conception of corporate responsibility. *Academy of Management Review, 32*(4), 1096–1120. https://doi.org/10.5465/amr.2007.26585837

Shrader, C. B., Blackburn, V. B., & Iles, P. (1997). Women in management and firm financial performance: An exploratory study. *Journal of Managerial Issues, 9*(3), 355–372.

Sial, M. S., Zheng, C., Cherian, J., Gulzar, M. A., Thu, P. A., Khan, T., & Khuong, N. V. (2018). Does corporate social responsibility mediate the relation between boardroom gender diversity and firm performance of Chinese listed companies? *Sustainability, 10*(10), 1–18. https://doi.org/10.3390/su10103591

Sirmon, D. G., Hitt, M. A., & Ireland, R. D. (2007). Managing firm resources in dynamic environments to create value: Looking inside the black box. *Academy of Management Review, 32*(1), 273–292.

Stiles, P. (2001). The impact of boards on strategy. *Journal of Management Studies, 38*(5), 627–650.

Suchman, M. C. (1995). Managing legitimacy: Strategic and institutional approaches. *The Academy of Management Review, 20*(3), 571–610. https://doi.org/10.2307/258788

Valls Martínez, M. C. (2019). Profitability, corporate social responsibility and gender in private healthcare in Spain. *Revista Española de Investigaciones Sociológicas, 168*(December), 111–128. https://doi.org/10.5477/cis/reis.168.111

Valls Martínez, M. C., & Cruz Rambaud, S. (2019). Women on corporate boards and firm's financial performance. *Women's Studies International Forum, 76*(102251), 1–11. https://doi.org/10.1016/j.wsif.2019.102251

Valls Martínez, M. C., & Ramírez-Orellana, A. (2019). Patient satisfaction in the Spanish national health service: Partial least squares structural equation modeling. *International Journal of Environmental Research and Public Health, 16*(24), 1–18. https://doi.org/10.3390/ijerph16244886

Valls Martínez, M. C., Martín Cervantes, P. A., & Soriano Román, R. (2021a). Public management resources of the NHS. In A. Farazmand (Ed.), *Global encyclopedia of public administration, public policy, and governance* (pp. 1–7). Springer. https://doi.org/10.1007/978-3-319-31816-5_4262-1

Valls Martínez, M. C., Ramírez-Orellana, A., & Grasso, M. S. (2021b). Health investment management and healthcare quality in the public system: A gender perspective. *International Journal of Environmental Research and Public Health, 18*(5), 1–25. https://doi.org/10.3390/ijerph18052304

Velte, P. (2017). Does board composition have an impact on CSR reporting? *Problems and Perspectives in Management, 15*(2), 19–35. https://doi.org/10.21511/ppm.15(2).2017.02

Ward, J. H. (1963). Hierarchical grouping to optimize an objective function. *Journal of the American Statistical Association, 58*(301), 236–244. https://doi.org/10.1080/01621459.1963.10500845

Webb, E. (2004). An examination of socially responsible firms' board structure. *Journal of Management and Governance, 8*(3), 255–277. https://doi.org/10.1007/s10997-004-1107-0

María del Carmen Valls Martínez is a PhD in Economics and Business. In 1995 she joined the Economics and Business Department in the University of Almería (Spain), where she is associate professor in Financial Economics and Accounting. Her current research interests include Financial Operations, Ethical Banking, Gender Economics Studies, Health Economics and Sustainability. She has published numerous books and book chapters in reputable publishers such as Springer, Dykinson, Pirámide, McGraw-Hill, etc. She is the author of a large number of articles in prestigious scientific journals, such as Corporate Social Responsibility and Environmental Management, Physica A: Statistical Mechanics and its Applications, European Research on Management and Business Economics, Mathematics, Plos One, Women's Studies International Forum, International Journal of Environmental Research and Public Health, etc. She is also an editor at Plos One and a guest editor at Frontiers in Public Health, Mathematics, Sustainability, etc.

Rafael Soriano Román holds a degree in Translation and Interpretation and a law degree. He is a practicing lawyer and a PhD student in Economics, Business and Law. He has written several book chapters for publishers such as Dykinson and Springer. He has also presented several papers at conferences. His current research interests are in sustainability, gender and market risk.

Mayra Soledad Grasso holds a PhD in Economics and Business Administration. Her main line of work is health economics, where she has published articles in prestigious journals such as International Journal of Environmental Research and Public Health, Mathematics and Healthcare. She has also presented several papers at international scientific conferences and is the author of several book chapters.

Pedro Antonio Martín-Cervantes earned his Ph.D. in Economic Science from the University of Almería. His research interests include the application of econometric methods in different fields such as corporate social responsibility, health economics, political analysis, sustainability, etc. He has published work in prestigious international journals (European Research on Management and Business Economics, Sustainability, the International Journal of Environmental Research and Public Health, among others). Many of his collaborations have been presented at national and international conferences, while also being included in book chapters from renowned publishers such as Springer Nature.

Saving Lives and Minds: Understanding Social Value and the Role of Anchor Institutions in Supporting Community and Public Health before and after COVID-19

Julian Manley, Craig Garner, Emma Halliday, Julie Lee, Louise Mattinson, Mick Mckeown, Ioannis Prinos, Kate Smyth, and Jonathan Wood

1 Introduction

The idea of corporate social responsibility (CSR) has a history stretching back to Bowen (1953) who asked the fundamental question that underpins all subsequent discussions to this very day, 'What responsibilities to society may businessmen reasonably be expected to assume?' (Bowen, 1953, p. xi). The archaic use of the term 'business*man*' aside, this question still holds today, as long as the society we are discussing is designed along capitalist and neoliberal lines. When working in a framework of *corporate* social responsibility, it is the corporation as a legal person that assumes responsibility, with all the patriarchal implications of dependency extending to a sense of charity that this entails. Although this may at first sight appear to be obvious, it is nevertheless important to point this out, since in much of the discussion of CRS (Carroll, 2021), this framework is taken for granted. There are signs that this securely rooted approach in capitalism has been shaken by recent events, not least the COVID-19 pandemic, and that the pandemic may be only the

J. Manley (✉) · M. Mckeown · I. Prinos
University of Central Lancashire, Preston, UK
e-mail: jymanley@uclan.ac.uk; mmckeown@uclan.ac.uk; iprinos1@uclan.ac.uk

C. Garner · J. Lee · L. Mattinson
Community Gateway Association, Preston, UK
e-mail: craig.garner@communitygateway.co.uk; julie.lee@communitygateway.co.uk; louise.mattinson@communitygateway.co.uk

E. Halliday
Lancaster University, Lancaster, UK
e-mail: e.halliday@lancaster.ac.uk

K. Smyth · J. Wood
Lancashire Teaching Hospitals NHS Foundation Trust, Preston, UK
e-mail: kate.smyth@LTHTR.nhs.uk; jonathan.wood@LTHTR.nhs.uk

© The Author(s), under exclusive license to Springer Nature Switzerland AG 2023   193
S. O. Idowu et al. (eds.), *Corporate Social Responsibility in the Health Sector*, CSR,
Sustainability, Ethics & Governance, https://doi.org/10.1007/978-3-031-23261-9_8

explicit manifestation of a systemic revolution that challenges how people work and how businesses function in a world that is facing multiple catastrophes. In his 'guarded' considerations of the future of CRS, Carroll (2021) suggests that even before the pandemic, there had been a shift away from a hard-core financial and economic focus of businesses to an approach that fulfils an evolutionary sequence identified by Visser (2011), who 'maintains that we have moved through the ages of greed, philanthropy, marketing, management, and now responsibility' (Carroll, 2021, p. 1270). Maybe such an optimistic stance on evolutionary responsibility must always be limited by the system. Indeed, some of the literature interprets changes in CSR as being profound and systemic as opposed to a constant refining of the capitalist system (e.g. see Mason, 2015; Blakeley, 2020, for a popular and political interpretation of systemic reactions to the economic collapse of 2008 and the recent pandemic, as well as Trebeck & Williams, 2019, for academic perspectives on the potential for systemic change beyond the capitalist paradigm). It is beyond the scope of this chapter to discuss such a radical possibility of change that explicitly rejects capitalism through a total reconceptualisation of systems of work and life that are encapsulated in terms such as 'degrowth' (Vandeventer & Lloveras, 2021) or 'post-growth' (Banerjee et al., 2021; Pansera & Fressoli, 2021). Nevertheless, such radical change as potential and possibility should be noted as a significant background to our discussion of the role of CRS or its redefinition as 'social value' in the context of the emergence of community wealth building projects in the UK, and in the specific case of this chapter, the context of the Preston Model (PM) (Manley & Whyman, 2021), which is the focus of our study. If we were to pursue an analysis of the meaning of 'post-growth' in the context of this chapter's focus on social value, we would adopt the 'cooperative economics' of Novkovic and Webb (2014). This is because our chapter considers the substantial shift in organisational strategy in two anchor institutions in Preston, UK, under the influence of the PM, which has its roots in the theoretical and practical applications of cooperative principles and values, whether through the creation and promotion of actual cooperative businesses or simply through an emphasis on cooperation as a vague, quasi-philosophical, yet valuable, approach to society and economics, which provides the motivational inspiration and drive for the PM to continue to thrive. Within our consideration of these two case studies—the first from the Lancashire Teaching Hospitals NHS Foundation Trust and the second from Community Gateway Association (CGA)—this chapter asks to what extent the lively social value engagement by these two anchor institutions could positively enhance health outcomes for communities in Preston.

The two in-depth interviews that form the basis of the responses below are taken from interviewees holding senior positions in our two anchor institution case studies in this chapter. They have been selected from a series of interviews belonging to a wider research project, in progress at the time of writing, which seeks to assess the public health impact of the Preston Model and community wealth building initiatives in the UK.

2   The Preston Model: A Brief Summary

2.1   Overview

The Preston Model of community wealth building has been in development since 2012 and envisages a significant change in the local economy for the people of Preston through the encouragement of a broadly interpreted adoption of cooperative principles and values leading to increased participation in local democratic involvement in communities (Prinos & Manley, 2022) and by efforts to generate and retain local economic wealth. The procurement habits of large institutions in Preston (known as 'anchor' institutions, meaning that they are rooted in place and are unlikely to move) have been changing so that more and more local suppliers and small businesses are given opportunities to trade with these anchor organisations. The PM seeks to go beyond standard 'regeneration' projects and takes a holistic and systemic position as its motivation. Therefore, the city has witnessed the creation of the Preston Cooperative Development Network (https://prestoncoopdevelopment.org/), to encourage the development of cooperatives in Preston; the Preston Cooperative Education Centre (https://prestoncoopeducationcentre.org/), to encourage the change of mindset and culture that can provide new opportunities for mutual support and a cooperative economy for Preston; a cooperative and community bank (https://nwmutual.co.uk/), set to provide financial support and opportunities for developing innovative local businesses in the region; and the creation of a project research and development group supported by the Mondragon cooperatives consultancy organisation to support the design of a socio-economic ecosystem for Preston (https://prestonmodel.net/). There are many indications that the PM is a positive force for good in the city. Through this redirection of local spend, a huge amount of financial capital has been relocated in Preston instead of being spent elsewhere. In a nutshell, Preston has become a wealthier city, and through economic wealth comes well-being and prosperity, as indicated in the results of various surveys and reports published since 2017 (for a full account, see Manley & Whyman, 2021).

2.2   Sustainability

Recent developments have placed sustainability, climate action and social value at the forefront of the future plans for the PM (PCC, 2021; Charnley-Parry et al., 2022; Our Central Lancashire).

According to Carroll (2021, p. 1273) 'sustainability' for businesses is little more than a renaming of CRS. In this development, we see a sense of CSR evolution within the capitalist frame, summed up by Carroll in the following sequence:

Corporate Social Responsibility (CSR1) → Corporate Social Responsiveness (CSR2) → Corporate Social Performance (CSP) → Corporate Social Impacts (CSI).

(Carroll, 2021, p. 1274)

The flow, starting in the 1950s and leading to the present day, goes from a moral 'responsibility' taken on board by companies in the first instance, to a 'response' that suggests action, to 'performance' that demonstrates results, to 'impacts' that can be measured and evaluated. This may represent the limits of CSR1&2/CSP/CSI within the boundaries of the neoliberal system. This emphasis on real action that can be measured also reflects a possible shift in company strategy from prioritising shareholders only (and therefore economic profit above all) to a concept that includes paying due attention to 'stakeholders' in the businesses, with 'stakeholders meaning any person who is in any way affected by the business. According to Carroll, the COVID-19 pandemic has provided significant impetus for this view:

> To a significant extent, the stakeholder view has been bolstered during the COVID-19 pandemic as companies have been striving hard to be visible manifestations of social concern.
> (Carroll, 2021, p.1268)

2.3   Anchor Institutions

The 'anchor institutions' in the two case studies that make up this chapter are not companies within a capitalist framework. Indeed, the big institutions in Preston and in the city that inspired the idea of 'anchor institutions Cleveland, Ohio, are in one way or another the providers of public services, such as hospitals, local government authorities, universities, housing associations and suchlike. However, such institutions are firmly embedded in the capitalist system and, especially in the UK and the USA, are expected to make profits, provide value for money and spend "taxpayers' money" wisely'. They therefore adhere, in the main, to the same system of values and socio-economic perspectives as private companies (Alperovitz et al., 2010). The push in the city of Preston, UK, to change the procurement habits of such institutions was largely driven by the local city council and supported by progressive experts in procurement policies. What this drive revealed was a paradoxical situation where these institutions were aware of and keen to commit their organisations to financial profitability—in the sense of being located in a market, with an emphasis on a narrow view of effectiveness and efficiency, and the introduction new public management into public services which have blurred the differences between public and private organisations—while at the same time proud of their status as providers of services to the communities they serve, and therefore readily able to understand and embrace the possibilities for change that came about with the Public Services (Social Value) Act in the UK (2012) (SV Act). Indeed, many anchor institutions in Preston are quick to claim that the work of the Preston Model was already inherent in their daily practices even before the PM came into being (Prinos, 2021; Prinos & Manley, 2022). The Social Value Act

> required Preston City Council and other local, public sector institutions to think differently about service contracts, above the Official Journal of the European Union (OJEU)

thresholds. Instead of just focusing upon compliance, cost and quality; they now needed to think across service design and tendering as to how suppliers could deliver wider social value.

(Urbact, 2018)

As in the case of CRS, 'social value' is connected by Preston City Council to sustainability:

'Social value' which can also be called 'sustainability' or 'added value' is the phrase that refers to additional activities which suppliers can deliver that go beyond the delivery of the good or service and which contribute towards achieving such goals. For example, social value can include the creation of jobs or apprenticeships, the provision of training, or activities that reduce carbon emissions.

(Preston City Council (n.d.))

The anchor institutions that are the subject of this chapter have long-standing commitments to social value, whether couched in CRS terms or as sustainability, which makes it less difficult to imagine changes in procurement from a principally financially focussed design to a social value strategy. As such, these anchors have been able to reimagine themselves as forces for the common good and have demonstrated, as part of the PM, an ability to provide mutual support among themselves, as well as providing impetus and motivation for change within the general socio-economic fabric of Preston.

An aspect of social value, particularly for these two anchor institutions—one that cares for the health and well-being of tenants through housing and another directly through health interventions—is public health. The public health impact of the Preston Model and similar community wealth building projects is, at time of writing, being researched by the Universities of Liverpool, Central Lancashire and Lancaster (NIHR, n.d.). This study is currently ongoing and could demonstrate Carroll's (2021) description of a shift from moral responsibility to action and impact. In the sections below, we firstly draw on interviews which have been conducted with professionals based in both anchor institutions that are the example institutions in this chapter, and secondly, we refer directly to case study reports from the anchor institutions themselves, providing a personal and professional response in the first case, and an institutional and aspirational description, in the second case.

3 Working in Two Anchor Institutions in Preston: Dealing with Health and Care through Social Value

3.1 The NHS

The NHS in Preston serves a community beyond the city, but has been clearly impacted by the development of changing procurement strategies that form part of the Preston Model. The following responses to interview have been provided by a respondent in a leadership position at the Lancashire Teaching Hospitals NHS

Foundation Trust. The move towards a more holistic vision of health and the prevention of illness combines a national trend with the Preston Model and the absolute necessity to take on board lessons learnt from the impacts of the pandemic. In particular, it became clear that outcomes were worse in areas of deprivation, which has raised awareness of the correlation between health inequalities and economic disadvantage. To take families out of poverty is also a positive health action.

At interview, our respondent from the NHS identifies the need for radical implementation of a social value policy with a moral value that must be actioned:

> ... we also need to invest in our local area whenever we can. I think there's a real moral argument and there would be something wrong if we didn't.
> (NHS respondent)

These realisations are not necessarily new, and indeed, a 2019 report by the Health Foundation creates a detailed case for the role of the NHS as an anchor institution (Reed et al., 2019), but the respect, attention and empowerment that are transferred when moving from theory to practice and from a totally clinically focussed approach to a systemic approach within the containing framework of the Preston Model are noticeable, and this makes a difference:

> I think hospitals are now starting to look at health inequalities and determinants of health in a way that they didn't previously. I think hospitals saw themselves and defined themselves in a very narrow way that was all about health care, whereas now, they are much more aware of wider factors and how they might be able to benefit their local community through the employment of local people and where they spend their resources. I think it is a change of emphasis.
> (NHS respondent)

When it comes to the Preston Model, there is a feeling among many interview respondents, including the responses in this chapter, that social value is a 'common sense' approach to living and working, and that this is implicitly pitted against 'economic sense'. Manley points this out as a simple and not-so-revolutionary way to 'challenge the system' (Manley & Whyman, 2021, p.19), and this idea is independently reiterated by our respondent:

> Colleagues in several different departments have started to look at the work they do differently, not what they do, but how they do it, to make sure that the local community gets better benefits than it did previously. Departments include human resources, waste disposal, actual clinical staff, procurement . . . . Before doing things, people consider whether they've maximised social value. It's still quite new in the Trust. It's common sense.
> (NHS respondent)

3.2   Community Gateway Association

Community Gateway Association was one of the very first anchor institutions to come out strongly in support of the Preston Model. Through a governance system that has been strongly influenced by cooperative governance, CGA seeks to improve the lives of tenants in multiple ways. Research has shown that there is a clear connection between adequate housing and physical and mental health (for a recent summary in the context of child health, see Nasim, 2022), although the wide scope of what constitutes housing and health (including factors associated with related matters, not just the actual housing itself, such as neighbourhood environment, family circumstances and the like) makes it difficult to precisely evaluate. The following responses to interview are taken from a respondent in a leadership role at CGA.

With regards to health, housing and the pursuit of social value, there is a constant balancing act between the organisation's primary task, that is to say providing housing, and the wider task associated with housing and health:

> That is the reality … we've got very clear corporate plan and objectives that we want to achieve, you know, but ultimately, we've got … 6700 homes and it sort of 10–15,000 individuals we're trying to serve so they've got a got to come first, everything's got to link back to that ultimately.
> (CGA respondent)

In the same way as the NHS clearly needs to focus on direct health interventions, CGA needs to focus on the roof above people's heads. However, in both cases there is a newer and growing realisation of the holistic nature of this task, which becomes significantly wider with the acknowledgement and promotion of social value actions and impact into the mix:

> I think I think the offering that the housing sector provides now is fundamentally different than it was when I started working in sectors 30 odd years ago, it's a much broader offering with a significantly more sort of social value…
> … there's it's more than just providing a home for somebody. It's a wide … offering.
> (CGA respondent)

Examples of this in CGA include education and training to provide tenants with increased opportunities for quality employment, including supporting the Preston Vocational Centre that provides work training skills for young people; encouraging the development of apprenticeship schemes; opening up the possibility of buying cheaper and yet higher quality food; and providing access to computers. A vision of 'community', however, is not restricted to tenants. As an anchor institution within the framework of the Preston Model, CGA also provides the living wage for contracted work which wherever possible is contracted to local people. Such contracts are also increasingly focussed on social value criteria. The spirit of this approach, and its translation into actions, is well encapsulated in comments from our respondent who emphasises local spending that supports communities through social value and employing local people to serve in the various departments of the housing association.

There is no doubt that the nuances, subtleties, complexities and the innovative nature of the drive for social value as a significant shift away from market-driven approaches, even if such values are clearly recognised as ethical values that are embedded in the work of the housing association, are a challenge. As our respondent points out, 'But you know, a lot a lot of it is easy to say and very, very hard to do, isn't it?'

In this context of newness and the potentially never-ending and overwhelming nature of an expansion of social value beyond the simple remit of housing, our respondent points to the cooperative need for sharing work, networking and learning from others:

> It's about trying to learn from others . . . see if we can work better together . . . there's a lot of organisations on in and around Preston trying to do really good things . . . the more we can understand who they are, what they do and how they do it, then . . . is that going to create opportunities where we can . . . tap into one another?
> (CGA respondent)

Ultimately, however, while there is no end to social value, especially in the sense that value of this kind is difficult to quantify, there is a background of fear that an anchor institution such as CGA may stretch itself too far in the pursuit of such social value. This points to the decline of the system and the encouragement of a very free market, which, in a neoliberal fashion, is relaxed about organisations such as CGA taking over the moral responsibilities embedded in social value:

> A lot of public services are getting squeezed and squeezed and squeezed and squeezed and therefore, you know, organisations like ours have tried to kind of do more and more and work close together, but the reality is, it just can't carry on indefinitely. So that's my fear.
> (CGA respondent)

This situation is always a threat to social value and community and leads in the end to the constant possibility of closing down or narrowing the focus of an organisation, understandably concerned about the 'nuts and bolts' of its mission.

4   Case Study Institutional Statements: Lancashire Teaching Hospitals NHS Foundation Trust and Community Gateway Housing Association

The following statements on the role of each of these anchor institutions in the promotion of social value are provided by the institutions in question.

4.1   NHS

Lancashire Teaching Hospitals (LTH) spends approximately £600 m per year, employs almost 9000 people and serves a local population of c340,000. In August 2021, the Board of LTH approved the creation of a Social Value Framework (SVF) linked to the hospital's status as an anchor institution. The framework is structured around four key areas:

1. People
2. Planet
3. Procurement
4. Place

Numerous departments in the hospital have embraced the SVF and are developing approaches to make it a reality, some of which are described below.

4.1.1 People

Recruitment Services

LTH takes social value (SV) seriously and works hard to stay connected to its communities, knowing that local employment supports better health outcomes, which ultimately reduces the impact on NHS services.

Partnership working with local agencies helps LTH achieve its priorities, by sharing ideas and good practice. Partners include local job centres, local authorities, colleges, third-party providers and the Shout Network which covers 600 local businesses. This helps LTH understand local demographics and the economic climate.

Some of the key recruitment service initiatives, which demonstrate SV, are as follows:

- Working with Preston City Council to analyse 34,000 application forms looking at applicant demographics to inform strategies for outreach work.
- Receiving data from Job Centre Plus to improve understanding of local labour market information and trends.
- Attending careers fairs, jobs events, etc., to engage with potential candidates and/or their families.
- Listening to feedback from local people about barriers to employment, issues with flexibility, etc.
- Running sessions on how to complete application forms and developing an associated e-learning resource.
- Commissioning pre-employment courses with a guaranteed interview at the end for hard-to-fill roles or mass vacancy roles.
- Giving lectures at local colleges/universities about NHS careers.
- Organising a leaflet drop throughout the area describing job opportunities at the hospitals.
- Advertising roles for mass recruitment, e.g. healthcare assistants, administrative roles and hotel services on two prominent outdoor digital screens.
- Activity on social media platforms with strong job-focussed accounts, encouraging people to share local job opportunities.
- Actively welcoming candidates from all backgrounds.

Educational Services

A focus of the work provides a broad range of career promotion activities and raises awareness of careers available in health care. LTH engages with communities from primary school age through all age groups, offering real-life experience through coordinated visits to the Life (Learning, Inspirations, Future Employment) Centre.

A broad range of widening participation programmes is offered, focussed on supporting people into employment and/or education. Examples include the following:

- Pre-employment programmes which work alongside other organisations' initiatives, e.g. Job Centres, Princes Trust, and Community Gateway.
- The Preston Widening Access Programme targeting students studying A levels who aspire to become doctors and who are disadvantaged in some way. This programme builds knowledge, skills and experience to support their application for medical school, with successful students guaranteed an interview at Manchester Medical School for the MBChB Medicine.
- The Work Familiarisation Programme targeting those with additional learning needs and/or disabilities to boost their confidence and relieve anxiety for hospital admission.
- Work experience, tailored to individual needs, provides people from various backgrounds with opportunities to supplement college activities, explore career options or look for career changes.
- Ready, Steady, Apply is a programme which helps people to develop their job application skills.

LTH's apprenticeships provide development opportunities that support career progression and skill development, including developing language skills for those where English is not their first language.

The Trust offers a range of placement opportunities for students and learners on formal education programmes and has a pastoral support portfolio targeted at helping learners with their health and well-being and providing a safe space for discussing issues.

LTH's educational portfolio provides accredited/non-accredited educational programmes that enable career progression and ongoing skill development, which supports staff retention.

Equality, Diversity and Inclusion

The Equality Diversity and Inclusion (EDI) Strategy contains several streams of work aligned to the SV agenda. These include the following:

- Working to ensure that LTH's workforce is proportionally representative of the community it serves, helping patients to feel cared for by people who understand

their needs and supporting individuals from disadvantaged communities to gain meaningful employment and achieve their career aspirations with the Trust.
• Improving the experience of work for colleagues with protected characteristics and providing a robust programme of education to reduce misperceptions, discrimination and sources of bias. Seven different Inclusion Ambassador Forums and Network Groups have been established including ethnicity, LGBTQ+, disability and carers.

An EDI Resource Centre has been developed as part of the learning management system. The Trust's mandatory training has ensured that 3500 employees have completed the Rainbow Badge training and Living Library events have been delivered. The training aims to create an inclusive workplace, deliver tailored care to patients and reduce bias and discrimination in the community.

4.1.2 Planet

Achieving Net Zero

The NHS has made a commitment to reach net-zero carbon (NZC) by 2040, with an 80% reduction to be achieved by no later than 2032.

The Trust is working with NHS Carbon & Energy Fund (CEF) to prepare proposals to reduce energy consumption and CO_2 emissions, including the following steps:

(a) Reducing heat loss through the building fabric and improving airtightness through improving insulation, re-glazing and other improvements.
(b) Ensuring the highest possible energy efficiency of heating, cooling and lighting by installing the most energy-efficient plant and equipment.
(c) Moving away from gas-fired heating systems to electrification of heat and power.
(d) Identifying opportunities to provide on-site renewables.

In addition to planning for the estate to be net zero, LTH has introduced the following sustainable services:

• Electric vehicle charging points.
• Monitoring water usage, identifying areas of excessive usage and reducing water consumption.
• Offering park-and-ride facilities to reduce the environmental impact of staff travel.
• Improving staff changing facilities to assist staff cycling and walking to work as an alternative to commuting by car.

A Clinical Perspective

Health care is a major contributor to climate change: the NHS is responsible for 5.4% of the country's carbon footprint.

Sustainable medicine recognises that health care must change to reduce resource use and carbon emissions.

Making A Green Emergency Department (ED) is a multidisciplinary 'green group' established in the ED of LTH aiming to:

- Raise awareness of the impact of the climate crisis on health.
- Reduce the department's carbon footprint.
- Campaign for institutional change.

Sustainability is now a standing item on the ED Directorate agenda, and progress is reviewed monthly. Progress includes reduced nitrous use, emissions from waste and improved energy use.

One benefit has been an increase in staff well-being, gained through a sense of purpose and determination to improve the department. The group links with local groups to increase cohesion with communities.

Some initiatives will require links to be made with external stakeholder agencies, e.g. improving links to primary care to streamline prescribing, referral pathways which encourage social prescribing, and creating a diagnostic stewardship framework to reduce unnecessary testing.

The Green Plan

LTH's 3-Year Green Plan Strategy will drive actions on sustainability and climate change and deliver SV through positive contributions to the local community. LTH will:

- Involve colleagues to deliver the Green Plan and embed wide-ranging actions within everyday working practices which will involve collaborating with community, voluntary sectors and educational institutions.
- Encourage medicines optimisation, the reduction in medical waste and the consideration of lower carbon alternative medicines.
- Commit to prevention approaches and reducing single-use plastics and PPE.
- Improve green spaces and ecological environments for both employees and patients.
- Increase levels of active travel and public transport, explore low-emission vehicle fleets and maximise efficiencies in the transport of goods and services.
- Improve energy efficiency across the hospital estate and reduce energy usage and waste reduction.

- Harness existing digital technology and systems to streamline service delivery.
- Reduce food waste and provide healthier, locally sourced, seasonal menus high in fruit and vegetables, and low in heavily processed foods.

Waste Management

The hospital's focus is on waste reduction and resource management, rather than disposal, including the following:

- Working to ensure that waste contracts are sustainable and meet Green Plan and NHS net-zero targets.
- Promoting waste management as a key focus of procurement decisions.
- Continuing to provide a clinical waste transfer facility for local authorities and small businesses in the north-west.
- Working with other local trusts, local authorities, charities and schools in relation to the reuse of surplus equipment and furniture.
- Continuing to participate in local and national waste networking groups and sharing good practice.

4.1.3 Procurement

The Lancashire Procurement Cluster (LPC)

LPC is a procurement service used by Blackpool Teaching Hospitals NHS Foundation Trust (FT), East Lancashire Hospitals NHS Trust and Lancashire Teaching Hospitals NHS FT.

SV and improving health outcomes for local communities are key to a new procurement approach, meaning that procurement teams are incentivised to deliver SV outcomes in equal measure to more traditional procurement objectives such as financial improvement and compliance. The Quadruple Aim Framework was established as follows:

- Quality of care—25%
- Population health and SV—25%
- Efficient and reducing cost—25%
- Staff joy—25%

Partnership working is vital for success, and the LPC is working with a diverse range of partners. A draft of the procurement policy utilises information from across Lancashire and South Cumbria (L&SC) partners, effectively creating a 'toolkit' of resources to support procurement professionals and NHS staff to embed SV into procurement opportunities. The policy includes the following:

- Findings of a review to identify common SV outcomes for organisations across L&SC.
- Common outcomes detailing the types of activities suppliers could deliver as part of a procurement process and the types of questions to ask in order to receive effective responses.
- A matrix detailing how outcomes are relevant to goods, services and works (essential for compliance with procurement regulations).

- How organisations can monitor the performance of suppliers against outcomes during the delivery of their contract.
- Key considerations when procuring goods, services and works.

The LPC is revising procurement activity by utilising spend to influence activities of the LPC and those suppliers it contracts. In addition, SV ambitions will be embedded into the way in which LPC recruits, incentivises and develops its teams.

4.1.4   Place

Lancashire Teaching Hospitals

The Trust has established a multidisciplinary Social Value Steering Group to develop its plans and actions. Consideration is being given to understanding the impact of this improved focus and the benefits it has for residents, staff and patients.

Central Lancashire

A Social Value Network at the place level has been established which includes numerous stakeholders from across Central Lancashire including the health; local authority; education; and voluntary, community, faith, and social enterprise sectors. The network uses the following definition of SV:

Social Value is the wider benefit gained by a local community from the delivery of public contracts or services.
  These benefits include
- employment
- training
- strengthened civil society
- improvements to the local environment
- mitigation of the climate risk

The Social Value Network meets bi-monthly, with local champions presenting and sharing their expertise. Actions and next steps are identified.

In addition, the network is looking into the development of an evaluation approach with measures against the pledges. This work is likely to include a combination of quantitative and qualitative research and is being supported by volunteers and by the University of Central Lancashire.

4.2   Community Gateway

Formed in 2005, Community Gateway (CGA) is a not-for-profit housing association based in Preston, owning over 6700 properties and employing 290 people working across the city. Eight out of ten Gateway colleagues live locally.

Craig Garner, Finance Director, explains the following: CGA is more than just a social housing landlord; we were founded on the Gateway model meaning we are tenant-led. It is our tenants who influence the wide range of additional services we offer to enhance their lives and all those who live in our homes and communities. Over the past few years, we have significantly increased the opportunities we offer for tenants to connect, engage and feed back to us to ensure our services are what they need and are delivered in the way that suits them.

Not only does this feedback provide crucial information for us to evolve our offer, but by participating in our various informal and formal boards and committees, attending a consultation event, or task and finish group, our customers develop self-confidence, grow their skills in presenting, public speaking or evaluating reports for example which in turn enhances their employability and personal growth as well as helping combat social isolation.

During the pandemic, the need to work closely and at pace with our partner organisations across Preston was essential to mobilise resources and get much-needed help and support to the most vulnerable—whether that be those shielding, socially isolated or in ill-health, families home-schooling children for the first time or those who suddenly found themselves out of work.

We immediately set up a Support Hub to take calls from customers and the whole organisation swung into action delivering nearly 3000 food parcels from our Purple Pantry food outlet, working with voluntary organisations like Recycling Lives, Fair Share and private providers such as Morrisons and James Hall & Co (SPAR UK) who supported our efforts. We brought our food outlet to people and made sure those struggling in our communities could still get access to affordable food.

Responding to concerned parents, we set up our first Uniform Bank to ensure children had clothes for the new school year and helped out with medication requests, benefits support and even dog-walking. We have now gone on to help others to provide the same range of services themselves in in local community centres and hubs.

At the same time, where we were unable to assist directly, we used our Support Hub and the additional 28,359 "let's chat" support calls we made to help or signpost people to other providers—for befriending services for example. It was true part-nership working. What we found was that sometimes the call itself went a long way to helping people feel less socially isolated, as did our sheltered accommodation newsletters, virtual events for VE Day and Christmas, interactive content on our website and customer magazine. In total, the Support Hub went on to help an incredible 866 people, and along with our SupportLine team's regular check-ins, we undertook nearly 50,000 visits and phone calls in 2020/2021.

Mary, a Gateway tenant said: 'I live on my own, so the calls kept me going and really were my lifeline. Having someone to talk to is worth its weight in gold'.

CGA teams also worked hard with community groups to enable them to become more sustainable and expand their services right at the heart of its estates. Corporate social responsibility is a key driver for the organisation and CGA teams were pleased to be able to support third sector groups and volunteers with professional expertise in marketing, access to grants and digital training to enhance not only their skills, but those of service users.

The organisation worked with Kids in the Kitchen and the Larder to produce online recipes and videos for customers, as well as deliver the ingredients themselves so that families could learn and cook together during lockdown. CGA's online offer was ramped up, with quizzes, employability skills, live Q and As and support for students online. Their charitable subsidiary Preston Vocational Centre continued its award-winning vocational training through blended learning.

Equally importantly, as a founding member of the 'Preston Model', CGA directly supports the regional economy by buying its goods and services from local companies, boosting the supply chain and creating more jobs for local people. In 2020/21, it spent over £9 m in the region on jobs, goods and services.

5 Discussion

5.1 The Anchor Institutions and the Preston Model: Coping with COVID-19

The COVID-19 pandemic has galvanised a turn from social value *intent* into social value *action* in Preston, with anchor institutions, such as the NHS and CGA, determined to build on the social capital that had been built up as a result of being involved and/or influenced by the Preston Model. It seems that the momentum has grown and that a permanent transformation of the way these anchors do business is on the table. For example, at the time of writing, the NHS in Lancashire is in the process of developing a Social Value Framework; CGA, in collaboration with Preston City Council, and supported by Homes England, has won two awards for a project—Making Homes from Houses—to turn empty houses into homes for those who need them, from homeless citizens to others on the housing waiting list (Blog Preston, 2022). Tellingly, Stephen Galbraith, Development Manager at CGA, points to the social value of the scheme as being as important as the economic benefit:

> By bringing these run-down homes up to scratch, not only are we improving the appearance of our neighbourhoods, but we are also helping to combat anti-social behaviour.
> (Blog Preston, 2022)

5.2 The Role of Social Value Post-COVID-19

Social value is an uncertain science, and to embrace its pursuit is to have faith in the systems and general values that form the vague basis for its conditions. In the world of business, the pursuit of social value can be viewed with desire, yet trepidation, especially if social value on the one hand and business objectives on the other are seen to be in possible conflict. Lashitew et al. (2021) even identify the skill in dealing with this potential conflict of interest as 'organisational ambidexterity' which 'relies on highly flexible managerial skills to devise the appropriate organizational designs through structures and processes that foster synergies and reduce frictions among social and financial goals' (Lashitew et al. 2021, no page number). Although this conflict may be less visible in the two anchor institutions that are the subject of this chapter, it is clearly going to be ever-present while our society and communities are structured around a neoliberal system that is focussed on financial profit before social value. In many ways, therefore, it is the system that needs changing, as much as the desire for 'doing good'. It could be argued that the reactions of anchor institutions to the pandemic was indeed a reaction to system change, that is to say a system changed by unforeseen circumstances of prolonged lockdowns and a sense of impending crisis. Clearly, there is a debate or struggle to understand what a post-pandemic society might look like, ranging from a 'back to normal approach' to 'things can never be the same' attitude. The Preston Model, which is a pre-pandemic development, is well placed to adapt to the latter of these tendencies. Of the 'four challenges' identified by the Preston Model Project Committee in consultation with a consultancy team from Mondragon in the Basque Country—home of the Mondragon cooperative system—one is building a 'social value business approach' (LKS Next, 2020, pp. 18–20). Along with Preston City Council's 'version 2' of the Preston Model, which identifies social value as one of its eight strands of action, the concept becomes an integral part of systemic development in the community wealth building project in Preston (Preston City Council, 2021).

5.3   The Indefinition of Social Value

A curiosity that emerges from our research is the range of different views as to what exactly social value entails. This is not necessarily negative or disadvantageous. It may be that social value is more akin to an attitude than a precise object of delivery, and although that does not sit comfortably in this world of 'SMART' (specific, measurable, achievable, relevant, time-bound) aims and objectives, perhaps it actually plays to the strengths of the different actors in their different organisations and circumstances. Therefore, for Community Gateway, social value must speak to the needs of the housing association's tenants, and this might place good housing and decent, living waged work before health outcomes, whereas the NHS might naturally emphasise health. Nevertheless, just as 'all roads lead to Rome', all social values lead to positive community outcomes, and a good job is likely to improve health outcomes, while good health is likely to be a good basis for seeking decent work (Caldwell et al., 2017).

For Community Gateway, then 'Social Value is the term used to describe the extra value created when we deliver a service that benefits our tenants and the wider community' (Community Gateway, n.d.). On the other hand, the development of a Central Lancashire Social Value Framework has initially adopted the definition found in Preston City Council's Community Wealth Building 2 document: "'Social value' is the wider benefit gained by a local community from the delivery of public contracts or services. These benefits include employment, training, a strengthened civil society, improvements to the local environment and mitigation of the climate risk' (Preston City Council, 2021, p. 9; James & Wood, 2021, slide 5), while the Public Health Foundation report on the role of the NHS as an anchor institution states that there is no definitive definition of the term, but then goes on to narrowly define it as referring 'to the wider societal benefits that can be gained from purchasing decisions' (Reed et al., 2019, p. 25), therefore locating social value in the domain of procurement. This is because this report acknowledges its position as being informed by the Centre for Local Economic Strategies (CLES) and the Democracy Collaborative, both of whom have been influential in developing the procurement policies for the Preston Model. The Social Value Portal, which was designed to support an understanding of the Social Value Act of 2012 (SV Act, 2012) and the measurement of social value nationally, spreads the value net far wider than this by describing five social value themes as part of its Themes–Outcomes–Measures (TOMS) design. These are as follows: (1) jobs (supporting the promotion of local skills and employment), (2) growth (the promotion of responsible regional business), (3) social (healthier, safer and more resilient communities), (4) environment (decarbonising and safeguarding our planet), and (5) innovation (social innovation). Clearly, then, the version of social value adopted in and around Preston is wired in to the development of the Preston Model, which provides a supporting framework for the different anchor institutions in Preston, although this might not be a suitable or even desirable framework for other localities in the UK.

5.4   Social Value, System Change and Democracy

For the interpretation of social value in its broadest sense, which is the sense that emerges from our research for this chapter, the systemic change that is inextricably linked with the development of social value as a community and organisational value—including but looking beyond social value as anchor institution procurement—is a shift towards participatory and citizen-led democracy. This is evidenced in various community facing processes which can serve to improve democracy in its broadest sense, even if this is not their primary purpose. For instance, coproduction processes within health, housing and welfare organisations, participatory community development and action research approaches, community asset mapping, citizen juries and assemblies all have potential to enhance democratic involvement and engagement of local citizens, and these can be successful in even the most unpropitious circumstances (see, e.g., Dzur, 2019). Yet, almost everywhere we look, democracy is in crisis, with denuded participation and faith in electoral processes accompanying new populist demagoguery and dog whistle politics that expose something of a lack of sophistication and resilience in our institutional democracy (Cossarini & Vallespin, 2019). As various commentators have pointed out, the legitimacy of democracy is at stake (Macdonald, 2021) and these crises of legitimacy are matched in an increasing lack of faith in our institutions of health and welfare and associated professional and administrative roles (Cahill, 2011, McKeown & Carey, 2015). That said, a desire for voice and influence on the part of citizens is arguably not diminished, rather it is the confidence in formal democratic processes and their ability to deliver that is lacking. A plethora of new social movements and citizen activism has emerged, precisely because opportunities for voice and influence have become so constrained (Tarrow, 2011). Part of the appeal of initiatives such as the Preston Model is the hope that aspects of the approach to community wealth building are implicitly democratising and that this can redeem the various legitimacy crises we find ourselves living through. A social value economy as part of a new democracy, and its benefits as described in this chapter, is embedded in the appeal of the Preston Model.

Even though electoral success in council elections in Preston seems to indicate democratic approval of the Preston Model, it is arguably only part, and an insufficient part, of the democratic appeal of community wealth building. The approaches to extending and sustaining economic justice, developing worker voice and control by building a cooperative ecosystem, and tackling stark inequities such as food poverty and migrant sanctuary are predicated upon a commitment to democratisation and citizen participation that goes far beyond intermittent involvement in elections. The work of CGA, with its cooperative approach to governance, and the NHS in Lancashire, with its shift from a narrow clinical approach to a social value approach, are indicative of the changes heralded by the Preston Model.

Therefore, one of the key challenges of the Preston Model is how to nurture and deepen democratic participation in ongoing deliberations across the anchor organisations, between community and social organisations in a spirit of inter-cooperation, and throughout the broader polity. If this can be realised, then it would be a key dimension of sought-after social value and also the means for bringing social value into being. There is some emerging evidence that more actively engaged citizens who can draw upon networks of social capital do better in terms of various health and well-being outcomes and that participation at local and neighbourhood levels also supports a more virtuous cycle of broader civic engagement, including electoral participation (Ziersch, 2005; Ziersch & Baum, 2004, Ziersch et al., 2005, 2009). Benefits are more likely to be seen at the community rather than the individual level, and the means by which positive outcomes can be achieved are sensitive to differences between places. In short, seeking and exercising democratic voice, influence and participation are healthy and generative, and the more we do it, the more we expect to do more of it. For this to be effective, there needs to be a fair distribution of effort so that individual community actors do not bear all of the burden, and community-wide initiatives must be rooted in understanding of place, local needs and demands. This is one of the vectors by which full success of the place-based Preston Model can be achieved and wider social value be brought about, in a self-sustaining cycle of cooperative, democratic voice and participation. The road to such success, however, will be complex and arguably requires sophisticated attention to means of support and mechanisms of action.

5.5 Comparison with Social Entrepreneurship Projects, Social Value and Public Health

Although it is beyond the scope of this chapter to present a detailed overview of how the Preston experience compares with social entrepreneurship, some references to other projects may help to illuminate similarities and differences. For further evidence linking the wider scope of better public health as a result of promoting social value projects, see Ashton et al. (2020).

This chapter began with Bowen's (1953) historical question to businessmen regarding the range of social responsibility that a business could reasonably expect to engage with. Bowen also introduced the term 'social entrepreneurship', which has gathered meaning and favour in recent years. On the face of it, such organisations are promoting social value ahead of financial benefit for the organisation, and, as such, they encompass similarities with the work of anchor institutions described in this chapter. The closest similarity can be found in one of the three definitions of social entrepreneurship in Martin and Osberg (2007):

> Forging a new, stable equilibrium that releases trapped potential or alleviates the suffering of the targeted group, and through imitation and the creation of a stable ecosystem around the new equilibrium ensuring a better future for the targeted group and even society at large.
> (https://ssir.org/articles/entry/%20social_entrepreneurship_the_case_for_definition#)

However, the two case studies described here are different in significant ways. First, they are large anchor institutions. They are rooted in the locality; therefore, they are comparatively sustainable and stable, and they have a community identity that immediately resonates with the social value associated with their work. Second, they are not institutions founded on social value. The NHS's primary mission is a clinical one, while Gateway's mission is housing. These institutions are not 'changemakers' in the spirit of social entrepreneurship as defined in Ashoka, for example (https://www.ashoka.org/en-gb/programme/ashoka-changemakers), although change is clearly part of a systemic change in communities that engage with community wealth building (Manley & Whyman, 2021). The increased emphasis on social value within the work of these institutions is closely connected to change in the sense of a changing systemic vision of how local actors can network and contribute to society. In Preston, this is known as the Preston Model. Unlike social entrepreneurship, the Preston anchor institutions do not rely on individual and innovative leaders ('entrepreneurs') but rather on a growing awareness of interconnectivity, community identity and system change.

6 Conclusion

As this chapter has outlined, the shift in emphasis from highlighting economic success to a growing interest in corporate social responsibility and more recently to social value is a change that is clearly evidenced in our two anchor institution case studies. System change models such as the PM represent a shift in local economic development, underpinned by cooperative principles and values that seek to generate and retain local economic wealth as well as enhance local democratic involvement of communities. Flowing from this, we suggest there may be a range of ways through which approaches underpinned by social value have benefits for health and well-being. Firstly, enhanced employment opportunities as well as improvements in working conditions and wages may lead to direct health benefits (Bambra et al., 2009). There is evidence to suggest, for example, that a greater sense of control in the workplace may reduce the risk of cardiovascular disease (Marmot et al., 1997). Secondly, by placing 'community' at the heart of economic development initiatives, it is possible that residents will feel a greater sense of control over their lives (Whitehead et al., 2016), as well as a strengthened sense of collective identity, promoting a more positive sense of place (Halliday et al., 2021). Finally, it is feasible that through a more effective system-wide response, involving mobilising resources and knowledge between organisations, 'social value' approaches can enable the creation of 'social capital' between cooperating social actors and the broader community. In turn, this may have helped buffer against the adverse economic impacts of the pandemic as well as having potential wider benefits such as contributing to cultural changes, shifts in organisational practices and individual attitudes. Nevertheless, when considering the role of community wealth building in tackling health inequalities, it is also important to be mindful of the limits of such initiatives, given that drivers of inequalities may be located within macro-global systems rather than within local areas (Naik et al., 2019). In this respect, although the signs to date are positive—all social and economic indices point to a sustained improvement in the overall circumstances of people in the Preston region—it remains to be seen to what extent models such as the PM can help mitigate against new and emergent system 'shocks', which have direct financial consequences (e.g. escalating living costs) for local communities already experiencing health and social inequalities.

The active interest of the two anchor institutions in this chapter—the Lancashire Teaching Hospitals NHS Foundation Trust and Community Gateway Association—in a systematic effort to build fairer economies, improve the social value of their various endeavours and locate themselves within a more cooperative ecosystem, all herald opportunities to deepen grassroots participatory democracy. The very particular democratic opportunities are represented by the development of cooperative enterprises and/or businesses, anchor institutions and other organisations that are turning to social value as a basic principle of work; inter-cooperation between these can hugely enhance such strengthening of community voice and democracy, with likely associated health and well-being benefits. This in turn has the potential to contribute to a renewal of more formal democratic processes and provide an antidote to various telling crises of legitimacy for health and social care services in particular and, notably, the wider polity.

While part of this social and economic transformation rests with a drive for social value changes within anchor institutions (Health Foundation 2019)—and moves beyond the basics of CSR—the evidence in this chapter suggests that the application of social value principles can signal a wider social and economic transformation.

Organisations such as the NHS in Lancashire and CGA are people-centred. The turn towards social value is also a turn towards what gives these organisations a sense of purpose. Our research suggests that the injection of social values into the fabric of anchor institutions and the communities they serve has the potential to result in healthier communities. The reality of this scenario will need to be tested in future research over time, but the possibilities of success appear to be within reach. Success, by means of healthier communities, would be both a hard-fought victory for public health and also a welcome expression of democratic transformation.

Acknowledgement

The authors would like to thank the following for their contributions to the NHS case study that forms part of this chapter: Sharon Robson, Nina Carter, Louisa Graham, Kerry Hemsworth, David Hounslea, Sarah James, Jacqueline Higham, Stefanie Johnson, Sian Fisher and Frances Balmer.
Disclaimer Julian Manley, Mick McKeown, Emma Halliday and Ioannis Prinos are funded by the National Institute for Health Research (NIHR) Public Health Research Programme, Grant Reference Number NIHR130808. The views expressed in this publication are those of the authors and not necessarily those of the National Institute for Health Research or the Department of Health and Social Care.

References

Alperovitz, G., Williamson, T., & Howard, T. (2010). The Cleveland model. *The Nation, 1*(1), 21–24.

Ashton, K., Schröder-Bäck, P., Clemens, T., Dyakova, M. A., & Bellis, M. A. (2020). The social value of investing in public health across the Life course: A systematic scoping review. *BMC Public Health, 20*, 597. https://doi.org/10.1186/s12889-020-08685-7

Bambra, C., Gibson, M., Sowden, A. J., Wright, K., Whitehead, M., & Petticrew, M. (2009). Working for health? Evidence from systematic reviews on the effects on health and health inequalities of organisational changes to the psychosocial work environment. *Preventive Medicine, 48*(5), 454–461.

Banerjee, S. B., City, M., Jermier, J. M., Peredo, A. M., Perey, R., & Reichel, A. (2021). Theoretical perspectives on organizations and organizing in a post-growth era. *Organization, 28*(3), 337–357.

Blakeley, G. (2020). *The Corona crash: How the pandemic will change capitalism.* Verso.

Blog Preston. (2022). Preston refurbished homes scheme wins two awards. Accessed June 1, 2022, from https://www.blogpreston.co.uk/2022/06/preston-refurbished-homes-scheme-wins-two-awards/.

Bowen, H. R. (1953). Social responsibilities of the businessman. .

Cahill, D. (2011). Beyond neoliberalism? Crisis and the prospects for progressive alternatives. *New Political Science, 33*(4), 479–492.

Caldwell, N. D., Roehrich, J. K., & George, G. (2017). Social value creation and relational coordination in public-private collaborations. *Journal of Management Studies, 54*(6), 906–928.

Carroll, A. B. (2021). Corporate social responsibility: Perspectives on the CSR Construct's development and future. *Business & Society, 60*(6), 1258–1278.

Charnley-Parry, I., Farrier, A. and Dooris, M. (2022) 'Climate resilience, social justice and Covid 19 recovery in Preston', Research Report. Accessed May 27, 2022, from Climate Resilience,

Social Justice and COVID-19 Recovery in Preston | Place Based Climate Action Network (pcancities.org.uk).

Community Gateway (n.d.) Accessed June 2, 2022, from https://www.communitygateway.co.uk/social-value.

Cossarini, P., & Vallespin, F. (Eds.). (2019). *Populism and passions: Democratic legitimacy after austerity*. Routledge.

Dzur, A. W. (2019). *Democracy inside: Participatory innovation in unlikely places*. Oxford University Press.

Halliday, E., Brennan, L., Bambra, C., & Popay, J. (2021). 'It is surprising how much nonsense you hear': How residents experience and react to living in a stigmatised place. A narrative synthesis of the qualitative evidence. *Health & Place., 68*, 102525.

James, S., & Wood, J. (2021). 'Developing a Central Lancashire social framework', Powerpoint presentation, for the social value Framework 'champions' group, November 2021.

Lashitew, A. A., Narayan, S., Rosca, E., & Bals, L. (2021). Creating social value for the 'base of the pyramid': An integrative review and research agenda. *Journal of Business Ethics*. Accessed June 1, 2022, from springer.com

LKS Next. (2020). 'Designing a cooperative entrepreneurship initiative for Preston: Challenges and strategic action lines A report presented to the University of Central Lancashire, Preston City Council and the Project Committee' Accessed June 1, 2022, from https://secureservercdn.net/1 60.153.137.14/dx0.b15.myftpupload.com/wp-content/uploads/2021/07/Final-Report-2-Pres ton-Ecosystem-Challenges-and-Action-Lines-.pdf.

Manley, J., & Whyman, P. B. (Eds.). (2021). *The Preston model and community wealth building*. Routledge.

Macdonald, T. (2021). Reviving democracy: Creating pathways out of legitimacy crises. European Journal of. *Political Theory*, 14748851211020625.

Marmot, M. G., Bosma, H., Hemingway, H., et al. (1997). Contribution of job control and other risk factors to social variations in coronary heart disease incidence. *The Lancet, 350*, 235–239.

Martin, R. L., & Osberg, S. (2007). Social Entrepreneurship: The case for definition. *Stanford Social Innovation Review*. Accessed September 1, 2022, from https://ssir.org/articles/entry/%20social_entrepreneurship_the_case_for_definition#.

Mason, P. (2015). *Postcapitalism: A guide to our future*. Penguin.

McKeown, M., & Carey, L. (2015). Democratic leadership: A charming solution for nursing's legitimacy crisis. *Journal of Clinical Nursing, 24*(3–4), 315–317.

Naik, Y., Baker, P., Ismail, S. A., et al. (2019). Going upstream—an umbrella review of the macroeconomic determinants of health and health inequalities. *BMC Public Health, 19*, 1678. https://doi.org/10.1186/s12889-019-7895-6

Nasim, B. (2022). Does poor quality housing impact on child health? Evidence from the social housing sector in Avon, UK. *Journal of Environmental Psychology*. https://doi.org/10.1016/j.jenvp.2022.101811. Journal pre-proof.

NIHR. (n.d.) 'Is Preston 'Building Back Better' post-Covid?' Accessed May 31, 2022, from https://arc-nwc.nihr.ac.uk/news/is-preston-building-back-better-post-covid/.

Novkovic, S., & Webb, T. (Eds.). (2014). *Co-operatives in a post-growth era*. Zed Books.

Our Central Lancashire (social value framework) (n.d.) Accessed May 27, 2022, from https://www.healthierlsc.co.uk/central-lancs.

Pansera, M., & Fressoli, M. (2021). Innovation without growth: Frameworks for understanding technological change in a postgrowth era. *Organization, 28*(3), 380–404.

Preston City Council/PCC. (2021). Accessed May 27, 2022, from https://www.preston.gov.uk/media/5367/Community-Wealth-Building-2-0-Leading-Resilience-and-Recovery-in-Preston-Strategy/pdf/CommWealth-ShowcaseDoc_web.pdf?m=637498454035670000.

Preston City Council. (n.d.). https://www.preston.gov.uk/article/3538/Making-Spend-Matter-Toolkit-Social-Value-Procurement-Frameworks-Frequently-Asked-Questions-FAQs-3.

Prinos, I. (2021). The Preston Model and co-operative development: A glimpse of transformation through an alternative model of social and economic organisation. In J. Manley & P. Whyman

(Eds.), *The Preston model and community wealth building, creating a socio-economic democracy for the future* (pp. 32–48). Routledge.

Prinos, I., & Manley, J. (2022). The Preston model: Economic democracy, cooperation and paradoxes in organisational and social identification. *Sociological Research*. https://doi.org/10.1177/13607804211069398

Reed, S., Göpfert, A., Wood, S., Allwood, D., & Warburton, W. (2019). Building Healthier communities: The role of the NHS as an anchor institution. *The Health Foundation*. Accessed May 31, 2022, from https://www.health.org.uk/publications/reports/building-healthier-communities-role-of-nhs-as-anchor-institution

SV Act. (2012). Accessed May 31, 2022, from https://www.legislation.gov.uk/ukpga/2012/3/enacted.

Tarrow, S. (2011). *Power in movement* (3rd ed.). Cambridge Univ. Press.

Trebeck, K., & Williams, J. (2019). *The economics of arrival*. Policy Press.

Urbact. (2018). Making spend matter. Accessed May 31, 2022, from https://urbact.eu/sites/default/files/media/final_transferability_study_-_making_spend_matter_-_october_2018.pdf.

Vandeventer, J. S., & Lloveras, J. (2021). Organizing degrowth: The ontological politics of enacting degrowth in OMS. *Organization, 28*(3), 358–379.

Whitehead, M., Pennington, A., Orton, L., et al. (2016). How could differences in 'control over destiny' lead to socioeconomic inequalities in health? A synthesis of theories and pathways in the living environment. *Health & Place, 39*, 51–61. https://doi.org/10.1016/j.healthplace.2016.02.002

Ziersch, A. M., & Baum, F. E. (2004). Involvement in civil society groups: Is it good for your health? *Journal of Epidemiology & Community Health, 58*(6), 493–500.

Ziersch, A. M., Baum, F., Darmawan, I. G. N., Kavanagh, A. M., & Bentley, R. J. (2009). Social capital and health in rural and urban communities in South Australia. *Australian and New Zealand Journal of Public Health, 33*(1), 7–16.

Ziersch, A. M., Baum, F. E., MacDougall, C., & Putland, C. (2005). Neighbourhood life and social capital: The implications for health. *Social Science & Medicine, 60*(1), 71–86.

Ziersch, A. M. (2005). Health implications of access to social capital: Findings from an Australian study. *Social Science & Medicine, 61*(10), 2119–2131.

Julian Manley is an academic researcher working in the University of Central Lancashire. His research is centred on the social and relational aspects of co-operatives and co-operation, with a special focus on the Preston Model and community wealth building. He is a founder member and Director of the Preston Co-operative Education Centre.

Craig Garner is the Executive Director of Resources at Community Gateway Association (CGA). He is born and bred in Preston and has over 30 years sector experience across a number of housing organisations across North West England. Craig joined CGA in 2018 and is a qualified accountant and chartered manager who is responsible for service functions including Finance, Information Technology & Digital Transformation, Governance and Business Assurance.

Emma Halliday is a Senior Research Fellow in the Faculty of Health and Medicine, Lancaster University. She has a background in history, and expertise in the use of qualitative methods for applied public health research. Her research focuses primarily on evaluating local government and community initiatives with potential to impact health inequalities.

Julie Lee is the Head of Income Management at Community Gateway Association (CGA). Julie has worked in income and debt management for over 30 years, of which the last 18 have been within the social housing sector. Julie joined CGA in 2007 and has been instrumental in setting up financial support services to benefit and improve the quality of life for customers within our local communities. She has developed many initiatives alongside partners to increase household income and tackle food poverty.

Louise Mattinson is the Executive Director of Customers & Communities at Community Gateway Association (CGA). She is responsible for CGA's Development, Repairs, Asset Management and front-line housing services. Louise joined CGA in 2006 and has over 25 years' experience within the Housing Sector across Preston and the Fylde Coast. Louise is passionate about Customer Service and ensuring all customers are listened to and treated as individuals.

Mick McKeown is Professor of Democratic Mental Health, University of Central Lancashire. He is also active in trade unions, linking union organising with cooperative development and education related to community wealth building. To this end, Mick is a founder member of Union Coops UK.

Ioannis Prinos is seeking to combine empirical research and social theory in understanding the socio-dynamics of unequal power relations and their consequences in terms of poverty, marginalisation, democracy, and inequality. His research is centred around the ways alternative paradigms of economic organisation focusing on cooperation and the pursuit of social value such as community wealth building, can help alleviate these issues. Currently, he is working with Dr. Julian Manley and Professor Mick Mckeown on assessing the overall impact of the Preston Model and on a research project funded by the NIHR, investigating the impact of community wealth building initiatives on health and wellbeing, with a focus on the socio-economic determinants of health, health inequalities, covid response and post-covid recovery.

Kate Smyth is a Non-Executive Director (NED) at Lancashire Teaching Hospitals NHS Foundation Trust (LTH) and is the NED lead for Anchor Institutions and Social Value. She is a chartered town planner and has worked in a number of local authorities developing local purchasing initiatives. She is a Lay Leader at the Yorkshire and Humber Patient Safety Translational Research Centre. Prior to joining LTH she worked for Calderdale CCG and was the vice-chair of Kirklees Neighbourhood Housing where she led on poverty and deprivation. She is co-chair of the Disabled NHS Directors Network.

Jonathan Wood is the Deputy Chief Executive and Director of Finance for Lancashire Teaching Hospitals Foundation Trust (LTH). He is the executive lead for Social Value at LTH. Jonathan qualified as an accountant having joined the NHS as a graduate trainee in the early 1990s. He has worked in a number of different NHS organisations across the North West and has worked in Lancashire since 2009.

Corporate Social Responsibility and Coping with COVID-19 Pandemic in the Global Health Service Institutions: The United Kingdom

Mohammed Ali and Courtney Grant

1 Introduction

1.1 Historical Context

There is a long history of the practical application of CSR in the UK, extending back hundreds of years (Idowu, 2009). This long history also includes socially responsible behaviour in the context of the healthcare system. A sense of social responsibility was fundamental to the forming of the UK National Health Service (Health Services Research UK, 2017), which came into effect when Health Secretary Aneurin Bevan launched the NHS at Park Hospital in Manchester as part of an overall plan to deliver a good standard of health care for all citizens (Denham, 2017). The forming of the NHS was one more step along a long-standing sociological pattern that moved towards creating a more equitable society (Denham, 2017).

Before the creation of the Church of England in the sixteenth century, religious orders (and most notably the monasteries) demonstrated a sense of social responsibility by caring for the most vulnerable members of society, including those who were infirm. However, this source of care was removed once the Church of England was formed because of the ex-communication from the Catholic Church (Greengross et al., 1999).

While hospitals such as St Bartholomew's Hospital in London existed in the seventeenth century, it was the eighteenth century that gave rise to a boom in hospital provisions known as the voluntary hospital movement (Waddington,

M. Ali (✉)
Guy's and St Thomas' NHS Foundation Trust/OUR IMPACT, London, UK
e-mail: mali@ourimpactldn.com

C. Grant
Transport for London, London, UK

© The Author(s), under exclusive license to Springer Nature Switzerland AG 2023
S. O. Idowu et al. (eds.), *Corporate Social Responsibility in the Health Sector*, CSR, Sustainability, Ethics & Governance, https://doi.org/10.1007/978-3-031-23261-9_9

2014). By the nineteenth century, there were hundreds of voluntary hospitals located across Britain (Waddington, 2014).

The nineteenth century also saw local authorities demonstrating social responsibility in the context of health care. The nation became disturbed by the high rates of maternal and infant mortalities. Local authorities, upon realizing that action needed to be taken, provided free maternal and child welfare services, which fell within the context of local authorities concentrating their efforts on environmental reform (Honigsbaum, 1990). Local authorities also provided health care for individuals with infectious diseases, as well as for people with mental illnesses and physical disabilities (Greengross et al., 1999).

While local authorities were responsible for 59% of hospital beds, the voluntary sector was responsible for the others, which were financially underpinned by charity (Gorsky, 2008). Specialists would provide medical care in these voluntary hospitals. These hospitals would typically put their efforts towards caring for those who were not in need of care on a long-term basis, and the specialists who worked there would also often have profitable private practices at other locations (Greengross et al., 1999).

It was clear that Victorian London depended on the voluntary hospital movement, which was underpinned by the social responsibility of philanthropists, who aimed to reduce the levels of misery experienced by members of society (Waddington, 2014). They believed in the need to contribute towards 'the common good' (Waddington, 2014). It was upon this philosophy that London's hospitals were built, and this ethos extended from the early eighteenth century and into the twentieth century (Waddington, 2014).

Reverend Andrew Reed established a foundation in 1813 to provide education and support to vulnerable children and orphans (The Andrew Reed Foundation, 2022). Among Reed's philanthropic contributions is the founding of the Royal Hospital for Incurables, now known as the Royal Hospital for Neuro-disability. In addition, he founded the London Orphan Asylum, the Royal Earlswood Hospital, and Reed's School (Cook, 2004).

By the twentieth century, the healthcare space in which philanthropy occupied now began to become occupied by the state (Moon, 2010). After 1914, philanthropy became undermined by tax rises, and as a result, voluntary hospitals began to suffer funding problems. In order to address the shortfall, they began to elicit financial contributions from the working class (Gorsky, 2008).

After the Second World War, more than 50 nations attended the United Nations Conference on International Organization (United Nations, 2022). During the next 2 months, these nations drafted and signed the United Nations Charter, creating the United Nations in the hope of preventing another world war (United Nations, 2022).

In the United Kingdom, the National Health Service Act of 1946 was passed, the act helped to establish the NHS on 5 July 1948 (Mann & Saeed, 2021). The NHS became a government-run, population-wide service, which was free at the point of use and was funded generally by taxation (Greengross et al., 1999). The aftermath of the Second World War also helped create a wider context of other reforms in the UK, including the modernizing of schools and social security systems (Shapiro, 2010).

Moreover, the Alma-Ata Declaration of 1978 emerged as a major landmark of the twentieth century, emphasizing the importance of primary health care in reaching the goals of health for all (World Health Organization, 2022). Additionally, the declaration reaffirmed the broader definition of health as '*a state of complete physical, mental, and social well-being, and not merely the absence of disease or infirmity, is a fundamental human right and that the attainment of the highest possible level of health is a most important world-wide social goal whose realization requires the action of many other social and economic sectors in addition to the health sector*' (World Health Organization, 2022). Therefore, the responsibility for health falls on more actors across a wider sector. In 2019, the United Nations Declaration on Universal Health Coverage was adopted by world leaders at the United Nations General Assembly (World Health Organization, 2019). The declaration reemphasized the importance of global health equity in that '*Health is a human right, where universal means universal; whereby all people have access and coverage to quality comprehensive health services, as well as interventions to address social determinants of health*' (Mitchell, 2019).

One of the significant reforms the NHS underwent was through the Social Care Act of 2012. The responsibilities of managing public health services were transferred to local authorities, while the commissioning was carried out by NHS Clinical Commissioning Groups (CCGs), which replaced Primary Care Trusts (Gadsby et al., 2017). In addition, the Health and Wellbeing Board was created. The board brought leaders from health and care systems together to improve community health and well-being and reduce health inequality (Gadsby et al., 2017). Furthermore, the Social Value Act 2012 required the public sector to ensure that the money it spends on services provides maximum economic, social and environmental value for communities (Public Health England, 2015). This would allow the private sector to become providers and investors in local borough initiatives through CSR (Public Health England, 2015). It has been implied that the NHS is the central jewel in the crown of corporate responsibility in the UK (Hastings, 1960).

1.2   Current Context

The COVID-19 pandemic impacted different aspects of people's lives within the United Kingdom, including specific health implications. COVID-19 has placed unparalleled demands on the UK healthcare system, including on the amount of available space for intensive care units and on equipment (Flynn et al., 2020). Evidence shows that COVID-19 has disproportionately impacted ethnic minority communities, with biological and social risk factors increasing the risk of ethnic minority groups contracting COVID-19 compared to non-ethnic minority individuals (Phiri et al., 2020). Bangladeshis had the highest risk of death caused by COVID-19, twice that of white Britons (Public Health England, 2020a). The increased risk of death was caused primarily by wider determinants of health, such as crowded multigenerational housing, occupational risks, and a lack of access to health care (Public Health England, 2020b). These risks become even more apparent considering that age, chronic comorbidities, and smoking are some of the biggest risk factors for COVID-19 deaths (Dessie & Zewotir, 2021).

It has been suggested that quick and decisive action was needed to lower the mortality risk of people from ethnic minority communities and to ascertain a better understanding of the reasons why there is a greater risk for these groups (Aldridge et al., 2020). As a result of the pandemic, organizations in the third sector have had to significantly change the way they deliver their services to health service institutions (Cottom, 2021). Despite being at risk, ethnic minorities played an important role in managing COVID-19's health effects in the UK (Phiri et al., 2020).

Doctors and other front-facing healthcare professionals from ethnic minorities were disproportionately impacted in terms of deaths (Phiri et al., 2020). Moreover, during the pandemic, the well-being of health professionals was affected. The data from the *NHS Staff Survey of 2020 and NHS Workforce Race Equality Standard* (WRES) found health staff from ethnic minorities to have consistently reported being discriminated against, harassed, and victimized at work (Phiri et al., 2020). Furthermore, poor fitting of personal protective equipment (PPE) with head coverings and beards has increased anxiety among Muslims and Sikh health workers due to the increased risk of developing COVID-19 infection when they come into contact with patients (Phiri et al., 2020). Also faced with the prospect of putting their lives at risk to save the lives of patients, health professionals were placed under extremely high pressure that may have negatively affected their mental health. Feelings of guilt, shame, and meeting high expectations may contribute to this pressure (Phiri et al., 2020). Therefore, the incorporation of ethical and moral imperatives into health operations, processes, and policies was required in tackling the concerns highlighted above.

In addition to the direct impact of COVID-19, the virus is likely to have increased the level of mortality and morbidity among individuals not infected with COVID-19 but who have not been able to get care for other conditions during the pandemic, both over the short- and medium-term period (Flynn et al., 2020). For example, weight management is important for preventing severe outcomes, such as cardiovascular disease, stroke, and heart failure. The UK healthcare system has seen fewer patients seeking emergency care during the pandemic (Flynn et al., 2020).

The pandemic has also led to a quadrupling of food insecurities among families in the UK, with associated increases in the level of food bank use and free school meal applications (Moore & Evans, 2020). High demand and low supply of food led to shortages during the COVID-19 pandemic, which was exacerbated by consumer stockpiling, slow food supply chain logistics, high staff absenteeism due to COVID-19-related illnesses, increased food waste, and outbreaks of COVID-19 in food processing centres (Wentworth, 2020). Moreover, the World Health Organization (WHO) reports that the reduction in physical activity levels of people and restrictions on access to specific foods have contributed to obesity (Flynn et al., 2020). In the UK, deprivation contributes to 63% of the population being either overweight or obese (Flynn et al., 2020). Obesity and hunger are closely related, as deprived families turn to energy-dense and higher-calorie foods because they are more affordable than nutrient-dense foods (Fanzo, 2019). While healthy foods satisfy the stomach for longer, energy-dense foods digest more quickly, resulting in a high level of dopamine release and the tendency to crave food (Volkow et al., 2011). The situation is further compounded by poorer household members missing meals or eating less to ensure their families have enough food to eat. In such a situation, obesity and hunger can coexist in the same household, causing a 'double burden of malnutrition' (National Institute for Health Care and Research, 2020). Additionally, a child's risk of obesity can be determined even before birth if the foetus' long-term development is negatively impacted by poor nutritional diets and lifestyle choices during pregnancy (Harvard T.H. Chan School of Public Health, 2022).

Data show that food insecurities in the UK have impacted four million people, including 2.3 million children (The Food Foundation, 2021). More than twice as many people from ethnic minorities suffer food insecurity compared to their white British counterparts (The Food Foundation, 2021). It is important to understand that factors such as poverty can adversely contribute to enhancing health inequalities; for example, inflation rising to 10.1% (Elliott, 2022) has resulted in higher living costs, which disproportionately affect low-wage earners, as well as those who are dependent on Universal Credit (Mahon, 2022). In the UK, child poverty rates reach their highest in London and Manchester (Action for Children, 2021). The London constituency of Bethnal Green and Bow has the highest levels of child poverty and has some of the largest proportions of Bangladeshis in the whole of the United Kingdom (Action for Children, 2021). In the 10% of the most income-deprived British neighbourhoods, there are three times as many Bangladeshis than white British residents (GOV.UK, 2020). With this in mind, it is important to note that malnutrition is a cause of poverty, and poverty causes malnutrition (Fleetwood, 2020). This impacts a person's health, and as such affects the healthcare institutions. Food poverty places people at a higher level of risk of chronic diseases, including cardiovascular disease, diabetes, and hypertension (Harrington et al., 2009).

Before 2020, 1.3 million children in the UK were classified as being entitled to free school meals (Sandhu, 2022). However, not all children who were living in food poverty qualified for free school meals. Additionally, half of the children eligible for free school meals did not receive any. This exacerbated child hunger once the pandemic began (Hefferon et al., 2021). A longitudinal study found that long-term child hunger can increase the risk of depression and suicidal ideation during adolescence, as well as the development of adverse health conditions, such as asthma and iron deficiency (Ke & Ford-Jones, 2015).

A COVID-19 pandemic of unprecedented proportions could also explode into a mental health crisis. Survey data from the *Coronavirus and the Social Impacts on Great Britain* suggest that 69% of respondents said that COVID-19 negatively impacted their well-being, with uncertainty, anxiousness, and boredom being common contributing factors (Marshall et al., 2020).

Furthermore, social distancing and lockdowns are likely to have contributed to an increased prevalence of mental health conditions among children and adults (Flynn et al., 2020). In fact, social isolation had a significant impact on children's mental health, especially since their mental health was already in crisis before the pandemic occurred (Hefferon et al., 2021). According to a study conducted before the pandemic, the UK had been identified as a country with high levels of austerity, resulting in negative effects on child health outcomes and social determinants of child health (Rajmil et al., 2020).

Therefore, these existing problems have been amplified by the pandemic. For example, Young Minds conducted a survey, which showed that 83% of adolescents with mental health needs found that the pandemic exacerbated their mental health issues. In addition to this, children between the ages of 4 and 10 suffered attentional and behavioural problems. Referrals for mental health services have increased, including an increase in the use of mental health platforms that are online-based. Moreover, it has been found that the risk of future depression is linked to the experience of loneliness during lockdown (Hefferon et al., 2021).

2 The Four Major Aspects of CSR

In the United States, Howard R. Bowen wrote a book called 'Social Responsibilities of the Businessman', which sparked modern CSR concepts (Carroll, 2016). Bowen suggests that large organizations have a societal responsibility that comes with power and decision-making ability that impacts the lives of the citizens they serve (Carroll, 2016). This led to the development of Archie Carroll's pyramid of CSR, one of the most popular models of CSR today. Carroll's pyramid of CSR has four components established through empirical research studies (Carroll, 2016). Carroll believes that corporations should take on economic, legal, ethical, and philanthropic responsibilities simultaneously with their business activities (Carroll, 2016). This section looks at how UK institutions have managed to cope, and are still coping, with their health obligations towards their citizens.

2.1 Economic Responsibilities

Companies that practice economic social responsibility pay their employees a fair salary and ensure their consumers receive good value for their money. Core to the accepted definition of CSR is the fact that businesses should engage in behaviours that embody economic responsibility with the aim of helping society to meet its expectations (Gutierrez-Huerter, 2020). A way in which this expectation is being met within the UK economy is through the expansion of social enterprises. Social enterprises are businesses with social, charitable, or community goals as their core mission. In 2018, social enterprises brought around £60 billion to the UK economy, which equates to three per cent of gross domestic product and brought in employment by five per cent (Financier Worldwide, 2020). Furthermore, social enterprises played a pivotal role during the pandemic, as more than 7000 of them supported frontline health and social services for deprived communities (Social Enterprise UK,

2020). People who are homeless face health disadvantages and have a higher risk of death than the general population due to multi-complex issues including poor living conditions, trauma, and drug and alcohol dependence (Public Health Scotland, 2021). Social enterprises like Beam work with civic institutions like Lambeth Council to help COVID-19-affected communities experiencing homelessness through work-related skills training, travel assistance, and childcare support (Lambeth Council, 2020).

As small- and medium-sized businesses suffered due to the COVID-19 restrictions, the UK Government awarded emergency grants, funds, and loans to help small businesses recover (Charity Bank, 2021). As part of the government's commitment during the coronavirus (COVID-19) outbreak, the voluntary, community, and social enterprise organizations received £750 million to provide vital support (Department for Digital, Culture, Media & Sport and Office for Civil Society, 2020). Funding was available to skill providers via the COVID-19 Emergency Recovery Support Fund. These funds were distributed to small- and medium-sized businesses through college-based programmes, private skill providers, charities, local authorities, and adult education centres (Mayor of London, 2020). With the support of the Department for Digital, Culture, Media & Sport, the National Lottery Community Fund established the COVID-19 Community-led Organizations Recovery Scheme (The National Lottery Community Fund, 2020). Its objective is to establish a number of intermediary organizations authorized to distribute funds of £200 million efficiently to communities disproportionately affected by COVID-19 (The National Lottery Community Fund, 2020).

Examples include John Lewis who provided £1 million in funding to support local communities through their Waitrose stores, which was used to support local communities and initiatives in supporting new parents, facilitating crafts classes, and supplying care packages for elderly and vulnerable people (John Lewis Partnership, 2020). Moreover, Tesco Bags of Help COVID-19 Communities Fund awarded £4 million to 800 organizations across the UK that provided support to vulnerable groups. A cash donation of £500 was awarded to charities, voluntary sector organizations, community organizations, and schools to establish food banks, telephone services, and an online patient service (Tesco PLC, 2020). It has been found that CSR initiatives have exceeded the usual measures in the area of charity. For example, UK banks have implemented CSR initiatives such as health support programmes (Das & Cirella, 2020). An aid package of £100 million was provided by Barclays Bank. One half of the funding was used to support charities working with vulnerable people impacted by COVID-19 and its underlying socio-economic factors. The second half of the funding will match donations from Barclays Bank employees to charities that are helping during the COVID-19 pandemic (Barclays Bank, 2020).

Additionally, well-being and recovery programmes played a significant role during the COVID-19 pandemic. The symptoms of mental illness were exacerbated during the period under the UK lockdown. The prevalence of moderate or severe depressive symptoms among adults in Great Britain was 21% from January to March

2021 (Baker, 2021), and intentional self-harm was the leading cause of death among people aged 5–34 in 2020 (Flinders, 2022).

The London School of Hygiene and Tropical Medicine, LSHTM, one of the UK's top universities in public health, designed an evidence-based well-being intervention for students called *Pentacell*. Alumni of LSHTM facilitated online meetings for up to five students through this programme, which promotes belonging, inclusion, and social connection (London School of Hygiene Tropical Medicine, 2021). Having initiatives like the Pentacell programme is important for the economy because essentially universities are places that produce the cadre of skilled workers of tomorrow. Furthermore, the COVID-19 pandemic highlights how important public health work is for the economy and population health because public health prevents disease, and a healthy population means a productive and reliable workforce.

In 2021, LSHTM's Vaccine Confidence Project partnered with YouTube and Klick Health to produce a series of videos addressing misinformation and hesitancy regarding the COVID-19 vaccination, known as community immunity. The video content targeted people between the ages of 18 and 30 with clear, easy-to-understand information and answers to their hesitancy to the COVID-19 vaccine (Little Black Book, 2021). Evidence indicates that the COVID-19 vaccine uptake actually enhanced the UK's economy because the vaccine improved public health, life expectancy, and human capital (Inman, 2021).

However, the pandemic lockdowns did actually have a negative impact on the economy. British unemployment rates soared to 4.9% in the 3 months leading up to October 2020 and redundancies topped 370,000 for the first time in more than a decade (Strauss, 2020). *The Social Enterprise Report* provides examples of social enterprise employment and skill providers of employment, apprenticeships, and work experience to the most disadvantaged communities (Social Enterprise UK, 2020). Furthermore, the international public speaking club Toastmasters International helped members develop their soft skills in communication, leadership, and organizational skills. Through the practice of these soft skills, people gain confidence, especially people who are afraid of public speaking or who experience impostor syndrome. Due to the lockdown, Toastmaster clubs switched to virtual and became a place for socializing, networking, and developing skills. In addition, Toastmasters International runs corporate clubs for companies such as Amazon, Google, and Microsoft, which help fund the development of their employees' leadership and communication skills (Toastmasters International, 2019). Communication skills can help people succeed in job interviews, negotiations with clients, and promotions. By doing so, profitability can increase, and the workforce can become more productive; thus, unemployment levels would decline.

2.2 Legal Responsibilities

A part of Carroll's pyramid of CSR also consists of businesses' legal obligations enacted by a society which encompass the standards that businesses are bound to meet pursuant to UK laws and regulations (Idowu et al., 2015).

As of 23 March 2020, the UK Government ordered all shops selling non-essential goods to close in order to contain the spread of COVID-19 (Johnson, 2020). While this was stipulated in law, businesses ceasing operations during this time lost trade in order to serve the greater public interest. Despite the difficulty of many businesses, the government provided a great deal of assistance. For example, employers were reimbursed for the employment costs incurred by furloughed employees, certain businesses were able to receive business rates relief, and small- and medium-sized businesses were eligible to apply for recovery loans (UK Government, 2020).

As a result of testing positive for COVID-19, government guidance was provided for workers to self-isolate (Department for Business, Energy & Industrial Strategy, 2020). The self-isolating workers at retailers like Tesco, Sainsbury, and Asda were paid full wages during their time off sick (Hawthorne, 2020). Many businesses also provide staff with personal protective equipment, protective screens, and hand sanitizers in order to prevent the spread of COVID-19 (Hawthorne, 2020). Additionally, government guidance allowed extremely vulnerable workers to shield and work from home (Department of Health and Social Care and Ministry of Housing, Communities & Local Government, 2020). Several retailers are now compensating shielded workers with full remuneration (Hawthorne, 2020). Finally, from 24 July 2020, the government mandated the general public to wear face coverings in enclosed public spaces (Department of Health and Social Care and The Rt Hon Matt Hancock MP, 2020), while companies like Morrison and Asda were rolling out free masks to visitors (Butler, 2021).

CSR-related legal aspects were also explored through the Health and Social Care Act 2008 (Regulated Activities) (Amendment) (Coronavirus) (No. 2) Regulations 2021 which saw frontline health and social care workers in England required to take mandatory COVID-19 vaccinations (Legislation.Gov.UK, 2021). It was intended to provide the best possible protection to the most vulnerable patients against COVID-19. However, it was withdrawn on 15 March 2022 (Department of Health & Social Care, 2022).

By law, the Equality Act 2010 protects workers from discrimination, harassment, and victimization, and employers are required to provide equal work opportunities and maintain good relations among workers of different groups (Equality and Human Rights Commission and Government Equalities Office, 2013). 25 May 2020 marked a turning point for race relations worldwide with the murder of George Floyd (BBC News, 2020). Floyd's death underlines the structural inequalities and institutional racism that exist in society today. With Floyd's passing, individuals, groups, and institutions took the opportunity to make a difference through reflection, denouncing, and distancing themselves from racism in all its forms. Many corporations and small and medium enterprises (SMEs) have responded with equality, diversity, and inclusion programmes. Black employees are heavily underrepresented in senior management roles in the UK. As a result, NatWest has a racial equality pledge which looks to promote more black staff into senior positions by 2% (NatWest, 2021). Ethnic minority business owners have been disproportionately affected by COVID-19, especially because of business losses related to lockdowns and in some cases businesses being ineligible for government funding (Jeraj, 2020).

Corporate giants such as Google and Timeout London have partnered up with Black Pound Day (which is a network and marketplace for black-owned businesses) to contribute to the economic recovery of black-owned businesses (Time Out London, 2020). Goldman Sachs partnered with Mentivity to launch the Raising Aspirations Project, which is a programme aimed at inspiring and motivating young Black British people to achieve excellence during the COVID-19 lockdown (Mentivity, 2021).

Additionally, neurodiverse people were disproportionately affected by COVID-19. The National Autistic Society reported that the COVID-19 lockdown negatively affected the mental health of 9 out of 10 autistic people, with 85% reporting elevated levels of anxiety (National Autistic Society, 2020). Thus, the quality of life of people who were under lockdown and could not manage their pre-existing conditions would be compromised. The technology company Recite Me provided a free accessible and inclusive landing page for businesses that allowed important information concerning COVID-19 to be accessed through their toolbar, which was helpful in making the information more easily accessible to people with diverse capabilities, including neurodiverse individuals (Recite Me, 2021). The Office of National Statistics found that only 21.7% of autistic people were employed in the UK in 2021 (Cusack, 2021). A number of companies have embraced inclusive strategies, such as Microsoft, Ernst Young, and SAP (Rudy, 2020). Microsoft has a dedicated Neurodiversity Hiring Program aimed at attracting and developing neurodiverse candidates, as well as having an inclusive and accessible workplace (Microsoft, 2022).

2.3   Ethical Responsibilities

A sense of genuine concern clearly falls within the ethical remit of CSR (Idowu, 2009). CSR of this type is centred on fair treatment. This means showing stakeholders that non-financial issues must be taken into account as well (Idowu, 2009). CSR has the advantage of not being constrained to a single definition, which allows for true development and innovation to take place through its practice and meaning, rather than how other disciplines view it (Idowu, 2009).

Human Factors is a scientific discipline that examines how environmental, organizational, and individual characteristics affect the way people work (Health and Safety Executive, 2022). Human Factors and CSR are intimately connected due to workplace ergonomics and the responsibility that comes with it, both of which are linked to CSR (Kurz et al., 2019). For example, an inadequate workspace design can exert physical and mental strain on employees, and it becomes an employer's ethical responsibility to create solutions.

Human Factors can help CSR in achieving goals related to sustainable development. It has been suggested that Human Factors can help to maximize the benefits of the economic, environmental, and social dimensions of CSR because of the fact that Human Factors is a discipline that operates within a sociotechnical context (Brown

& Legg, 2011). Human Factors can therefore help to address changes that take place within complex sociotechnical contexts (Brown & Legg, 2011).

Ergonomics can help to address the key challenges of CSR and sustainability (Holman et al., 2020). Sustainability is about ensuring that the ability to meet current needs does not compromise the ability to meet future needs, whether this is with regard to natural resources, social resources, and economic sustainability (Dul et al., 2012).

As such, it is important for companies to focus on aspects that go beyond financial performance. It has been suggested that Human Factors can play a role in developing programmes that promote social responsibility by optimizing performance and well-being (Dul et al., 2012).

As a number of Human Factors interventions are framed in the context of organizations, it has been suggested that the general sustainability paradigm is utilized at an organizational level (Zink & Fischer, 2013). The pandemic has seen companies resist unethical business practices and has also seen companies being very proactive with respect to getting involved in different activities of a CSR nature. This has been the case in particular with respect to offer help that is immediate, where direct assistance can be provided to help deal with the demands on public services. If companies are mindful of CSR, it is clear that there are opportunities to provide support (He & Harris, 2020). As an example, some manufacturing companies in the United Kingdom have changed their manufacturing lines to create goods that are novel to them, such as hand sanitizer, personal protective equipment, and PPE, with a number of companies donating these goods rather than seeking profit (He & Harris, 2020). This is a clear example of CSR.

The unprecedented nature of the pandemic has meant that companies have needed to respond in a rapid manner. Some companies have switched the type of products that they produce on their production lines, shifting towards the development of products that are essential to responding to the pandemic. For example, some companies have started to produce items such as hospital gowns, protective face masks, ventilators, and hand sanitizer. However, as companies start to produce products unfamiliar to their production lines, there is the risk that effectiveness may be reduced because of their lack of familiarity with the products that they are now developing. It could in fact be the case that acting in a socially responsible manner could inadvertently complicate the process of procurement on the part of healthcare authorities. For example, very early on in the pandemic in the UK, it was found that some design proposals for ventilators were unworkable, which meant that they could not be implemented and there was a need for increased ventilator capacity in the UK (Gutierrez-Huerter, 2020).

Wherever people are involved, Human Factors comes into play, especially the health and safety aspects of systems (Kumar & Sahu, 2016). The promotion of well-being and the promotion of health are central to CSR (Derenevich et al., 2021). Related to this definition, Human Factors aims to optimize well-being (Dul et al., 2012).

The COVID-19 pandemic has placed unprecedented pressure on the NHS, which has included the need for clinicians specializing in one discipline to transfer over to unfamiliar domains to help save lives. The potential implications of such moves are that they can increase the risk of errors and mistakes. It is well established that unfamiliarity can substantially increase the risk of error.

This is where Human Factors can add value, as this is the discipline that applies knowledge of human capabilities and limitations to the design of workplaces and systems. In the United Kingdom, the Chartered Institute of Ergonomics and Human Factors (CIEHF) has worked closely with the NHS to produce Human Factors guidance to help cope with the COVID-19 pandemic. These initiatives have supported various aspects of CSR. Human Factors is about optimizing both well-being and system performance (Dul et al., 2012).

The CIEHF has led a number of initiatives during the COVID-19 pandemic. CIEHF members have volunteered their time to provide expert guidance in order to help deal with the challenges associated with the pandemic (Chartered Institute of Ergonomics and Human Factors, 2020a). For example, they provided guidance to help ventilator manufacturers develop ventilators that are usable and safe. Their expert panel was approached by various firms that form part of the Ventilator Challenge UK group (Loughborough University, 2020). These initiatives have helped the NHS in the areas of education, staff well-being, skill development, and customer care, all of which are crucial aspects of CSR. This knowledge has helped the NHS to cope with the pandemic. Furthermore, the CIEHFs vaccinating a nation publication was translated into various languages, including Hindi and Bengali (Chartered Institute of Ergonomics and Human Factors, 2020a), to support India's response to the pandemic following surging rates of COVID-19 cases.

The creation of the NHS Nightingale hospitals increased the NHS's capacity to deliver acute care (House of Lords Public Services Committee, 2020). The local government re-housed 15,000 rough sleepers (House of Lords Public Services Committee, 2020).

It is important to draw on the skills and knowledge of stakeholders from the community and voluntary services. Through using a place-based approach, an increase in civic action has been witnessed across UK communities. For example, millions of citizens have given their time to support the NHS on a voluntary basis (House of Lords Public Services Committee, 2020).

The pandemic has also given rise to the forming of various mutual aid groups. For example, the UK has over 4000 of these groups. These groups have been helping the most vulnerable people in society, by performing services such as bringing food to people's homes as well as collecting and dropping off medicines, and also providing health advice (House of Lords Public Services Committee, 2020).

Another domain of CSR's ethical branch may also include companies participating in environmentally friendly practices in order to develop a social responsibility towards the environment. One of the key challenges during the COVID-19 pandemic is how to effectively manage clinical waste generated as a result of the pandemic (Flynn, 2020). Increased levels of clinical waste have been experienced, both as a result of COVID-19 and the resumption of other NHS services. These factors in turn increase the cost for the NHS and exacerbate the challenges already faced by the waste management industry. For example, NHS clinical waste contractors have reported substantial levels of clinical waste of an infectious nature. Although this waste is intended for orange colour-coded bags, the NHS has found that these bags are being used for different forms of waste, which is contributing to collections being missed, and, in turn, has adverse financial implications for the NHS. One of the key factors behind this is the fact that it is 45% cheaper to recycle waste than it is to send waste for heat recovery. This makes it clear how important it is to correctly segregate the waste and recycle what can be recycled (NHS Property Services, 2020). Similar efforts have been achieved in biomedical waste of needles and syringes. Companies such as Bio Systems Sharps Management Solution have partnered with the NHS in a bid to reduce plastic waste through their supply of reusable sharps containers as a sustainable alternative to single-use containers (Stericycle, 2021).

One of the other impacts of COVID-19 has been the effect of healthcare laundry on the NHS. As a result of this, the Textile Services Association (TSA), NHS Improvement (NHSI), and various departments of the government created the Interim Healthcare Laundry Certification (IHLC) scheme, which lets specialist hospitality companies help to address this challenge. The number of participating laundries is over 30 (Gard, 2021).

One of the other challenges has been the amount of disposable, single-use PPE items (such as single-use face masks) that have been used during the pandemic. This has culminated in the coining of the term 'PPE pollution' (Gard, 2021). Therefore, schemes such as the IHLC could support moving from single-use PPE to washable PPE.

Moreover, given that the UK Government has a target to shift the landscape towards using more reusable PPE and to lower the levels of clinical waste, schemes such as the IHLC scheme will help the UK Government achieve this aim (Gard, 2021). Therefore, such initiatives clearly display CSR.

2.4   Philanthropic Responsibilities

Companies or wealthy individuals participate in this type of CSR by donating resources to give back to society. The COVID-19 pandemic has exacerbated a number of social issues that are linked to a person's health, including food insecurity, homelessness, and mental health challenges. Given that many of these issues are interlinked, it is important to acknowledge that solving them requires a similar approach. A *systems thinking* approach helps to look at how different health systems interact and influence each other by exploring the deeper interconnections, the synergies, and the gaps that exist within health systems (World Health Organization, 2009). Furthermore, the systems thinking model places people at the centre of the system since people are what initiate, drive, and influence the interactions between different health systems (World Health Organization, 2009). The COVID-19 pandemic has given rise to greater collaborations between public services, charities, and the private sector. Through approaches such as Making Every Contact Count, community members, organizations, and health specialists can identify and provide information and referrals to members regarding broader health and well-being issues (National Institute for Health and Care Excellence, 2022). Making Every Contact Count has been used to prevent homelessness by taking a collaborative approach to addressing wider issues, such as troubled childhoods and adolescent health, drug and alcohol use, crime, unemployment, and debt (Hill, 2012). As a result, philanthropy is also important for large organizations as reflected in their activities, such as providing or supporting food banks (Gutierrez-Huerter, 2020).

Manchester United Footballer Marcus Rashford has led a significant philanthropic movement during the COVID-19 pandemic in order to tackle food poverty, which in turn has a positive impact on health, thereby helping to reduce the demands on health service institutions. Rashford has spearheaded the Child Food Poverty Task Force, which has lobbied the UK Government to support children who are most vulnerable to experiencing food poverty, by putting into place recommendations from the National Food Strategy policy (The Food Foundation, 2021).

Although free school meal vouchers have helped to alleviate the problem, there are still nonetheless a substantial number of children who are still unable to ascertain a healthy diet, which places them at risk of not growing and developing in an optimum way and will make it more of a challenge for them to effectively participate in a school setting and within society as a whole (The Food Foundation, 2021). This problem is likely to have public health implications.

Rashford has raised donations worth £20 million through contributions from supermarkets, all of which are going towards addressing the problem of UK-based food poverty. Rashford also featured at the apex of the Sunday Times Giving List, becoming the youngest ever person to do so (ITV, 2021).

Rashford wrote a blog where he asked healthcare stakeholders to sign up for the Healthy Start Scheme created by the UK Government (Rashford, 2021). However, despite Rashford lobbying and the fact that free school meal vouchers were continued by the UK Government, there were a number of additional recommended actions that were not taken up with respect to increasing free school meal eligibility. Moreover, it was recommended that a Child Food Poverty Taskforce should be developed, but this recommendation was also not taken forward. The Child Poverty Action Group has undertaken modelling, which shows that positive benefits can be achieved even with modest increases to child benefits, such as the suggestion that child poverty can be lowered by 5% with a £10 weekly addition to child benefits (Hefferon et al., 2021). This will lead to positive health benefits for the affected groups. Several recommendations for child food poverty mitigation are outlined in the National Food Strategy (Moore & Evans, 2020).

Children in Need ran the event the Big Night, which formed part of their overall strategy to help children and young people who have been directly impacted by COVID-19. Children in Need are setting up a funding structure that will allow philanthropists to directly contribute to the cause. The needs that they are actively seeking to address include the need to tackle poverty (and in turn help to address the health inequalities that this situation generates), the need to address emotional well-being and mental health issues, and the need to improve access to essentials for children who are disabled or seriously ill. This is important because disabled or seriously ill children have also been negatively impacted by the pandemic (The Beacon Collaborative, 2020).

It is clear that a great deal of charitable giving has occurred. However, a great deal of this has been in-kind giving, whereas what is needed more acutely is cash giving. The challenge with the latter is that, if it is to be done effectively, then it will take substantial resources to organize. Given that a lot of regular fundraising has been frozen during the pandemic, the lack of cash giving, and the number of resources it takes to arrange, will have the implication of not effectively addressing the financial shortfalls experienced by organizations in the third sector (Gutierrez-Huerter, 2020). As such, if companies can afford to continue to support these organizations, then they should continue to provide cash donations and then re-evaluate the situation once the crisis has been contained (Gutierrez-Huerter, 2020). A Public Health England report suggested that through the use of strategies from the field of CSR, healthy, sustainable communities and places can be created and developed as a result (Public Health England, 2015).

3   Practical Implications

This section looks at how UK health service institutions could benefit from broadening their capability, which would help them to cope with a future pandemic of this nature in the future, at the same time allow them to maintain their health responsibilities with respect to supporting other regular health matters.

Based on the current literature review, the following best practices and implementation strategies are effective in embedding CSR into the UK health service institutions.

Having better coordination with specialist disciplines can help determine what is workable and usable which can avoid the problems that occurred with ventilator design where firms ventured into areas that fell outside of their expertise. Therefore, creating a centralized database of domain experts that can be matched up in case of a crisis would be beneficial, as well as supporting health organizations in achieving long-term sustainable change, such as dealing with future pandemics.

It would also be beneficial to capture valuable good practice procedures to ensure that positive change is captured and can be reused for future crises. For example, the CIEHF produced guidance titled, *Achieving Sustainable Change*, which supports healthcare workers in capturing best practices during the COVID-19 pandemic, with the aim of bringing about strategic and policy change. Positive change should be documented as early as possible to reduce the risk of it not being documented at all (Chartered Institute of Ergonomics and Human Factors, 2020b).

Furthermore, corporations should continue investing in sector-wide approaches to create long-term sustainable projects for the most vulnerable communities affected by COVID-19. In particular, people who lived in the most deprived communities were at the greatest risk of COVID-19-associated deaths and remained vulnerable to underlying health problems (Marmot et al., 2020). Investing in the short term would only address the symptoms and not the underlying causes. So, investing long term in the aforementioned communities will lead to their growth in social, economic, and cultural terms (Marmot et al., 2020). Furthermore, a sector-wide approach focusing on the wider social determinants of health that tackles health inequalities will help to develop a more equitable public health system (Marmot et al., 2020).

Previous CSR work has shown that when engaging stakeholders, the use of a fragmented approach can reduce the likelihood of encouraging stakeholder engagement that is genuine (Lynch-Wood & Williamson, 2005). There are more areas where Human Factors can help with future planning for pandemics. It would be beneficial to systematically map out all the areas where a Human Factors intervention could help to reduce complexity. For example, preparing vaccine materials for patients is a challenging task and is an area where the ease with which this information can be read is key (Moore & Millar, 2021). The Human Factors literature can help to address readability issues.

It is important to recognize both the private and the third sector as key providers of public services. If they are to help with the delivery of services, then they need to be provided with support at a level that is appropriate (House of Lords Public Services Committee, 2020). The UK health service institutions would benefit from engaging locally with communities, as this will help to gain their support for initiatives because incorporating their ideas will make the outcomes more appropriate to the local needs of the people who are directly impacted by these initiatives (Vertigans & Idowu, 2017).

It is clear that it is necessary to develop strategies that look to the long term, which engage local authorities on the goal to lower the extent to which there is a reliance on the provision of emergency food resources, and that there is instead a shift towards a food system in the UK that is sustainable in the long term (Moore & Evans, 2020). This should result in positive health benefits in the long term and help to make the UK more resilient to future pandemics by helping to ensure that any such pandemics do not further exacerbate food inequalities, or at the very least minimize the extent to which they exacerbate such food inequalities to the greatest extent possible.

Once the COVID-19 pandemic has ended, there will still remain a need for global health cooperation. If UK health service institutions collaborate with international partners, then this will provide an opportunity to share data, share resources, and learn from other stakeholders. This will help the UK and their international partners to be better prepared for future pandemics (Brown, 2022).

4 Conclusion

As illustrated in this chapter, CSR has been embedded within the UK healthcare system to varying extents throughout the last few hundred years. Some aspects of CSR occurred during the time that pre-dated the implementation of the NHS, as evidenced by philanthropic-funded voluntary hospitals, demonstrating a level of social responsibility. This would indicate that the UK has a long history of having contributed towards the development of CSR in the context of health care. The COVID-19 pandemic has given rise to more social responsibility aimed at bolstering UK healthcare institutions' ability to address the challenges imposed on them.

However, the situation has given rise to companies venturing into areas outside of their area of expertise, and several errors have occurred as a result. Systems thinking and learning lessons from other industries in terms of how to mitigate against such errors can help with the planning of future waves of the COVID-19 pandemic and for future pandemics.

Disclaimer Courtney Grant is employed by Transport for London and is a Fellow, a Chartered Member, and a Council Member of the Chartered Institute of Ergonomics and Human Factors. Mohammed Ali is employed by the National Health Service. The perspectives of the authors are not necessarily those of Transport for London, the Chartered Institute of Ergonomics and Human Factors, or the National Health Service.

References

Action for Children. (2021). Where is child poverty increasing in the UK?. Retrieved April 19, 2022, from https://www.actionforchildren.org.uk/blog/where-is-child-poverty-increasing-in-the-uk/.

Aldridge, R., Lewer, D., Katikireddi, S., Mathur, R., Pathak, N., Burns, R., Fragaszy, E., Johnson, A., Devakumar, D., Abubakar, I., & Hayward, A. (2020). Black, Asian and minority ethnic groups in England are at increased risk of death from COVID-19: Indirect standardisation of NHS mortality data. *Wellcome Open Research, 5*, 88. https://doi.org/10.12688/wellcomeopenres.15922.2

Baker, C. (2021). Mental health statistics: Prevalence, services and funding in England. House of Commons Library. Retrieved April 3, 2022, from https://commonslibrary.parliament.uk/research-briefings/sn06988/.

Barclays Bank. (2020). Barclays launches £100 million COVID-19 Community Aid Package. Barclays. Retrieved April 4, 2022, from https://home.barclays/news/press-releases/2020/04/barclays-launches-p100-million-covid-19-community-aid-package/.

BBC News. (2020). George Floyd: What happened in the final moments of his life. Retrieved April 18, 2022, from https://www.bbc.co.uk/news/world-us-canada-52861726.

Brown, C., & Legg, S. (2011). Chapter 3 human factors and ergonomics for business sustainability. *Critical Studies on Corporate Responsibility, Governance and Sustainability, 3*, 59–79. https://doi.org/10.1108/S2043-9059(2011)0000003011

Brown, G. (2022). *Seven ways to change the world: How to fix the most pressing problems we face.* Simon & Schuster UK.

Butler, S. (2021). Tesco and John Lewis will ask customers and staff to wear face masks. Retrieved September 12, 2022, from https://www.theguardian.com/business/2021/jul/15/tesco-john-lewis-ask-customers-staff-face-masks-england.

Carroll, A. (2016). Carroll's pyramid of CSR: Taking another look. *International Journal of Corporate Social Responsibility, 1*(1). https://doi.org/10.1186/s40991-016-0004-6

Charity Bank. (2021). COVID-19 emergency funding for charities and social sector organisations. Retrieved April 2, 2022, from https://www.charitybank.org/news/covid-19-emergency-funding-for-charities-and-social-sector-organisations.

Chartered Institute of Ergonomics and Human Factors. (2020a). COVID-19: Human factors response. Retrieved August 26, 2021, from https://covid19.ergonomics.org.uk.

Chartered Institute of Ergonomics and Human Factors. (2020b). Achieving sustainable change—capturing lessons from Covid-19. Retrieved August 30, 2021, from https://www.linkedin.com/pulse/achieving-sustainable-change-capturing-lessons-from-covid-19-sujan.

Cook, G. C. (2004). Caring for "incurables": The 150th anniversary of the Royal Hospital for neuro-disability. *Putney. Postgraduate medical journal, 80*(945), 426–430. https://doi.org/10.1136/pgmj.2003.017673

Cottom, S. (2021). COVID-19 resilience in the third sector. *Health services insights., 14*, 11786329211013547. https://doi.org/10.1177/11786329211013547

Cusack, J. (2021). Autistic people still face highest rates of unemployment of all disabled groups. *Autistica*. Retrieved April 18, 2022, from https://www.autistica.org.uk/news/autistic-people-highest-unemployment-rates/

Das, D., & Cirella, G. (2020). *Corporate social responsibility practices in the United Kingdom., 9*, 51–54.

Denham, M. (2017). Victorian philanthropists, philosophers and activists. *British Geriatrics Society*. Retrieved August 26, 2021, from https://www.bgs.org.uk/resources/victorian-philanthropists-philosophers-and-activists/

Department for Business, Energy & Industrial Strategy. (2020). [Withdrawn] If you need to self-isolate or cannot attend work due to coronavirus. Retrieved September 12, 2022, from https://www.gov.uk/guidance/if-you-need-to-self-isolate-or-cannot-attend-work-due-to-coronavirus.

Department for Digital, Culture, Media & Sport and Office for Civil Society. (2020). Financial support for voluntary, community and social enterprise (VCSE) organisations to respond to coronavirus (COVID-19). Retrieved September 12, 2022, from https://www.gov.uk/guidance/financial-support-for-voluntary-community-and-social-enterprise-vcse-organisations-to-respond-to-coronavirus-covid-19.

Department of Health & Social Care. (2022). [Withdrawn] Coronavirus (COVID-19) vaccination as a condition of deployment for the delivery of CQC-regulated activities in wider adult social care settings. Retrieved September 12, 2022, from https://www.gov.uk/government/publications/vaccination-of-workers-in-social-care-settings-other-than-care-homes-operational-guidance/coronavirus-covid-19-vaccination-as-a-condition-of-deployment-for-the-delivery-of-cqc-regulated-activities-in-wider-adult-social-care-settings.

Department of Health and Social Care and The Rt Hon Matt Hancock MP. (2020). Face coverings mandatory in shops, supermarkets, shopping centres and enclosed transport hubs from Friday. Retrieved September 12, 2022, from https://www.gov.uk/government/news/face-coverings-mandatory-in-shops-supermarkets-shopping-centres-and-enclosed-transport-hubs-from-friday.

Department of Health and Social Care and Ministry of Housing, Communities & Local Government. (2020). Plans to ease guidance for over 2 million shielding. Retrieved September 12, 2022, from https://www.gov.uk/government/news/plans-to-ease-guidance-for-over-2-million-shielding.

Derenevich, M. G., Bitencourt, R. S., Junior, O. C., & Wu, V. (2021). Macroergonomics as a way for social responsibility: A study in a university hospital. In F. W. Leal, U. Tornado, & F. Frankenberger (Eds.), *Integrating social responsibility and sustainable development*. World Sustainability Series, Springer. https://doi.org/10.1007/978-3-030-59975-1_41

Dessie, Z. G., & Zewotir, T. (2021). Mortality-related risk factors of COVID-19: A systematic review and meta-analysis of 42 studies and 423,117 patients. *BMC Infectious Diseases, 21*, 855. https://doi.org/10.1186/s12879-021-06536-3

Dul, J., Bruder, R., Buckle, P., Carayon, P., Falzon, P., Marras, W. S., Wilson, J. R., & Van der Doelen, B. (2012). A strategy for human factors/ergonomics: Developing the discipline and profession. *Ergonomics, 55*(4), 377–395.

Elliott, L. (2022). After UK inflation again outpaces forecasts, a 13% peak looks optimistic. Retrieved August 17, 2022, from https://www.theguardian.com/business/2022/aug/17/after-inflation-again-outpaces-forecasts-13-per-cent-peak-looks-optimistic-bank-of-england.

Equality and Human Rights Commission and Government Equalities Office. (2013). Equality Act 2010: Guidance. Retrieved September 12, 2022, from https://www.gov.uk/guidance/equality-act-2010-guidance#:~:text=Print%20this%20page,Overview,strengthening%20protection%20 in%20some%20situations.

Fanzo, J. (2019). If you're poor in America, you can be both overweight and Hungry. Retrieved September 14, 2022, from https://www.bloomberg.com/opinion/articles/2019-07-17/how-hunger-and-obesity-coexist-in-america.

Financier Worldwide. (2020). The impact of social entrepreneurship on economic growth. *Financier Worldwide*. Retrieved April 3, 2022, from https://www.financierworldwide.com/the-impact-of-social-entrepreneurship-on-economic-growth#.YkmEty3MJhE/

Fleetwood, J. (2020). Social justice, food loss, and the sustainable development goals in the era of COVID-19. *Sustainability, 12*, 5027. https://doi.org/10.3390/su12125027

Flinders, S. (2022). Mental health. Retrieved April 3, 2022, from https://www.nuffieldtrust.org.uk/news-item/mental-health-indicator-update.

Flynn, D., Moloney, E., Bhattarai, N., Scott, J., Breckons, M., Avery, L., & Moy, N. (2020). COVID-19 pandemic in the United Kingdom. *Health Policy and Technology, 9*(4), 673–691. https://doi.org/10.1016/j.hlpt.2020.08.003

Flynn, G. (2020). Dealing with Covid-19 PPE waste—the next big question for the NHS. *Anenta*. Retrieved August 26, 2021, from https://www.anentawaste.com/dealing-with-covid-19-ppe-waste.html/

Gard, R. (2021). Laundry industry steps in to support healthcare sector. *Wearwell*. Retrieved August 26, 2021, from https://shop.wearwell.co.uk/blogs/news/laundry-industry-steps-in-to-support-healthcare-sector/

Gadsby, E., Peckham, S., Coleman, A., Bramwell, D., Perkins, N., & Jenkins, L. (2017). Commissioning for health improvement following the 2012 health and social care reforms in England: What has changed? *BMC Public Health, 17*(211). https://doi.org/10.1186/s12889-017-4122-1

Gorsky, M. (2008). Hospital governance and community involvement in hospital governance in Britain: Evidence from before the National Health Service. *International Journal of Health Services, 38*(4), 751–771. https://doi.org/10.2190/hs.38.4.j

GOV.UK. (2020). People living in deprived neighbourhoods. Retrieved March 13, 2022, from https://www.ethnicity-facts-figures.service.gov.uk/uk-population-by-ethnicity/demographics/people-living-in-deprived-neighbourhoods/latest.

Greengross, P., Grant, K., & Collini, E. (1999). *The history and development of the UK National Health Service 1948–1999* (pp. 4–6). DFID Health Systems Resource Centre. Retrieved April 17, 2022, from https://assets.publishing.service.gov.uk/media/57a08d91e5274a31e000192c/The-history-and-development-of-the-UK-NHS.pdf

Gutierrez-Huerter, G. (2020). COVID-19 poses new challenges for corporate social responsibility efforts. King's College London. Retrieved August 26, 2021, from https://www.kcl.ac.uk/news/covid-19-poses-new-challenges-corporate-social-responsibility-efforts.

Harrington, J., Lutomski, J., Molcho, M., & Perry, I. (2009). Food poverty and dietary quality: Is there a relationship? *Journal of Epidemiology & Community Health, 63*(16). https://doi.org/10.1136/jech.2009.096701p

Harvard T.H. Chan School of Public Health. (2022). Prenatal and early life influences. Retrieved September 14, 2022, from https://www.hsph.harvard.edu/obesity-prevention-source/obesity-causes/prenatal-postnatal-obesity/.

Hastings, S. (1960). Aneurin Bevan: An appreciation of his services to the health of the people. Socialist Health Association. Retrieved August 26, 2021, from https://www.sochealth.co.uk/national-health-service/the-sma-and-the-foundation-of-the-national-health-service-dr-leslie-hilliard-1980/aneurin-bevan-and-the-foundation-of-the-nhs/aneurin-bevan-an-appreciation-of-his-services-to-the-health-of-the-people-pamphlet-published-by-the-socialist-medical-association-c-1960/.

Hawthorne, E. (2020). How are supermarkets protecting and supporting their staff during the coronavirus crisis?. Retrieved September 12, 2022, from https://www.thegrocer.co.uk/people/how-are-supermarkets-protecting-and-supporting-their-staff-during-the-coronavirus-crisis/603512.article.

He, H., & Harris, L. (2020). The impact of Covid-19 pandemic on corporate social responsibility and marketing philosophy. *Journal of Business Research, 116*, 176–182. https://doi.org/10.1016/j.jbusres.2020.05.030

Health and Safety Executive. (2022). Human factors/ergonomics—introduction to human factors. Retrieved September 12, 2022, from https://www.hse.gov.uk/humanfactors/introduction.htm.

Health Services Research UK. (2017). A brief history of NHS politics 1948–2030 [Video]. Retrieved from https://www.youtube.com/watch?v=u3nvsAMrriA.

Hefferon, C., Taylor, C., Bennett, D., Bennett, D., Falconer, F., Campbell, M., Williams, J. G., Schwartz, D., Kipping, R., & Taylor-Robinson, D. (2021). Priorities for the child public health response to the COVID-19 pandemic recovery in England. Archives of disease in childhood. *BMJ Journals, 106*, 553–538. https://doi.org/10.1136/archdischild-2020-320214

Hill, T. (2012). Making every contact count: A joint approach to preventing homelessness. Local Government Information Unit. Retrieved March 14, 2022, from http://lgiu.org/wp-content/uploads/2012/09/Making-every-contact-count-a-joint-approach-to-preventing-homelessness.pdf.

Holman, M., Walker, G., Lansdown, T., & Hulme, A. (2020). Radical systems thinking and the future role of computational modelling in ergonomics: An exploration of agent-based modelling. *Ergonomics, 63*(8), 1057–1074. https://doi.org/10.1080/00140139.2019.1694173

Honigsbaum, F. (1990). The evolution of the NHS. *BMJ, 301*(6754), 694–699. https://doi.org/10.1136/bmj.301.6754.694

House of Lords Public Services Committee. (2020). A critical juncture for public services: Lessons from COVID-19. 1st Report of Session 2019–21. Ordered to be printed 10 November 2020 and published 13 November 2020. Published by the Authority of the House of Lords. HL Paper 167.

Idowu, S., Capaldi, N., Fifka, M., Zu, L., & Schmidpeter, R. (2015). Dictionary of corporate social responsibility. *CSR, Sustainability, Ethics & Governance, 137*. https://doi.org/10.1007/978-3-319-10536-9

Idowu, S. O. (2009). The United Kingdom of Great Britain and Northern Ireland. In S. O. Idowu & W. L. Filho (Eds.), *Global practices of corporate social responsibility* (pp. 11–35). Springer.

Inman, P. (2021). UK economy rebounds in March after rapid Covid vaccine rollout. Retrieved August 30, 2021, from https://www.theguardian.com/business/2021/may/12/uk-economy-rebounds-in-march-after-rapid-covid-vaccine-rollout.

ITV. (2021). Marcus Rashford tops Sunday Times Giving List in record-breaking year for donors. Retrieved August 30, 2021, from https://www.itv.com/news/2021-05-21/marcus-rashford-tops-sunday-times-giving-list-in-record-breaking-year-for-donors.

Jeraj, S. (2020). UK businesses run by ethnic minorities are "particularly exposed" to Covid-19's impact. *New Statesman*. Retrieved April 18, 2022, from https://www.newstatesman.com/spotlight/2020/08/uk-businesses-run-by-ethnic-minorities-are-particularly-exposed-to-covid-19s-impact

John Lewis Partnership. (2020). John Lewis Partnership announces new measures to support customers and the vulnerable. Johnlewispartnership.co.uk. Retrieved April 3, 2022, from https://www.johnlewispartnership.co.uk/media/press/y2020/jlp-announces-new-measures-to-support-customers.html.

Johnson, B. (2020). Prime Minister's statement on coronavirus (COVID-19): 23 March 2020. Retrieved September 9, 2022, from https://www.gov.uk/government/speeches/pm-address-to-the-nation-on-coronavirus-23-march-2020.

Ke, J., & Ford-Jones, E. (2015). Food insecurity and hunger: A review of the effects on children's health and behaviour. *Paediatrics & Child Health, 20*(2), 89. https://doi.org/10.1093/pch/20.2.89

Kumar, J., & Sahu, M. (2016). Corporate ergonomic responsibility—a tool for sustainability of corporate social responsibility. *International Journal of Engineering Trends and Technology, 41*(4), 210–215.

Kurz, L., Jost, L., Roth, K., & Ohlhausen, P. (2019). Focusing sustainable human resource management—framework for sustainability management in research organizations. In W. L.

Filho (Ed.), *Social responsibility and sustainability: How businesses and organizations can operate in a sustainable and socially responsible way* (pp. 57–73). Springer Nature Switzerland AG.

Lambeth Council. (2020). Love Lambeth. Retrieved September 12, 2022, from https://love. lambeth.gov.uk/beam-supporting-lambeth-residents-into-employment/.

Legislation.Gov.UK. (2021). The Health and Social Care Act 2008 (Regulated Activities) (Amendment) (Coronavirus) (No. 2) Regulations 2021. Retrieved September 12, 2022, from https:// www.legislation.gov.uk/ukdsi/2021/9780348228861.

Little Black Book. (2021). Klick health and the vaccine confidence project discuss community immunity in latest films. Retrieved April 3, 2022, from https://www.lbbonline.com/news/klick-health-and-the-vaccine-confidence-project-discuss-community-immunity-in-latest-films.

London School of Hygiene & Tropical Medicine. (2021). Pentacell: LSHTM's systemic student wellbeing initiative. *LSHTM*. Retrieved April 3, 2022, from https://www.lshtm.ac.uk/aboutus/alumni/blogs/2021/pentacell-lshtms-systemic-student-wellbeing-initiative

Loughborough University. (2020). Rapid manufacture of hospital ventilators could cost lives if not properly designed, according to new guidelines. Retrieved August 26, 2021, from https://www.lboro.ac.uk/media-centre/press-releases/2020/april/rapid-manufacture-of-ventilators-could-cost-lives/.

Lynch-Wood, G., & Williamson, D. (2005). A contextual review of CSR policy and law in the UK. *Mountbatten Journal of Legal Studies, 9*(1&2), 4–20.

Mahon, L. (2022). Black families more likely to go hungry due to 'food insecurity'. Retrieved August 17, 2022, from https://www.voice-online.co.uk/news/uk-news/2022/03/15/black-families-more-likely-to-go-hungry-due-to-food-insecurity/.

Mann, H., & Saeed, M. Z. (2021). Britain's National Health Services (NHS)—the 8th wonder of the world author. https://doi.org/10.13140/RG.2.2.14266.21446.

Marmot, M., Allen, J., Goldblatt, P., Herd, E., & Morrison, J. (2020). Build Back Fairer: The COVID-19 Marmot review. The pandemic, socioeconomic and health inequalities in England. Build Back Fairer: The COVID-19 Marmot review. The pandemic, socioeconomic and health inequalities in England, pp. 196–204.

Marshall, L., Bibby, J., & Abbs, I. (2020). Emerging evidence on COVID-19's impact on mental health and health inequalities. Retrieved August 17, 2022, from https://www.health.org.uk/news-and-comment/blogs/emerging-evidence-on-covid-19s-impact-on-mental-health-and-health.

Mayor of London. (2020). COVID-19 emergency recovery support fund (ERSF). Retrieved September 11, 2022, from https://www.london.gov.uk/coronavirus/covid-19-emergency-recovery-support-fund-ersf#acc-i-62018.

Mentivity. (2021). Raising aspirations project. Retrieved April 18, 2022, from https://www.mentivity.com/raisingaspirationsproject.

Microsoft. (2022). Neurodiversity hiring. Microsoft.com. Retrieved April 18, 2022, from https://www.microsoft.com/en-us/diversity/inside-microsoft/cross-disability/neurodiversityhiring.

Mitchell, C. (2019). *PAHO/WHO: Un declaration on universal health coverage: PAHO advocates for health system transformation.* Retrieved March 20, 2022, from https://www3.paho.org/hq/index.php?option=com_content&view=article&id=15447%3Aun-declaration-on-universal-health-coverage-paho-advocates-forhealth-system-transformation&Itemid=0&lang=en#gsc.tab=0

Moon, J. (2010). An explicit model of business-society relations. In A. Habisch, J. Jonker, M. Wegner, & R. Schmidpeter (Eds.), *Corporate social responsibility across Europe* (pp. 51–65). Springer.

Moore, J. B., & Evans, C. (2020). Tackling childhood food poverty in the UK (Policy Brief No. 4). Policy Leeds. University of Leeds. https://doi.org/10.5518/100/54.

Moore, J., & Millar, B. (2021). Improving COVID-19 vaccine-related health literacy and vaccine uptake in patients: Comparison on the readability of patient information leaflets of approved

COVID-19 vaccines. *Journal of Clinical Pharmacy and Therapeutics, 46*. https://doi.org/10. 1111/jcpt.13453

National Autistic Society. (2020). Left stranded: Our new report into the impact of coronavirus. *National Autistic Society*. Retrieved April 18, 2022, from https://www.autism.org.uk/what-we-do/news/coronavirus-report

National Institute for Health Care and Research. (2020). 20/48 Food insecurity—health impacts and mitigation. Retrieved September 14, 2022, from https://www.nihr.ac.uk/documents/2048-food-insecurity-health-impacts-and-mitigation/24905.

National Institute for Health and Care Excellence. (2022). Making every contact count. Retrieved March 14, 2022, from https://stpsupport.nice.org.uk/mecc/index.html.

NatWest. (2021). The legacy of George Floyd. NatWest Business Hub. Retrieved April 18, 2022, from https://natwestbusinesshub.com/articles/the-legacy-of-george-floyd.

NHS Property Services. (2020). How to dispose of waste correctly. Retrieved August 26, 2021, from https://www.property.nhs.uk/news-insight/insights/how-to-dispose-of-waste-correctly/.

Phiri, P., Delanerolle, G., Al-Sudani, A., & Rathod, S. (2020). COVID-19 and the Black, Asian and Minority Ethnic (BAME) communities: An antagonistic relationship. https://doi.org/10.2196/preprints.22581.

Public Health England. (2015). Local action on health inequalities Using the Social Value Act to reduce health inequalities in England through action on the social determinants of health. Retrieved August 26, 2021, from https://assets.publishing.service.gov.uk/government/uploads/system/uploads/attachment_data/file/460713/1a_Social_Value_Act-Full.pdf.

Public Health England. (2020a). Disparities in the risk and outcomes of COVID-19. Retrieved March 14, 2022, from https://assets.publishing.service.gov.uk/government/uploads/system/uploads/attachment_data/file/908434/Disparities_in_the_risk_and_outcomes_of_COVID_August_2020_update.pdf.

Public Health England. (2020b). Beyond the data: Understanding the impact of COVID-19 on BAME groups. Retrieved March 14, 2022, from https://assets.publishing.service.gov.uk/government/uploads/system/uploads/attachment_data/file/892376/COVID_stakeholder_engagement_synthesis_beyond_the_data.pdf.

Public Health Scotland. (2021). Homeless people. Retrieved September 12, 2022, from http://www.healthscotland.scot/population-groups/homeless-people#:~:text=can%20be%20damaged.-,Homeless%20people%20have%20a%20much%20higher%20risk%20of%20death%20from,abuse%2C%20drug%20use%20and%20violence.

Rajmil, L., Hjern, A., Spencer, N., Taylor-Robinson, D., Gunnlaugsson, G., & Raat, H. (2020). Austerity policy and child health in European countries: A systematic literature review. *BMC Public Health, 20*(1). https://doi.org/10.1186/s12889-020-08732-3

Rashford, M. (2021). Marcus Rashford: Every child deserves the best chance in life, and here is how health professionals can help. Retrieved August 26, 2021, from https://blogs.bmj.com/bmj/2021/08/04/marcus-rashford-every-child-deserves-the-best-chance-in-life-here-is-how-health-professionals-can-help/.

Recite Me. (2021). Recite Me continues to host free inclusive landing pages during COVID-19. Recite Me. Retrieved April 18, 2022, from https://reciteme.com/news/recite-me-continues-to-host-free-inclusive-landing-pages-during-covid-19.

Rudy, L. (2020). Top 10 employers seeking autistic employees. [online] Verywell Health. Retrieved April 18, 2022, from https://www.verywellhealth.com/top-autism-friendly-employers-4159784.

Sandhu, S. (2022). 1.3 million children living in poverty are not entitled to free school meals, say campaigners. Retrieved September 14, 2022, from https://inews.co.uk/news/uk/free-school-meals-children-living-poverty-not-entitled-benefit-campaign-780320.

Shapiro, J. (2010). The NHS: The story so far (1948-2010). *Clinical Medicine, 10*, 336–338. https://doi.org/10.7861/clinmedicine.10-4-336

Social Enterprise UK. (2020). Social enterprise and COVID-19. *Social Enterprise*, 1–15. Retrieved April 2, 2022, from https://www.socialenterprise.org.uk/wp-content/uploads/2020/05/Social-Enterprise-COVID-19-research-report-2020.pdf

Stericycle. (2021). Corporate social responsibility report. Retrieved August 30, 2021, from https://www.stericycle.com/content/dam/stericycle/global/documents/Stericycle-Corporate-Social-Responsibility-Report-2021.pdf.

Strauss, D. (2020). UK unemployment rate rises to 4.9% in 3 months to October. Ft.com. Retrieved April 3, 2022, from https://www.ft.com/content/5bf2aaf5-c3ad-4a9a-9083-feccd8becac6.

Tesco PLC. (2020). Tesco bags of help COVID-19 communities fund review. [online] Tesco PLC. Retrieved April 4, 2022, from https://www.tescoplc.com/updates/2020/tesco-bags-of-help-covid-19-communities-fund-review/.

The Andrew Reed Foundation. (2022). Andrewreedfoundation.org. Retrieved March 20, 2022, from https://www.andrewreedfoundation.org/2110/the-andrew-reed-foundation.

The Beacon Collaborative. (2020). Thematic funding spotlight: Children and young people. Retrieved August 29, 2021, from https://www.beaconcollaborative.org.uk/covid-guide/children-and-young-people/.

The Food Foundation. (2021). *A crisis within a crisis: The impact of Covid-19 on household food security* (p. 18). The Food Foundation. Retrieved from https://foodfoundation.org.uk/sites/default/files/2021-10/FF_Impact-of-Covid_FINAL.pdf

The National Lottery Community Fund. (2020). COVID-19 community-led organisations recovery scheme. Retrieved September 12, 2022, from https://www.tnlcommunityfund.org.uk/funding/programmes/covid-19-community-led-organisations-recovery-scheme.

Time Out London. (2020). Time Out London and Google partner to create the first ever edition focused on black businesses, black voices and black change makers. Retrieved April 18, 2022, from https://www.timeout.com/about/latest-news/time-out-london-and-google-partner-to-create-the-first-ever-edition-focused-on-black-businesses-black-voices-and-black-change-makers-110320.

Toastmasters International. (2019). Toastmasters popularity grows among fortune 500 companies. Prnewswire.com. Retrieved April 3, 2022, from https://www.prnewswire.com/news-releases/toastmasters-popularity-grows-among-fortune-500-companies-300866101.html.

UK Government. (2020). COVID-19 financial support for businesses. Retrieved September 11, 2022, from https://www.gov.uk/government/collections/financial-support-for-businesses-during-coronavirus-covid-19.

United Nations. (2022). History of the United Nations. Retrieved March 20, 2022, from https://www.un.org/en/about-us/history-of-the-un.

Vertigans, S., & Idowu, S. O. (2017). Corporate social responsibility in times of crisis: An introduction. In S. O. Idowu, S. Vertigans, & S. A. Burlea (Eds.), *Corporate Social responsibility in times of crisis: Practices and cases from Europe, Africa and the World* (pp. xi–xviii). Springer.

Volkow, N., Wang, G., & Baler, R. (2011). Reward, dopamine and the control of food intake: Implications for obesity. *Trends in Cognitive Sciences, 15*(1), 37–46. https://doi.org/10.1016/j.tics.2010.11.001

Waddington, K. (2014). Victorian philanthropy and doctor self-interest gave rise to modern hospitals. Retrieved August 30, 2021, from https://theconversation.com/victorian-philanthropy-and-doctor-self-interest-gave-rise-to-modern-hospitals-24371.

Wentworth, J. (2020). Effects of COVID-19 on the food supply system. Retrieved August 16, 2022, from https://post.parliament.uk/effects-of-covid-19-on-the-food-supply-system/.

World Health Organization. (2009). *Systems thinking for health systems strengthening* (pp. 30–33). World Health Organization.

World Health Organization. (2019). WHO welcomes landmark UN declaration on universal health coverage. Retrieved August 30, 2021, from https://www.who.int/news/item/23-09-2019-who-welcomes-landmark-un-declaration-on-universal-health-coverage.

World Health Organization. (2022). Declaration of Alma-Ata. Retrieved March 20, 2022, from https://www.who.int/teams/social-determinants-of-health/declaration-of-alma-ata.

Zink, K. J., & Fischer, K. (2013). Do we need sustainability as a new approach in human factors and ergonomics? *Ergonomics, 56*(3), 348–356. https://doi.org/10.1080/00140139.2012.751456

Mohammed Ali is a Global and Public Health Professional, Public Speaker, Filmmaker and Author. He has a BA (Hons) in Accountancy and Business, and an MSc in Global Health Policy from the London School of Hygiene and Tropical Medicine. He is the founder of digital think tank organisation OUR IMPACT. He specialises in applying participatory action research into public health programs for marginalized communities. He is also the Immediate Past President for Sutton Speakeasy (Toastmasters International), a Sounding Board Member for the London School of Hygiene and Tropical Medicine and a facilitator for the university's multi-award winning Pentacell initiative.

Courtney Grant has a BA (Hons) in Psychology, and an MSc in Human-Computer Interaction with Ergonomics. He is a Fellow, a Chartered Member, and a Council Member of the CIEHF. He is also a Registered European Ergonomist. He was also a member of the CIEHF COVID-19 Expert Panel. He is also the Human Factors and Stroke Lead for the Patients' Forum Ambulance Services (London).

Responsible Innovation During the COVID-19 Pandemic: A Case Study from Türkiye

Gizem Aras Beger, Gönenç Dalgıç Turhan, and Gülen Rady

1 Introduction

The recent widespread research interest in corporate social responsibility (CSR) (Shahzad et al., 2020) and its growing importance for society, highlighted during the COVID-19 pandemic, has stimulated new perspectives. While CSR definitions vary, the fundamental idea concerns the firm's commitment to act responsibly in terms of its triple bottom line (Hamza & Jarboui, 2020). According to the widely recognized definition proposed by Carroll (1999), CSR refers to business behavior "that it is economically profitable, law-abiding, ethical, and socially supportive." Currently, "the most important shift lies in the purpose of CSR," which needs to be linked to innovation (Szutowski & Ratajczak, 2016). As Carroll (2021) puts it, "It is necessary for CSR-proactive firms to give serious consideration to innovation-thinking."

In recent years, the relationship between CSR and innovation has received much scholarly attention (Belas et al., 2021). This has led to a debate as to whether CSR drives innovation (Halme & Korpela, 2014) or innovation drives responsibility (Szutowski & Ratajczak, 2016). Others argue that social responsibility is an important driver of innovation for business organizations (Ortiz-Avram et al., 2018; Voegtlin & Scherer, 2017). Voegtlin and Scherer (2017), however, claim that there is a clear theoretical connection between CSR and innovation although, with some exceptions, the link between them has not yet been sufficiently established. While some claim the relationship is reciprocal (Hlioui & Yousfi, 2020), empirical

G. Aras Beger (✉) · G. Rady
The Department of Business Administration, Yasar University, Izmir, Türkiye
e-mail: gizem.beger@yasar.edu.tr

G. Dalgıç Turhan
Karyatech Electrical Equipment, Izmir, Türkiye

© The Author(s), under exclusive license to Springer Nature Switzerland AG 2023
S. O. Idowu et al. (eds.), *Corporate Social Responsibility in the Health Sector*, CSR, Sustainability, Ethics & Governance, https://doi.org/10.1007/978-3-031-23261-9_10

studies indicate that the relationship generally flows from CSR to innovation (Wagner, 2010; Ratajczak & Szutowski, 2016). Innovation is generally conceptualized as a type (Halme & Laurila, 2009), stage (Ratajczak & Szutowski, 2016), part (Tsai et al., 2012), or critical facets (MacGregor & Fontrodona, 2008) or pillars of CSR (see FCC Innovation, 2020). Even more, one of the recent example worldwide at the point of urgent need for responsible innovation (RI), COVID-19 -related efforts also stand out as a CSR response to a grand societal challenge especially in healthcare sector (Voegtlin et al., 2019).

Based on a detailed literature review, the present study first explains RI theoretically in terms of its components. It then discusses how RI was able to assist during the COVID-19 pandemic through practices, processes, and outcomes in Türkiye's healthcare sector by considering RI needs/necessities/requirements. By focusing on the case of the Turkish pharmaceutical company, Abdi İbrahim, the study provides valuable insights into how healthcare companies especially have tackled the pandemic using RI. The case clearly demonstrates how each component of RI plays an important role and shows how rapid and effective responses could be developed for future pandemics. Finally, the case emphasizes the crucial importance of inclusion in successful RI.

2   Responsible Innovation

In terms of CSR, "RI means taking care of the future through collective stewardship of science and innovation in the present" (Stilgoe et al., 2013, p. 1570). RI can be characterized in multiple ways as a strategy, discipline, field of study, product, process, or outcome (De Hoop et al., 2016; Koops, 2015). Stilgoe et al. (2013) offer a four-dimensional RI framework of anticipation, reflexivity, inclusion, and responsiveness. *Anticipation* asks "what if" questions, envisioning risks, impacts, outcomes, dangers, or hidden opportunities to RI (Stilgoe et al., 2013; Aymerich-Franch & Fosch-Villaronga, 2020; Owen et al., 2012). *Reflexivity* refers to rethinking and reviewing the organization's prevailing assumptions, norms, and values and reflecting them upon ongoing RI processes (Zhang et al., 2019; Aymerich-Franch & Fosch-Villaronga, 2020; De Hoop et al., 2016). *Inclusion* refers to the meaningful and broad involvement of the voices of the general population and diverse stakeholders through invitation and inclusive processes that go beyond just stakeholder engagement (Genus & Stirling, 2018; Reber, 2018; Owen et al., 2013; Zhang et al., 2019; Brand & Blok, 2019). *Responsiveness* establishes an outward-looking engagement rather than only responding to a traditionally inward-looking code of conduct. That is, it enables an understanding of "what it is for innovation to be socially responsive" (Groves, 2017, p. 373). Going beyond a classical CSR perspective, it seeks new solutions that respond instantly to change.

The COVID-19 pandemic is one of the most serious global health crises of recent decades. As of February 2022, global cases and deaths were officially estimated at 404 million and 5.5 million, respectively (Coronavirus Resource Center, 2022).

Therefore, the situation highlights the innovative regulations in a more socially responsible way especially for the health care. In this context, the four RI dimensions outlined above need closer attention. Although the dimensions differ, they are not separated by distinct boundaries. Rather, the boundaries are blurred at some points.

2.1 Anticipation

Anticipation focuses heavily on activities like establishing national and global databases to ensure that the expected and unexpected risks and consequences of COVID-19 can be measured. It also requires designing technologies for tracking, analyzing, and evaluating critical data. Such anticipatory practices, which relate to all health aspects of COVID-19, rely on "looking at what is already there to clarify which path to take" (Alami et al., 2020, p. 3). Important examples of activities for predicting future pandemic scenarios include COVID-19 spread forecasting, trend prediction, and government healthcare capacity planning.

2.2 Reflexivity

Reflexivity generally refers to "taking a step back and considering the innovation activities from a broader perspective" (Bruynseels, 2021, p. 122) or "holding a mirror up to one's own activities, commitments and assumptions" (Stilgoe et al., 2013, p. 1571). During the pandemic, reflexivity took various forms, such as "the development and engagement with best practice guidelines, codes of conduct and international standards." However, its persistence once adopted is doubtful (Rose et al., 2021, p. 308). These orientations have been embodied in governmental policies through innovating reflexive responses, such as "instituting information campaigns and intensive testing, enforcing social distancing and quarantining; tracking infection chains; and closing borders, and putting restrictions on traveling etc." (Scherer & Voegtlin, 2020, p. 31).

2.3 Inclusion

During the pandemic, inclusion was not only related to stakeholder involvement but also emphasized inclusive dialog in multilevel governance, science, technology, and innovation (Koch, 2020). During the struggle against COVID-19, inclusive efforts can be traced to various prominent actions. These include establishing coronavirus scientific committees comprising academics and other experts from multiple disciplines (e.g., Science Advisory Board) (Erdem, 2020) or expanding health, social, and educational services to all groups needing protection from COVID-19, including the most vulnerable groups (e.g., the elderly, the disabled, and chronic patients) (Turkey 3RP [Regional Refugee and Resilience Plan] Country Chapter, 2021). Another common practice is the "inclusion of relevant information about all affected social groups" (Leslie, 2020, p. 25) to disseminate knowledge about COVID-19 and ease COVID-19 detection.

2.4 Responsiveness

By focusing more on being proactive rather than just active in the fight against COVID-19, responsiveness concerns the immediate steps taken to overcome pandemic challenges. Examples include quarantine or isolation practices, various lockdowns like closing borders to travelers from high-risk countries, strict flight restrictions, the transition to remote and online learning, teaching, and working, mandatory use of face masks, and digital COVID-19 tracing apps. Other responsive steps include vaccine development and mandatory vaccinations, production of test kits, and capacity planning by hospitals for patients with serious complications requiring a ventilator (Ibeh et al., 2020).

3 COVID-19 Innovation in the Health Sector in the Republic of Türkiye

Before examining responsible innovation practices, processes, and outcomes during the COVID-19 pandemic in Türkiye, it is imperative to acknowledge the impact of the pandemic within the Turkish context. As of February 2022, the pandemic caused 13,265,374 confirmed cases and 91,646 deaths in Türkiye (World Health Organization, 2022). Türkiye's first COVID-19 case was announced on March 11, 2020, by the Ministry of Health on the same day that the World Health Organization (WHO) declared COVID-19 as a global pandemic. Since then, to contain the spread of the virus, the authorities introduced various mitigating measures to prevent the spread of COVID-19, such as lockdowns, mask mandates, school closures, and restrictions on public and private gatherings, and domestic and international travel.

Under these circumstances, it became extremely important for institutions to implement socially responsible practices and processes to help alleviate the pandemic's devastating effects. These organizations should also evaluate the current situation by identifying the people's needs/necessities/requirements to create a better future for humanity. For instance, Açıkgöz and Günay (2020) analyzed potential threats and opportunities in a post-pandemic world. They concluded that the pandemic would provide some economic and political opportunities to some countries if they manage to mitigate its impacts. Regarding Türkiye specifically, they said, "It seems that the Turkish economy will be destroyed in the short term like the many other countries but if The Republic of Türkiye can manage to control the virus soon, this would bring sustainable growth with the accelerating rise in manufacturing exports, tourism revenue and foreign investment in the short term." Thus, like other countries, Türkiye should not be content with several standard precautions. Rather, it also needs to exploit the opportunities by speeding up socially responsible actions, particularly in the healthcare sector, and developing new strategies to adjust to the post-pandemic world.

During the pandemic, such developmental orientations have deepened in almost every sector, with the health sector being the most affected. As mentioned earlier, one important socially responsible way to progress in today's world is through innovation. Scherer and Voegtlin, 2018, argue that "without responsible innovation, it is impossible to meet the grand challenges and to achieve sustainable development." Regarding health care, the rapid change of events during the pandemic confirmed that responsible innovation plays a particularly critical role in fighting globally against COVID-19. The pandemic has been an important catalyst for the healthcare sector, which has long been interested in innovation. According to McKinsey (2020), 90% of healthcare executives believe that the pandemic will fundamentally change their business, while 85% predict lasting changes to customer preferences.

Since the start of the pandemic, the healthcare sector in Türkiye has tried to develop ways to handle the crisis in a socially responsible manner. As an important element of social responsibility implementations, RI has indeed become one of the main means of coping with this healthcare crisis. These practices, processes, and outcomes of RI in Türkiye's healthcare sector are evaluated terms of needs/necessities/requirements.

3.1 Practices and Needs

To meet society's developing needs, Türkiye's government implemented several different practices, as summarized below.

In response to the developing needs of society during the pandemic, all sectors, including health care, have experienced high levels of digitization and digital innovation. Many mobile phone applications were developed and implemented, either for global use (e.g., Apple COVID-19, Corona Checker) or country-specific use (e.g., Canada COVID-19, Hayat Eve Sığar) (Kelleci et al., 2020). One of the most prominent examples of RI in Türkiye is the Ministry of Health's application "Hayat Eve Sığar" ("Life fits the home") which was launched in April 2020. This mobile application informs citizens about the pandemic to minimize its risks and prevent its spread. It tracks the infected people, risky areas, and tracks vaccination reports (Hayat Eve Sığar, 2020). The app was rolled out April 2020, which shows the importance of responsiveness as an RI dimension.

Given rising COVID-19 cases in neighboring countries, such as Iran, Iraq, Greece, and Bulgaria, Türkiye's government introduced measures to mitigate the risk of virus spread. The growing number of cases globally increased the public's need for scientific and trustworthy knowledge. In January 2020, a Science Advisory Board with 26 participants was formed as a consulting forum regarding the pandemic (Erdem, 2020). This board was instrumental in informing the Turkish public about the virus and providing scientific guidance to avoid virus contact. Inclusion of subject matter experts in consultancy and decision making was a strong indication of the inclusion dimension in responsible innovation.

Another important RI in healthcare practice in Türkiye was the establishment of pandemic hospitals. The pandemic undoubtedly increased pressure on the healthcare system across the country and increased the workload of healthcare workers, including doctors, nurses, and hospital staff. To mitigate the risk of overload, hospitals were defined as pandemic hospitals if they had at least two clinicians with infection disease specialty, clinical microbiology, and pulmonology specialties and had third-degree intensive care beds (Erdem, 2020; T.C. Sağlık Bakanlığı, 2020). This is a good example of both contingency planning in health care and out-of-the-box thinking to solve an emerging problem.

To prevent the virus spreading, several days before the first case was officially recorded, Türkiye's higher education council (Yuksekogretim Kurulu—YÖK) warned education institutes to prepare for digital and remote education (YÖK, 2020). YÖK highlighted five important dimensions of remote education: legislation, infrastructure, human resources, content, and application (Sözen, 2020).

3.2   Processes and Necessities

According to DeSalvo et al. (2021), an effective public health response to a pandemic depends on developing foundational capabilities, such as being prepared and responsive, assessing conditions, communicating properly, and partnering with communities.

As the COVID-19 pandemic developed into a healthcare crisis, the public's greatest need was to get regular access to truthful information from the government authorities. The process for this in Türkiye was that the Science Advisory Board formally convened twice a week while also meeting regularly with MoH staff to discuss pandemic-related issues and reach a consensus on critical and emerging issues (Keskinkilic et al., 2021). Following each science advisory board meeting, Minister of Health, Fahrettin Koca, held regular press conferences to pass on the latest data and information to the public. In addition, Turkish policy makers regularly met with WHO and other international bodies to share experiences and best practices.

In addition, to improvements to assessment and surveillance, starting from the early days of the pandemic, the practice of filiation and case tracking was implemented by the Ministry of Health. This process enabled strict monitoring of the contacts of anyone who may have been exposed to the virus and were subject to isolation (Demirtaş & Tekiner, 2020). In addition to contact tracing, health departments organized and invested in testing facilities to keep under control the growing number of cases.

It is important to note that Türkiye has one of the most detailed and elaborative health coverage systems, which helped during a healthcare crisis like the COVID-19 pandemic. According to a presidential decree published in April 2020, all healthcare costs related to COVID-19 were made free of charge for Turkish citizens and residents (Presidential decree, 2020). The flexibility and urgency around public policy development helped to meet the health needs of millions of citizens and residents.

3.3 Outcomes and Requirements

The abovementioned measures and the pandemic in general impacted Turkish society deeply, socially, economically, and environmentally. According to Aytaç and Kurtdaş (2015), health is no longer a personal issue but political, societal, economic, and technological. The pandemic has taken a major toll on the social lives and well-being of all humanity. According to Bostan et al. (2020), the Turkish community responded intensely to the pandemic and exerted maximum effort to protect themselves from the pandemic, implying that the crisis substantially impacted Turkish people. Generally, the pandemic had the greatest social impact on societies with more poverty, vulnerabilities, and inequalities. Globally, the pandemic exacerbated existing gender inequality, with implications for Türkiye as too (Alon et al., 2020).

According to the World Bank (2020), the pandemic threatens years of hard-won benefits while billions of jobs are threatened worldwide. As early as the summer of 2020, the World Bank (2020) warned that the pandemic could trigger the most challenging global economic depression since the Second World War. Similarly, OECD Secretary-General, Angel Gurría, warned: "Many countries would fall into recession and countries would be dealing with the economic fallout of the Covid-19 pandemic for years to come" (BBC, 2020). The pandemic's economic effects are indeed being experienced globally in terms of higher unemployment, business closures, demand and supply shocks, and declining world financial markets. These developments have in turn damaged the healthcare industry.

Due to urgent measures taken, Türkiye managed the early days of the chaotic pandemic environment reasonably successfully while also demonstrating elements of RI in its healthcare sector. The main requirement from the public was to control the spread of COVID-19 and eliminate the burden on the healthcare system. Despite some debatable outcomes, these initial requirements were met to the extent possible.

Responding to the pandemic requirements was a learning curve for most countries, including Türkiye. Even though some practices were not followed diligently or had limited public engagement, overall pandemic management and the resilience of Türkiye's healthcare sector were strong. Overall, the country was able to cope with this healthcare crisis.

Although the government signaled to the public that a "new normal" had arrived in May 2021, a dynamic process and contingency plans based on up-to-date developments, public behavior, and engagement will continue (Keskinkilic et al., 2021).

4 The Case of Abdi İbrahim: The Fight Against COVID-19

COVID-19 pandemic dominated almost all aspects of economic and social life throughout 2020 and 2021 (Brammer et al., 2020). Besides the measures taken to protect public health by slowing the spread of coronavirus, the preliminary actions were undoubtedly carried out in the health sector. As a part of healthcare services, the pharmaceutical industry played a crucial role during the outbreak in terms of research and development to develop and manufacture tests, vaccines, and other new products for treating and immunizing individuals against coronavirus infection. As a pioneering company in the Turkish pharmaceutical sector since 1912, Abdi İbrahim Pharmaceuticals implemented various RI practices during the pandemic (Abdi İbrahim, 2022a, 2022b, 2022c, 2022d, 2022e). These practices were also in line with the triple bottom-line approach (Elkington, 1994) and the economic, legal, ethical, and philanthropic components of CSR (Carroll, 1991).

4.1 Embedding the Dimensions of RI Within the Practices of Abdi İbrahim Pharmaceuticals

As explained previously in detail, the four IR dimensions have unique characteristics, yet cannot be completely differentiated by definite boundaries as they are also intertwined and mutually reinforcing (Stilgoe et al., 2013). That is, they sometimes may not be represented under a single dimension. Therefore, as shown in Table 1, a range of RI examples exist that can include multiple RI dimensions I as well as reflecting only the dominant orientations.

The following section presents Abdi İbrahim's most admirable and inspiring projects and practices during the pandemic to show how the four RI dimensions are reflected behaviorally in innovation processes.

4.1.1 Drug Development and Vaccine Production for COVID-19 Treatment

Following the COVID-19 outbreak, there were two main issues on the agenda: determining ways of protection from the virus and finding effective medical treatments for infected individuals. The development of new drugs and vaccines was therefore vital, and Abdi İbrahim Pharmaceuticals deployed its resources to provide

Table 1 Examples from Abdi İbrahim of the four RI dimensions

Projects and practices in Abdi İbrahim	Relevant RI dimensions
Drug development and vaccine production	Anticipation Inclusion Responsiveness
Biotechnology investments	Anticipation
"Healthy at Home" project	Reflexivity
Viral Agenda	Inclusion
"We Heal the Future" competition	Responsiveness

full support for new drug development (Abdi İbrahim, 2020a, 2020b). At this stage, the company focused on chloroquine, first discovered in 1934 and used as the active ingredient for treating malaria. Coincidentally, the company had already started production of a chloroquine-based drug in anticipation of a potential malaria epidemic (Gazete Duvar, 2020). Such foresight about future health issues and shaping production activities based on "what if" questions provides clear examples of the anticipation dimension of RI.

Abdi İbrahim Pharmaceuticals was already licensed to produce the chloroquine drug when the COVID-19 pandemic started. During the following days, research indicated that the drug could be used to treat COVID-19, so its production was prioritized (Abdi İbrahim, 2020a, 2020b). Although the company was the assigned manufacturer of the drug in Türkiye, their suffered from a materials shortage due globally increasing demand. In response, supported by the Ministry of Foreign Affairs and the Ministry of Trade, the company used good international relationships with India and China. Through these inclusive practices, it successfully imported large quantities of chloroquine, specifically 134 kg from India and 50 kg from China (Gazete Duvar, 2020).

The company then produced 1.6 million tablets for the treatment of COVID-19, which it donated to the Turkish Ministry of Health in 2020 (Abdi İbrahim, 2020a, 2020b). Regarding the cost of this donation, company chairman Nezih Barut said, "We never calculated this. We, as Abdi İbrahim, measure our own value by the contributions we make to our country" (Gazete Duvar, 2020). This way of thinking exemplifies the responsiveness dimension of RI.

In January 2021, the Ministry of Health licensed Abdi İbrahim Pharmaceuticals to manufacture and fill vials of both inactive and mRNA-based anti-coronavirus vaccines (Abdi İbrahim, 2021) in collaboration with Acıbadem Labcell (a cell laboratory and cord blood bank) (Dünya, 2021). This exemplifies the inclusion dimension of RI.

Abdi İbrahim Pharmaceuticals played a key role in the fight against the pandemic in Türkiye by producing and delivering drugs and manufacturing vaccines. In doing so, it collaborated with various national and international companies and the Turkish authorities. Overall, therefore, Abdi İbrahim Pharmaceutical's motivations, collaborations, efforts, and responses to public health and values, and its RI enabled it to achieve crucial medical achievements during the pandemic.

4.1.2 Biotechnology Investments

Biotechnology, which has become a common term since the 1980s, can be defined as "the integration of natural sciences and engineering sciences in order to achieve the application of organisms, cells parts thereof and molecular analogues for products and services" (IUPAC, 1992, p. 148). Accordingly, medicines produced using biotechnologies are called biopharmaceuticals or bio-drugs. These have been used since the mid-2000s (Walsh, 2002).

In 2018, in line with biotechnological developments, Abdi İbrahim Pharmaceuticals established Türkiye's largest biopharmaceuticals manufacturing facility, called "AbdiBio Biopharmaceutical Production Facility" to create greater long-term value by producing and exporting high value-added innovative products (Abdi İbrahim, 2022b). Nezih Bulut explained the motivation behind this investment by considering the current situation of the Turkish pharmaceutical industry. Currently, 80% of raw materials for drugs manufactured in Türkiye are imported, making the industry dependent on international markets for raw materials (Tıraş, 2020). Abdi İbrahim Pharmaceuticals therefore developed a corporate strategy focused on biotechnological investments to invent new molecules and produce bio-drugs instead of spending paying for imported raw materials (Dünya, 2021).

Envisioning the opportunities and potential impacts of biotechnology highlights the anticipation dimension of RI. This anticipation also led to inclusive initiatives during the pandemic as it became more difficult to import raw materials. For example, Abdi İbrahim Pharmaceuticals collaborated with Sabancı University and Mersin University (Dünya, 2021) and acquired a 28.5% stake in the Swiss biotech firm OM Pharma (Abdi İbrahim, 2020b). Hence, as Stilgoe et al. (2013) suggest, one dimension of RI can lead to another dimension. That is, although anticipation was the dominant orientation during the company's biotechnology investment initiative, it also reinforced the inclusion dimension, as revealed in Abdi İbrahim's strategic partnerships and collaborations to strengthen its competitiveness in biotechnology.

4.1.3 "Healthy at Home" Project

"Healthy at Home" (Evine Sağlık) was Abdi İbrahim Pharmaceutical's internal communication project that won the 2020 Golden Stevie Award in the category of "The Most Innovative Use of Human Resources Technology during the Pandemic" (Anadolu Ajansı, 2020). For this project, the company reconsidered its activities and values in response to changing working conditions following the COVID-19 outbreak. It then launched its successful RI project on March 30, 2020. The main purpose was the socialization and self-development of colleagues via "learning from one another" (Abdi İbrahim, 2022c). By reconsidering its own values, norms, and principles and reflecting them in its practices, processes, and outcomes, Abdi İbrahim Pharmaceuticals took a socially responsible step toward fighting the pandemic.

The dominant orientation of the project was the reflexivity dimension. In addition to volunteer colleagues, internal and external experts participated in the project's digital communication platform with 45-min podcasts (Anadolu Ajansı, 2020). Inviting and listening to diverse communities of professionals on health, psychology, and communication exemplified reflective behavior within the inclusion dimension of RI.

4.1.4   Viral Agenda

The Viral Agenda (Viral Gündem) was Abdi İbrahim Pharmaceutical's 2021 Bronze Stevie Award-winning communication project in the category of "Communications/ Public Relations Campaign-Covid-19 Related Information" (within the "Covid-19 Response" category) (Medikalteknik, 2021). The project provided an online communication platform for sharing reliable and evidence-based medical information with the public regarding developments in the pandemic. A team of doctors working in different specialties within the company's Medical Directorate scanned hundreds of COVID-19-related publications every week and prepared an agenda with the most accurate content. This Viral Agenda was delivered to 4600 employees and target audiences every week via email, LinkedIn, Twitter, Facebook, Instagram, and the company's official website. This project prevented the spread of unscientific or unverified information by providing accurate information (Anadolu Ajansı, 2021).

The Viral Agenda, prepared by expert doctors from diverse specialties and ensuring the public dissemination of the most accurate knowledge about COVID-19, exemplifies the inclusion dimension of RI.

4.1.5   We Heal the Future Competition

This competition is named after Abdi İbrahim Pharmaceuticals' following organizational commitment: *"We promise to work to heal the future in order to leave a more liveable and greener world for future generations"* (Abdi İbrahim, 2022d). The competition was designed within the framework of the company's "Social Innovation Program in Health and Life Sciences" and announced on April 28, 2021. The competition had the following seven objectives: (1) support social entrepreneurs in health and medicine; (2) strengthen and increase social benefit-oriented ideas; (3) understand and enhance the potential of social entrepreneurship; (4) increase integration into the start-up ecosystem; (5) contribute to the development of special creative solutions for the problems in health and medicine; (6) increase the number of entrepreneurs working in the health and pharmaceuticals with a focus on social benefit; and (7) provide opportunities to social benefit-oriented entrepreneurs and social entrepreneurial candidates in health and pharmaceuticals in developing their ventures (Abdi İbrahim, 2022e). The competition winners were announced on February 28, 2022, and provided with monetary rewards together with support on training, communication, incubation, and mentoring (Anadolu Ajansı, 2022).

Based on the competition's aims and support provided to the successful social entrepreneurs, it considered both innovation and science from a responsible perspective and with an outward-looking engagement. The competition thus precisely exemplifies the idea expressed by De Hoop et al. (2016) as going beyond "window-dressing" to empower innovative and social benefit-oriented business ideas. These aspects fundamentally relate to the responsiveness dimension of RI. In addition, Abdi İbrahim Pharmaceuticals established partnerships for this competition that brought together experts in social innovation, health, and life sciences. Hence, it also exemplifies inclusion.

5   Conclusion

Through a case study, this chapter demonstrated the importance of RI as one of the key pillars of CSR in the fight against the COVID-19 pandemic. Due to the urgent transformations imposed by the pandemic, the healthcare sector became one of the most important pioneers in overcoming the crisis. Although most countries experienced various challenges in this struggle, the process involved trial and error. Although not always sufficient, considering the various measures taken, Türkiye demonstrated various RI elements in the health sector from the first days of the pandemic. Generally, Türkiye succeeded to a certain extent in managing this health crisis in a socially responsible way through a wide range of RI practices, processes, and outcomes.

The case of Abdi İbrahim's response to COVID-19 reveals how RI works through its components and helped countries cope with the pandemic. In particular, given its involvement in various collaborations and partnerships with other players, the case demonstrates that the inclusion component of RI is an important driver for socially responsible companies. For instance, especially for producing new drugs and vaccines, the company's domestic and international collaborations and partnerships exemplified inclusion. By expanding its vaccine production capability and by importing large quantities of raw materials to manufacture drugs, Abdi İbrahim highlighted the importance of inclusion in this health crisis. The company's collaborations with universities in biotechnology investments, the participation of volunteer colleagues, and internal and external experts in its "Healthy at Home" Project, the inclusion of expert doctors in preparing the "Viral Agenda," and finally its partnerships in the "We Heal the Future Competition" demonstrated the critical importance of inclusion in all cases.

Given that the pandemic is still affecting countries worldwide, it seems clear that similar crises will continue to put pressure on global health services. Consequently, RI will continue to offer tremendous opportunities, especially for solving future health-related challenges.

References

Abdi İbrahim. (2020a, April 3). *In providing full support to the fight against the pandemic, Abdi İbrahim commences production of drug proposed by authorities for COVID-19 treatment.* Retrieved February 2, 2022, from https://www.abdiibrahim.com.tr/en/press/press-releases/in-providing-full-support-to-the-fight-against-the-pandemic-abdi-ibrahim-commences-production-of-drug-proposed-by-authorities-for-covid-19-treatmen

Abdi İbrahim. (2020b, September 18). *Historic move from Abdi İbrahim.* Retrieved February 2, 2022, from https://www.abdiibrahim.com.tr/en/press/press-releases/abdi-ibrahim-acquires-28-5-percent-stake-in-swiss-biotech-firm-om-pharma

Abdi İbrahim. (2021, January 7). *Abdi İbrahim granted permission to manufacture Covid-19 vaccine.* Retrieved February 2, 2022, from https://www.abdiibrahim.com.tr/en/press/press-releases/abdi-ibrahim-granted-permission-to-manufacture-covid-19-vaccine

Abdi İbrahim. (2022a). *Bir Bakışta Abdi İbrahim.* Retrieved February 2, 2022, from https://www.abdiibrahim.com.tr/hakkimizda/bir-bakista-abdi-ibrahim

Abdi İbrahim. (2022b). *AbdiBio production center.* Retrieved February 3, 2022, from https://www.abdiibrahim.com.tr/en/corporate/facilities/turkey/abdibio-production-center

Abdi İbrahim. (2022c). *Healthy at home.* Retrieved February 3, 2022, from https://www.abdiibrahim.com.tr/en/career/get-to-know-the-healers-of-life-better/healthy-at-home

Abdi İbrahim. (2022d). *Our commitment.* Retrieved February 10, 2022, from https://www.abdiibrahim.com.tr/en/sustainability/our-commitment

Abdi İbrahim. (2022e). *We heal the future: Social innovation program in health and life sciences.* Retrieved March 7, 2022, from https://www.wehealthefuture.com

Açıkgöz, Ö., & Günay, A. (2020). The early impact of the Covid-19 pandemic on the global and Turkish economy. *Turkish Journal of Medical Sciences, 50*(SI-1), 520–526.

Alami, H., Mörch, C., Rivard, L., Rocha, R., & Silva, H. (2020). *Can we innovate responsibly during a pandemic?* Retrieved February 10, 2022, from https://www.docdroid.com/juJtgpG/can-we-innovate-responsibly-during-a-pandemic-pdf#page=3

Alon, T., Doepke, M., Olmstead-Rumsey, J., & Tertilt, M. (2020). The impact of the coronavirus pandemic on gender equality. *Covid Economics Vetted and Real-Time Papers, 4*, 62–85.

Anadolu Ajansı. (2020). *Abdi İbrahim Stevie Awards'ta iki altın ödül aldı.* Retrieved February 5, 2022, from https://www.aa.com.tr/tr/sirkethaberleri/saglik/abdi-ibrahim-stevie-awardsta-iki-altin-odul-aldi-/658971

Anadolu Ajansı. (2021). *Abdi İbrahim'e Stevie Uluslararası İş Ödülleri'nden 2 ödül.* Retrieved February 5, 2022, from https://www.aa.com.tr/tr/sirkethaberleri/saglik/abdi-ibrahime-stevie-uluslararasi-is-odullerinden-2-odul/666788

Anadolu Ajansı. (2022). *Abdi İbrahim İyileştiren Fikirler Yarışması'nda kazanan sosyal girişimler belli oldu.* Retrieved March 7, 2022, from https://www.aa.com.tr/tr/sirkethaberleri/saglik/abdi-ibrahim-iyilestiren-fikirler-yarismasinda-kazanan-sosyal-girisimler-belli-oldu/671257

Aymerich-Franch, L., & Fosch-Villaronga, E. (2020). A self-guiding tool to conduct research with embodiment technologies responsibly. *Frontiers in Robotics and AI, 7*, 22.

Aytaç, Ö., & Kurtdaş, M. Ç. (2015). Sağlik-Hastaliğin Toplumsal Kökenleri Ve Sağlik Sosyolojisi. *Fırat Üniversitesi Sosyal Bilimler Dergisi, 25*(1), 231–250.

BBC News. (2020). *Global economy will suffer for years to come, Says OECD.* Retrieved March 24, 2020, from https://www.bbc.com

Belas, J., Çera, G., Dvorský, J., & Čepel, M. (2021). Corporate social responsibility and sustainability issues of small-and medium-sized enterprises. *Corporate Social Responsibility and Environmental Management, 28*(2), 721–730.

Bostan, S., Erdem, R., Öztürk, Y. E., Kılıç, T., & Yılmaz, A. (2020). The effect of COVID-19 pandemic on the Turkish society. *Electronic Journal of General Medicine, 17*(6), em237.

Brammer, S., Branicki, L., & Linnenluecke, M. K. (2020). COVID-19, societalization, and the future of business in society. *Academy of Management Perspectives, 34*(4), 493–507.

Brand, T., & Blok, V. (2019). Responsible innovation in business: A critical reflection on deliberative engagement as a central governance mechanism. *Journal of Responsible Innovation, 6*(1), 4–24.

Bruynseels, K. (2021). Responsible innovation in synthetic biology in response to COVID-19: The role of data positionality. *Ethics and Information Technology, 23*(1), 117–125.

Carroll, A. B. (1991). The pyramid of corporate social responsibility: Toward the moral management of organizational stakeholders. *Business Horizons, 34*(4), 39–48.

Carroll, A. B. (1999). Corporate social responsibility: Evolution of a definitional construct. *Business and Society, 38*(3), 268–295.

Carroll, A. B. (2021). Corporate social responsibility (CSR) and the COVID-19 pandemic: Organizational and managerial implications. *Journal of Strategy and Management, 14*(3), 315–330.

Coronavirus Resource Center. (2022). *COVID-19 dashboard*. Retrieved February 10, 2022 from https://coronavirus.jhu.edu/data

De Hoop, E., Pols, A., & Romijn, H. (2016). Limits to responsible innovation. *Journal of Responsible Innovation, 3*(2), 110–134.

Demirtaş, T., & Tekiner, H. (2020). Filiation: A historical term the COVID-19 outbreak recalled in Turkey. *Erciyes Medical Journal, 42*(3), 354–3588.

DeSalvo, K., Hughes, B., Bassett, M., Benjamin, G., Fraser, M., Galea, S., & Gracia, J. N. (2021). Public health COVID-19 impact assessment: Lessons learned and compelling needs. *National Academy of Medicine Perspectives*. Retrieved February 28, 2022, from https://www.ncbi.nlm.nih.gov/pmc/articles/PMC8406505/

Dünya. (2021). *Abdi İbrahim Yönetim Kurulu Başkanı Nezih Barut: Dünya ilaç sektörünün DNA'sı kökten değişiyor*. Retrieved February 3, 2022, from https://www.dunya.com/ekonomi/abdi-ibrahim-yonetim-kurulu-baskani-nezih-barut-dunya-ilac-sektorunun-dnasi-kokten-degisiyor-haberi-618934

Elkington, J. (1994). Towards the sustainable corporation: Win–win–win business strategies for sustainable development. *California Management Review, 36*(2), 90–100.

Erdem, İ. (2020). Koronavirüse (Covid-19) Karşı Türkiye'nin Karantina ve Tedbir Politikaları. *Electronic Turkish Studies, 15*(4), 377.

FCC Innovation. (2020). *Green in the grey: An update on Corporate Social Responsibility in 2020*. Retrieved December 14, 2021, from https://www.fccinnovation.co.uk/blog/corporate-social-responsibility-2020

Gazete Duvar. (2020). *Abdi İbrahim: Korona için Favipiravir'i üretebiliriz*. Retrieved February 1, 2022, from https://www.gazeteduvar.com.tr/saglik/2020/05/04/abdi-ibrahim-korona-icin-favipiraviri-uretebiliriz

Genus, A., & Stirling, A. (2018). Collingridge and the dilemma of control: Towards responsible and accountable innovation. *Research Policy, 47*(1), 61–69.

Groves, C. (2017). Review of RRI tools project, http://www. rri-tools. eu. *Journal of Responsible Innovation, 4*(3), 371–374.

Halme, M., & Korpela, M. (2014). Responsible innovation toward sustainable development in small and medium-sized enterprises: A resource perspective. *Business Strategy and the Environment, 23*(8), 547–566.

Halme, M., & Laurila, J. (2009). Philanthropy, integration or innovation? Exploring the financial and societal outcomes of different types of corporate responsibility. *Journal of Business Ethics, 84*(3), 325–339.

Hamza, S., & Jarboui, A. (2020). CSR: A moral obligation or a strategic behavior? In: *Corporate social responsibility* (pp. 1–15).

Hayat Eve Sığar. (2020). *HES Güvenli Alan ve Hes Kodu*. Retrieved February 16, 2022, from https://hayatevesigar.saglik.gov.tr/

Hlioui, Z., & Yousfi, O. (2020). CSR and innovation: Two sides of the same coin. In *Corporate social responsibility*. IntechOpen.

Ibeh, I. N., Enitan, S. S., Akele, R. Y., Isitua, C. C., & Omorodion, F. (2020). Global impacts and Nigeria responsiveness to the COVID-19 pandemic. *International Journal of Healthcare and Medical Sciences, 6*(4), 27–45.

IUPAC. (1992). Compendium of chemical terminology (the "Gold Book"). In A. D. McNaught & A. Wilkinson (Compiled). *Blackwell scientific publications, Oxford (1997)* (2nd ed.). Online version (2019–) created by S. J. Chalk. ISBN 0-9678550-9-8. https://doi.org/10.1351/goldbook

Kelleci, M., Tel, H., & Kısaoğlu, Ö. (2020). *Covid-19 Pandemisi ile Kullanıma Giren Akıllı Telefon Uygulamaları.* 4. Uluslararası Hemşirelik Ve İnovasyon Kongresi, 66.

Keskinkilic, B., Shaikh, I., Tekin, A., Ursu, P., Mardinoglu, A., & Mese, E. A. (2021). A resilient health system in response to Coronavirus disease 2019: Experiences of Turkey. *Frontiers in Public Health, 8,* 871.

Koch, S. (2020). Responsible research, inequality in science and epistemic injustice: An attempt to open up thinking about inclusiveness in the context of RI/RRI. *Journal of Responsible Innovation, 7*(3), 672–679.

Koops, B. J. (2015). The concepts, approaches, and applications of responsible innovation. *Responsible Innovation, 2,* 1–15.

Leslie, D. (2020). Tackling COVID-19 through responsible AI innovation: Five steps in the right direction. *Harvard Data Science Review.* Retrieved February 11, 2022, from https://hdsr.mitpress.mit.edu/pub/as1p81um/release/3

MacGregor, S. P., & Fontrodona, J. (2008, July). *Exploring the fit between CSR and innovation.* Working Paper WP-759. Center for Business and Society.

McKinsey. (2020). *Innovation in a crisis: Why it is more critical than ever.* Retrieved February 15, 2022, from https://www.mckinsey.com/business-functions/strategy-and-corporate-finance/our-insights/innovation-in-a-crisis-why-it-is-more-critical-than-ever

Medikalteknik. (2021). *Abdi İbrahim wins two more international awards.* Retrieved February 5, 2022, from https://www.medikalteknik.com.tr/abdi-ibrahim-wins-two-more-international-awards/

Ortiz-Avram, D., Domnanovich, J., Kronenberg, C., & Scholz, M. (2018). Exploring the integration of corporate social responsibility into the strategies of small-and medium-sized enterprises: A systematic literature review. *Journal of Cleaner Production, 201,* 254–271.

Owen, R., Macnaghten, P., & Stilgoe, J. (2012). Responsible research and innovation: From science in society to science for society, with society. *Science and Public Policy, 39*(6), 751–760.

Owen, R., Stilgoe, J., Macnaghten, P., Gorman, M., Fisher, E., & Guston, D. (2013). A framework for responsible innovation. In R. Owen, J. Bessant, & M. Heintz (Eds.), *Responsible Innovation: Managing the responsible emergence of science and innovation in society* (Vol. 31, pp. 27–50). Wiley.

Presidential Decree. (2020). *Addition to the presidential decree of 13 April 2020 No: 2399.* Retrieved September 26, 2020, from https://www.resmigazete.gov.tr/eskiler/2020/04/20200414-16.pdf

Ratajczak, P., & Szutowski, D. (2016). Exploring the relationship between CSR and innovation. *Sustainability Accounting. Management and Policy Journal, 7*(2), 295–318.

Reber, B. (2018). RRI as the inheritor of deliberative democracy and the precautionary principle. *Journal of Responsible Innovation, 5*(1), 38–64.

Rose, D. C., Lyon, J., de Boon, A., Hanheide, M., & Pearson, S. (2021). Responsible development of autonomous robotics in agriculture. *Nature Food, 2*(5), 306–309.

Scherer, A. G., & Voegtlin, C. (2020). Corporate governance for responsible innovation: Approaches to corporate governance and their implications for sustainable development. *Academy of Management Perspectives, 34*(2), 182–208.

Shahzad, M., Qu, Y., Javed, S. A., Zafar, A. U., & Rehman, S. U. (2020). Relation of environment sustainability to CSR and green innovation: A case of Pakistani manufacturing industry. *Journal of Cleaner Production, 253,* 119938.

Sözen, N. (2020). Covid 19 sürecinde uzaktan eğitim uygulamaları üzerine bir inceleme. *Avrasya Sosyal ve Ekonomi Araştırmaları Dergisi, 7*(12), 302–319.

Stilgoe, J., Owen, R., & Macnaghten, P. (2013). Developing a framework for responsible innovation. *Research Policy, 42*(9), 1568–1580.

Szutowski, D., & Ratajczak, P. (2016). The relation between CSR and innovation. Model approach. *Journal of Entrepreneurship, Management and Innovation, 12*(2), 77–94.

T.C. Sağlık Bakanlığı. (2020). *Sağlık Hizmetleri Genel Müdürlüğü. Sayı: 14500235-403.99/, konu: pandemi hastaneleri.* Retrieved February 21, 2022, from https://im.haberturk.com/images/others/2020/03/20/pandemi_genelge.pdf

Tıraş, H. H. (2020). Türkiye'de İlaç Sektörünün Gelişimi; Bir Durum Değerlendirmesi. *Journal of Economics and Research, 1*(1), 42–59.

Tsai, H., Tsang, N. K., & Cheng, S. K. (2012). Hotel employees' perceptions on corporate social responsibility: The case of Hong Kong. *International Journal of Hospitality Management, 31*(4), 1143–1154.

Turkey 3RP [Regional Refugee and Resilience Plan] Country Chapter. (2021). *Needs.* Retrieved February 19, 2022, from https://www.unhcr.org/tr/wp-content/uploads/sites/14/2021/03/3RP-Turkey-Country-Chapter-2021-2022_EN-opt.pdf

Voegtlin, C., Georg Scherer, A., Stahl, G. K., & Hawn, O. (2019). Grand societal challenges and responsible innovation. *Journal of Management Studies, 59*, 1–28.

Voegtlin, C., & Scherer, A. G. (2017). Responsible innovation and the innovation of responsibility: Governing sustainable development in a globalized world. *Journal of Business Ethics, 143*(2), 227–243.

Wagner, M. (2010). Corporate social performance and innovation with high social benefits: A quantitative analysis. *Journal of Business Ethics, 94*, 581–594.

Walsh, G. (2002). Biopharmaceuticals and biotechnology medicines: An issue of nomenclature. *European Journal of Pharmaceutical Sciences, 15*(2), 135–138. https://doi.org/10.1016/s0928-0987(01)00222-6

World Bank. (2020). *Saving lives, scaling-up impact and getting back on track World Bank Group COVID-19 Crisis Response Approach Paper.* Retrieved February 16, 2022, from https://reliefweb.int/sites/reliefweb.int/files/resources/World-Bank-Group-COVID-19-Crisis-Response-Approach-Paper-Saving-Lives-Scaling-up-Impact-and-Getting-Back-on-Track.pdf

World Health Organization. (2020). *WHO Statement regarding cluster of pneumonia cases in Wuhan, China.* Retrieved January 10, 2020, from https://www.who.int/china/news/detail/09-01-2020-who-statement-regarding-cluster-of-pneumonia-cases-in-wuhan-china

World Health Organization. (2022). *Coronavirus dashboard.* Retrieved February 15, 2022, from https://covid19.who.int/

YÖK. (2020). *Basın açıklaması.* Retrieved February 23, 2022, from https://www.yok.gov.tr/Sayfalar/Haberler/2020/YKS%20Ertelenmesi%20Bas%C4%B1n%20A%C3%A7%C4%B1klamas%C4%B1.aspx

Zhang, S. X., Choudhury, A., & He, L. (2019, July). Responsible innovation: The development and validation of a scale. *Academy of Management Proceedings, 1*, 12437. Academy of Management.

Gizem Aras Beger, completed her Bachelor's degree in Business Administration at Anadolu University-Faculty of Economics and Administrative Sciences under a rectorate success scholarship. She earned both master's and PhD degree with full scholarship in Business Administration from Yaşar University. Her research interests include corporate social responsibility, sustainability, responsible innovation, business ethics, and strategic management. She has been working as a research assistant at the Department of Business Administration/Yaşar University since 2018.

Gönenç Dalgıç Turhan is both a business professional on marketing and international trade and a researcher on business studies. She graduated from Faculty of Business, Department of International Relations of Izmir Dokuz Eylül University in 2001. She worked for 10 years in pension and life insurance sector at managerial positions. Meanwhile, she studied business administration, management and organization at Dokuz Eylül University (graduating with an M.A. in 2006). She conducted a study as a Visiting Researcher at Hohenheim University, Institute of Marketing and Management, Department of Business StartUps and Entrepreneurship between 2013 and 2014 and received her PhD on Business Administration in 2017 with a thesis on corporate sustainability from Yaşar University where worked as a part-time lecturer during 2015–2017. She has been working at Karyatech Electrical Equipments Industry and Trade Inc. Co. as Import Manager since 2016. Her research focuses on sustainability, innovation, and entrepreneurship.

Gülen Rady is a global marketing and commercial activation leader who has deep corporate experience in healthcare and medical technology industry across global markets. She is a recognized thought leader in value-based pricing and new product activation who has participated to many international speaking engagements related to pricing strategy and digital transformation. She has a B.A. degree in Economics from Bogazici University and an MBA from University of Virginia, Darden Graduate School of Business. She currently continues her PhD studies in Business in Yasar University.

Part II
CSR and COVID-19 Pandemic in Africa

Ghana
  Nigeria
  Zambia

Grappling with COVID-19: The Implications for Ghana

Sam Sarpong

1 Introduction

Ghanaians were alerted to a reported COVID-19 case in Nigeria on 27 February 2020 and told to take the necessary precautions before it emerged in the country. That was a time when people were on tenterhooks and the thought that the disease had finally arrived in the West African sub-region sent people into hysteria. Ghana confirmed its first cases of COVID-19 on 12 March 2022. The country's President, Nana Akufo-Addo, subsequently made a nation-wide address to assure the populace of the government's preparedness to contain the pandemic. The government immediately made available US$100 million to enhance Ghana's response plan and preparedness for the pandemic. It also ordered the suspension of all international travels of public officials except for critical assignments in the days following the detection of COVID-19 in the country.

The initial response showed a strong determination on the part of government to deal with the situation. Some Ghanaians, who hitherto had dismissed the pandemic as a fluke, were, meanwhile, left in no doubt about the enormity of the problem when cases began to soar in the country. The reality of the situation dawned on them when news about the demise of some renowned personalities from the disease reached them. Among the dignitaries who died were Ghana's former President, Jerry Rawlings, who reportedly succumbed to COVID-19 complications in 2020. Following from that, those who had dismissed the pandemic as a distraction began to shift steadily their position and also to observe the necessary protocols that had been put in place.

S. Sarpong (✉)
School of Economics and Management, Xiamen University Malaysia, Sepang, Selangor, Malaysia
e-mail: samsarpong@xmu.edu.my

© The Author(s), under exclusive license to Springer Nature Switzerland AG 2023
S. O. Idowu et al. (eds.), *Corporate Social Responsibility in the Health Sector*, CSR, Sustainability, Ethics & Governance, https://doi.org/10.1007/978-3-031-23261-9_11

The chapter presents a comprehensive review of the COVID-19 situation in Ghana. It provides an insight into the pandemic's impact on major areas like health, education and the socio-economic lives of the people. It also looks at the role played by both government and businesses since the inception of the pandemic and how societal preconceptions and considerations shaped people's opinions about the pandemic. The chapter largely relies on secondary data sources, namely reports from government COVID-19 addresses and documents, briefs from international agencies and media reports.

The chapter is structured as follows: first, it opens up with an overview of the pandemic situation in Ghana. Both the literature review and the methodology employed are subsequently discussed. It then delves into the misconceptions and myths surrounding the disease. From there, the chapter provides some details about Ghana's health system and its supportive role in the fight against COVID-19. It also portrays its challenges too. Following from this, it assesses the pandemic's impact on the educational system and society as a whole. The private sector's role in the fight against the pandemic is also explored. The lessons learnt from the pandemic and what ought to be done in the light of that are also incorporated into the work. This is followed by a conclusion.

2 Literature Review

Since the beginning of the COVID-19 outbreak in December 2019, a substantial body of COVID-19 literature has been generated. Many scholars have in recent times investigated and written about the COVID issue and its effect on a number of countries. The early COVID-19 medical literature originated primarily from Asia and focused mainly on clinical features and diagnosis of the disease, but many areas of potential research began to be published subsequently (Liu et al., 2020). Considerable efforts have since gone into various aspects of this pandemic to critically assess the way countries have fared in terms of dealing with it. Not only have COVID-19 publications increased exponentially (Palayew et al., 2020), but they have also covered a lot of methodologies too. Despite that, there are still knowledge gaps yet to be filled and areas for improvement for the global research community (Sarpong, 2021).

Some scholars have dealt with the issue through different approaches like clinical trials, surveys and the use of secondary sources. For instance, Silva and Mont'Alverne (2020) explored the social contexts in which the COVID-19 transmission occurs among vulnerable populations in Brazil through the use of online surveys. Sarpong (2021), meanwhile, looked at COVID-19's trajectory and sought to find out whether the lessons from it can be harnessed in the pursuance of the Sustainable Development Goals. Turcotte-Tremblay et al. (2021), on the other hand, examined the unintended consequences of COVID-19 mitigation measures. The rising burden of the ongoing COVID-19 epidemic in South Africa also motivated Reddy et al. (2021) to apply modelling strategies to predict COVID-19 cases and deaths. Witteveen (2020) dwelt on the induced economic hardship associated with COVID-19 in the UK. Meanwhile, the inequalities that the pandemic brought to the fore have also found space in a number of analyses by scholars. A contribution by Engzell et al. (2020), for instance, investigated the learning inequality during the COVID-19 pandemic in the Netherlands.

Many of these papers have given us much knowledge as to how COVID-19 has impacted differently on people and countries altogether. Quite significantly too have been the different methodologies that were employed in the course of these assessments. In some instances, data for some of the issues on COVID-19 have dwelt mainly on secondary sources.

3 Methodology

The methodology for this work comprises a review of the literature. The review covers a search, evaluation and synthesis of the relevant information on COVID-19 and how it has affected Ghana. It dwells extensively on journals, government statements, policy briefs as well as general observations. At a time when vast amounts of data are being collected and archived by researchers all over the world, the practicality of utilising existing data for research is becoming more prevalent (Andrews et al., 2012; Smith, 2008). Given the increasingly availability of previously collected data to researchers, it is important to see secondary data analysis as a systematic research method. In one commonly cited approach in social research, Bowen (2009) recommends first skimming the documents to get an overview, then reading to identify relevant categories of analysis for the overall set of documents and finally interpreting the body of documents.

The chapter, thus, dwelt upon these principles advanced by Bowen (2009). The chapter's quest was to have an insight into the COVID situation in Ghana. Looking out for data in this regard and also determining the kind of information that would be needed were quite vital for the success of this work. The decision was, therefore, made to utilise existing data to assist in unravelling some of the issues that have come about as a result of the pandemic.

4 Overview of the Pandemic in Ghana

Although Ghana's first two cases of COVID-19 were all imported, the pandemic quickly spread through the country, and within a week of the first cases, Ghana confirmed cases in individuals with no links to foreign travel. The majority of cases were in the environs of the two heavily populated cities of Accra and Kumasi. Immediate measures were, therefore, instituted to detect, contain and prevent the spread of the disease. The major measures instituted by the government at that time included:

- A lockdown of the major cities, Accra and Kumasi (from 30 March 2020). The government also included Kasoa, a thriving town in the central region of the country. (The lockdown was lifted on 20 April 2020.)
- Closure of schools, places of worship, restaurants and bars, and limitations on the number of people at gatherings.
- Use of face masks in all public spaces was made mandatory.
- Restrictions on travel (border closure) and public gatherings.
- Free water for domestic users for 6 months to ensure water security in households.

These measures were underpinned with education on the disease and its transmission as well as preventive measures such as personal hygiene. During the period of the lockdown, an enhanced surveillance in the form of active case search and contact tracing strategies were activated to early detect, isolate and treat all confirmed cases.

The government also banned all public gatherings including conferences, workshops, funerals, festivals, political rallies, church activities and other related events at the onset of the pandemic to reduce the spread of the virus. Beaches were also closed likewise basic schools, senior high schools and universities. People were also prevented from travelling beyond a certain radius. These measures were quite unprecedented. As such, some could not put up with this, as they had no experience with such restrictions. The military and other security agencies were, therefore, called in to ensure people did not flout the directives (Asante & Mills, 2020). To a very large extent, the general public adhered to the lockdown directives. The streets were quieter, non-essential shops were closed, and nightlife was virtually non-existent.

The lockdown imposed on the major cities was just for a short while, as the government was swayed to lift it after 3 weeks because of the severe impact on a large segment of Ghanaians. Indeed, the structure of the Ghanaian economy and the severe hardships that came with the lockdown provided the needed impetus for it to be lifted. Ghana has a large informal economy which was seriously impacted by the lockdown. Many traders, artisans and street vendors, among others, were not able to go to work or to buy their daily requirements for the period they were under lockdown (Akinwotu & Asiedu, 2020). Several people complained of severe hardships, and there was ample evidence of vulnerability in many households across the country (ILO, 2020). In several markets within Accra and Kumasi, law enforcement officers were seen reprimanding people because they ventured out of their homes in search of food during the lockdown period. Consequently, the lockdown was eased after the third week by the President during a televised address to the nation.

Post-lockdown measures were vigorously enforced to control the spread of the infection. These included personal hygiene measures, mandatory wearing of masks, ban on social gathering, social distancing, increase in the number of testing sites with enhanced surveillance and other response activities. On 30 March 2022, the President, Nana Akufo-Addo, in another address to the nation, announced new measures which included the reopening of the country's borders and the lifting of embargoes placed on public gatherings. He cited the rapidly declining infections and a relatively successful inoculation campaign as the reasons behind this.

Nana Akufo-Addo also announced that Ghana would start producing its own COVID-19 vaccines in 2024. He said a National Vaccine Institute would be established to lay out a strategy for the country to begin the first phase of the commercial production of the jabs (Reuters, 2022). To further strengthen the health delivery system, the president reiterated plans by the government to build 111 new hospitals across the country to serve the growing health needs of the population.

5   Current Developments on COVID-19

As of 12 March 2022, more than 12.7 million COVID-19 doses had been administered in Ghana. So far, Ghana has fully vaccinated around 21.4% of its 30 million inhabitants against coronavirus. It has indeed shown a capacity to rein in the disease at an early stage. It also became the first country to receive the COVID-19 vaccines from the COVAX Facility. COVAX is co-led by the Coalition for Epidemic Preparedness Innovations (CEPI), Gavi and the World Health Organisation (WHO), alongside key delivery partner, UNICEF. This is the only global initiative that is working with governments and manufacturers to ensure COVID-19 vaccines are available worldwide to both higher-income and lower-income countries.

At the onset of the pandemic, the country struggled to acquire personal protective equipment and medicines to manage COVID-19 cases. Testing was also an issue in the early days of the disease as testing facilities were limited. Since then, testing capabilities have been ramped up across the country. In addition, several dedicated COVID-19 treatment centres were set up, with the aim of having at least one centre in each of the 16 regions in the country. Ghana's preparations towards the pandemic and the measures it undertook received a lot of praises from near and wide in view of these measures.

The personal involvement of Ghana's president, Nana Akufo-Addo, in the fight against the pandemic has been acknowledged by both Ghanaians and international bodies. He has been very relentless in his periodic broadcasts to the nation detailing what Ghanaians should do in the wake of the pandemic. During his 2021 Christmas address, for instance, he urged Ghanaians to act and live responsibly and to adhere to the protocols of social distancing, enhanced hygiene and mask wearing (CGTN, 2021).

Ghana has since come a long way since the pandemic began. The pandemic has affected many people in diverse ways. After a strong and effective initial emergency response to the pandemic, the government's attempts at relaxing restrictions and reopening socio-economic activities were thwarted by the propagation of new variants in Ghana in early 2021 (World Bank, 2021). Numerous interventions have since been made to stop the spread of COVID-19 and its impact on the public. Some of these initiatives have been saddled with challenges, leading to severe hardships among vulnerable groups. Although some of these measures have helped to avoid a soar in the number of cases, there have been some misconceptions too about the disease which, to some extent, have acted against the COVID campaign.

6   The Social Reality, Misconceptions and Myths About COVID-19

There is still a belief among some Ghanaians that Africans have some immunity against COVID. Some people have remained steadfast in this belief despite the deaths of high-profile Africans. According to Padayachee and du Troit (2020), this came about because the initial cases were mostly recorded in western countries and people, therefore, thought Africans were not particularly susceptible to it. Some young people were also imbued with the notion that the condition only affected the elderly as the death rate soared among the elderly. These misconceptions served as risk attenuators among Ghanaians, especially the younger generation who felt they were somehow invulnerable from it. As the infection evolved in the country, another misconception emerged that the hot climate in Africa inhibited viral replication and transmission (Tabong & Segtub, 2021). The government had to release a statement debunking that.

It also came to a point when certain herbal practitioners even started prescribing remedies for the management of COVID-19. The use of local remedies such as neem tree (*Azadirachta indica*) and herbal preparations also came to the fore. Myths about the efficacy of locally manufactured gin (*akpeteshie*) and hydroxychloroquine as prophylaxis also led to abuse of such substances (Tabong & Segtub, 2021). Some people were of view that heavy drinking of the local gin could kill the virus in the blood stream. In general, all sorts of medications gained currency as people tried various means to have some protection against the COVID-19 infection. Some even still hold on to the belief that COVID-19 vaccines can make our body magnetic.

Many of these misconceptions have not yet been lost on the people, and as part of efforts to curtail and also to address these, the Ghana's Health Service, the Ministry of Information, FactSpace West Africa and UNICEF in March 2022, invited some of the country's popular social media influencers to a sensitisation workshop to discuss issues relating to this (UNICEF, 2022). The workshop was also centred on building partnerships in combating online misinformation on COVID-19. It was also focused on creating the needed awareness about the positive impact of vaccine administration in the country. It is worth noting that the preponderance of COVID-19 vaccine myths is causing some Ghanaians to forego vaccinations at a time when new and more transmissible coronavirus variants are spreading across the country and the world as a whole. The rollout of COVID-19 vaccines in the country was supposed to be highly patronised, but just as the supply of vaccines has been increasing in Ghana, so too has misinformation about their safety and efficacy been gaining grounds.

Ghana still has a poor health system, weak government capacity to manage a public health response, and limited water availability and related infrastructure that are crucial to health care (IFC, 2020).

7   Ghana's Health System

Incidentally, the health system in Ghana has faced a myriad of challenges in dealing with the pandemic. There have been weak coordination mechanisms, lack of/or inadequate quarantine and isolation facilities, irregular supply of laboratory supplies and other material and lack of adequate legislation in dealing with public health emergencies (World Bank, 2021). Besides, there is a shortage of adequately trained health workers. This has been compounded by an unequal distribution of health personnel. Many health personnel prefer to be in urban areas and also to work for urban hospitals, a situation that has led to fewer staff taking up roles in rural areas. The chronic lack of investment in healthcare infrastructure and equipment has made it harder for African nations, like Ghana, to retain skilled healthcare workers, provide essential medicines and reduce the mortality rates of perennial diseases like malaria (Human Rights Watch, 2020).

There are also high prices of drugs as well as limited coverage of health insurance in the country. Ghana's National Health Insurance Scheme (NHIS) budget is hugely underfunded so a lot of patients are resigned to spending their own money for their health care. The NHIS, which is publicly financed through taxes on goods, services and income, does not cover all disease conditions (Siaw-Frimpong et al., 2021). A notable fact about the healthcare needs and services in the country is that emergency preparedness, coverage of essential services, financial protection of the poorest and most vulnerable population and service accessibility and delivery remain very limited. Therefore, with the onset of the pandemic, many were those who felt that the health service in the country would crumble.

Indeed, the effect of the pandemic on the entire health system in Africa was feared and speculated by several scholars and institutions to be highly vulnerable to the pandemic (BBC, 2020; WHO, 2020). Much of the conversation that surrounded the potential impact of COVID-19 on Africa seemed to have stemmed from uninformed assumption. Doomsday predictions that Africa would have dead bodies on the streets did not really materialise as thought. However, a lot really went on among countries like Ghana to avert any possible disaster. Timely, strategic implementation of targeted emergency care solutions in many African countries has helped to avoid needless human suffering and deaths.

Ghana has impressed in the way it has handled the pandemic within its health systems. In the wake of the pandemic, the Ministry of Health of Ghana through the Ghana Health Service constituted a team of health experts to spearhead the initiation and implementation of strategies to combat the spread of the disease. As a standard practice, this team of experts provided regular updates to the government for onward communication to the citizenry.

Ghana has also received enormous support from multilateral agencies. Recently, the World Bank, through the Pandemic Emergency Financing Facility (PEF), provided funds to WHO to enhance the capacity of Ghana's health system to respond to COVID-19 (WHO, 2021). As a result, treatment facilities in all 16 regions of Ghana received critical medical supplies such as oxygen concentrators, patient monitors and arterial blood gas analysers. Some 360 multidisciplinary health staff were also trained to effectively manage COVID-19 patients in isolation, treatment facilities and at home. The World Bank funding helped to a great extent to ensure that a greater proportion of patients infected with COVID-19 and who required intensive care and ventilator support had good chances of survival.

Despite the aforementioned efforts, a disturbing issue that emerged from COVID situation was that the few healthcare workers at the forefront of the fight had to endure so much. Some experienced mental health issues due to stressful working conditions and the perceived fear of acquiring COVID-19 infection. They were extremely worried that they could potentially risk their families and loved ones for COVID-19 as they continued to be in hospital settings for clinical duty. Some were even shunned by their relatives for continuing to be in that setting and had to appear on TV to dispel any rumour and also assure the community of the safety of their work.

Aside from the state of health care in the country which has acted as a hindrance to effective health delivery, the pandemic has also caused a huge blow to the country's educational system.

8   COVID and Ghana's Educational System

Closures across learning institutions of all types remained one of the courses of action by the government to stem the tide of the pandemic. The long period of disruption resulted in serious consequences for the educational sector. Parents had to put up with their wards at home, and some had then lost their jobs and therefore experienced severe insecurities with their wards around them. The pandemic also had a negative toll on children's mental health as they were increasingly confined at home, spent less time with their friends and classmates and had limited possibilities for socialising with other children.

Beyond students, school closures also affected teachers and parents in diverse ways, although things are easing up now with the reopening of all educational institutions. In many instances, educators, often in collaboration with parents and guardians, had to maintain a distance learning environment with varying degrees of success. Many teachers faced frustrations relating to logistical difficulties, lack of support and training as they resorted to remote teaching. At the same time, the burden placed on parents to facilitate the learning of their primary or secondary school children stretched beyond available resources. For these parents and guardians, attention allocated to ensure children remain engaged amidst learning disruption, came at the cost of time spent at work, with a disproportionate impact on women, thus causing further pressure and stress in these households where a loss of income generates further pressure (UNESCO, 2020).

9   The Socio-economic Impact of COVID

The life-threatening health situation and the widespread disruptions in economic and social life arising from COVID-19 unleashed differential impact on communities in the country. Some sectors and some social groups were hard hit than by others. Women, youth and informal sector workers were the hardest hit. Prior to the pandemic, economic and social conditions in Ghana were challenging enough. The pandemic, meanwhile, added a fillip to the already precarious conditions some of these people were faced with. The swift and aggressive containment measures came at enormous economic cost to many people. Ghana's rapid growth (7% in 2017–2019) was halted by the pandemic (World Bank, 2022a). According to the World Bank, Ghana's economy continues to suffer the impacts of the pandemic as growth is yet to get back to pre-pandemic levels, and this could be compounded by the war in Ukraine. These developments have since given rise to high prices for several key commodities, including food and fuel, adding to prior inflationary pressures in Ghana.

The World Bank Group's (WBG) Board of Executive Directors has, meanwhile, discussed a new 5-year Country Partnership Framework (CPF) for Ghana for 2022–2026. The CPF is expected to support Ghana in its COVID-19 and medium-term development agenda. It is designed around three mutually reinforcing focus areas, namely enhancing conditions for private sector development and quality job creation; improving inclusive service delivery; and promoting resilient and sustainable development (World Bank, 2022b).

Whilst President Akufo-Addo has been widely commended by the World Health Organisation (WHO) for using effective COVID-19 measures, these same measures seem to strike a disproportionate cost on the working class and the poor who can least afford the burden. For instance, jobs in the informal sector, like petty trading, account for a large number of people; hence, it was improbable to permit such people to work from home. The shutdown stifled their livelihoods because of the nature of their work. They had virtually no savings or substantial wealth to fall back on to sustain them throughout the public health measures. Some firms had to close, sending workers home with no idea when or if they might return to work. Though things have since normalised in recent times, the considerable hardships that were experienced are still ongoing.

10   Economic Response Package and Support

Ghana was one of the first in the West African sub-region to implement a COVID-19 economic response package. The monetary measures included the Central Bank reducing monetary policy rate to 14.5% (lowest in 8 years) and the primary reserve requirement to provide liquidity to accelerate domestic investment. The fiscal stimulus consisted of tax forbearance and 3 months' absorption of the water bill for all households, full electricity bill rebate for over one million active 'lifeline customers'—poorest households—and 50% for non-poor households and businesses, based on March electricity consumption. There was also distribution of food supplies to the poorest communities. These measures were in place until October 2020.

To further lessen the economic impact of the pandemic on Ghanaians, Parliament approved a COVID-19 Alleviation Programme to support businesses and households. Ghana's efforts to combat the virus even caught the attention of the international community and donor agencies, prompting the IMF to approve a package as direct budget support to the government (IMF, 2020). Whilst governments around the world put up considerable funds to protect their citizens from income shortfalls and job losses, support in most African countries, including Ghana, was so limited (Lakemann et al., 2020).

Donations have, however, come from individuals, churches, the private sector and various entities towards the COVID-19 Fund set up by the government to mobilise funds to be used in complementing the government's efforts in addressing the pandemic issues. Relief projects have also been established by some organisations to provide support to vulnerable families by ensuring that basic necessities such as food and water are provided. The quest to help the government in the fight against the pandemic has been quite massive. Other donations have also found their way into hospitals, research centres, prisoners, the destitute, etc. Throughout the pandemic, the government partnered with the private sector to roll out economic reliefs and recovery programmes.

11   COVID-19 and Responsible Business

Companies are often seen to have the capacity and resources to promptly and systemically mobilise. They are thus seen as especially effective in executing disaster relief and recovery management (Ballesteros et al., 2017; McKnight & Linnenluecke, 2016). Hence, the involvement of the private sector in disaster relief and their engagement in society has become more common and important (Johnson et al., 2011; McKnight & Linnenluecke, 2016). In one way or the other, it is a means by which companies can also shore up their corporate social responsibilities (CSR).

Such activities range from responsive and episodic-based measures to proactive activities such as raising funds and partnering with relevant organisations in disaster relief (Johnson et al., 2011). Given the emergent situations that arise during disasters, companies' disaster relief CSR activities have particular importance to society, benefiting many people and local communities in need (Ballesteros et al., 2017; Johnson et al., 2011; McKnight & Linnenluecke, 2016).

Many organisations in Ghana were quick to offer their help in the course of the pandemic. Fidelity Bank, a Ghanaian-owned bank, for instance, donated GHC 1 million to support the building of the COVID-19 Infectious Disease and Isolation Facility, the first of this type of facility to be built in Ghana, that the government came up with. The bank also reduced interest rates on personal loans for qualifying customers to 17.5%, which included a relief option of a maximum 3-month repayment holiday. It also postponed loan repayments and restructured loans for existing borrowers hard hit by COVID-19. Besides, it waived interbank transfer fees on all digital transactions and mobile wallet transaction fees at the early stages of the pandemic. Some other banks pursued similar lines of engagement.

Ghana also received vaccines donated as part of the MTN Group's vaccination programme to African countries. Other donations in the form of handheld thermometers, sanitisers, goggles, face masks, protective gowns and other COVID-19 prevention equipment were made by a number of firms. The government set up a COVID-19 Trust Fund to which several companies in Ghana contributed to. The acts of corporate philanthropy saw firms donating money and other equipment to several public sector institutions too. The corporate donations augmented the government's effort in the fight against the pandemic. Many of the firms endeared themselves to Ghanaians as they showed a high level of responsibility. Some were able to ask their employees to work from home for a period.

During the pandemic, more active and responsible actions from the business sector had been required. The benevolent acts of the firms drew in a lot of public commendation for the way the firms cemented their role in society. The acts positioned them as socially responsible as they helped to address the challenges the country was facing. The companies' systemic donations during the disaster benefited their businesses by increasing stakeholders' awareness of their CSR activities and, accordingly, increased trust towards them and their reputation (see Qiu et al., 2021; Madsen & Rodgers, 2015). Through this means, they were able to leverage the risks surrounding COVID-19 as opportunities for their future sustainable businesses. Such a role implies that they can be seen as rescuers who can help solve the economic and social problems that the COVID-19 pandemic has caused.

The motive of companies' CSR activities has been argued to be mainly self-serving, ultimately pursuing certain economic benefits (Matten & Crane, 2005); however, many of the acts seen in the recent past suggest that the donations were not just for their own interests but rather for common economic and social benefits, with a good and genuine motive. As argued by McKnight and Linnenluecke (2016) and Johnson et al. (2011), respectively, the role of the private sector is important and effective during disasters. The private sector's capabilities of resources and experiences, as well as the mobilisation and transformation of resources, complemented government's efforts.

Although most businesses were challenged by the pandemic, they have continued to reach out to society and showed citizenship behaviours (Mahmud et al., 2021). The need for public support and the public's desire to see the involvement of such firms has made the companies to feel obligated to implement CSR activities so that they can be seen as good corporate citizens. Whilst society expects such responsible roles during difficult times, the performance of these roles is also a useful foundation from which to obtain positive business outcomes in the future (see Johnson et al., 2011).

12 Lessons Learnt

The COVID-19 pandemic continues to claim lives, wreak havoc to economies and disrupt livelihoods around the world. Both the private and public sectors in Ghana have been severely affected. Across the private sector, informal businesses, micro, small and medium enterprises (MSMEs), as well as large businesses are still under severe strain. For instance, the hospitality sector has suffered tremendously. The occupancy rates have been quite low, with many hotels closed and others still on the road to recovery.

The economic impact of COVID-19 has exposed major vulnerabilities within both the social setting and business operations in the country. Whilst some companies shielded their workforce from such impacts, many others had to lay off workers or reduce their working hours. There were enormous challenges for both small and large businesses. The government took the extraordinary step to support business with aid packages to struggling companies and workers (Adams, 2020).

The pandemic has exposed and exacerbated some ingrained social issues too, such as poverty and inequality. The poor have been at the receiving end as many remain jobless in view of the ramifications of the pandemic. Meanwhile, others are still struggling to access health care because they do not have the means to do so. The government has since stated that it is determined that health care, especially, will be within easy and affordable reach of every Ghanaian in foreseeable future. It has reiterated its desire to ensure that each district in the country has a state-of-the-art district hospital within the shortest possible time. The importance of pursuing this goal has been reinforced by the lessons from COVID-19.

In the informal sector, which supports the livelihoods of many citizens, many small businesses are reporting the loss of revenues across board. Meanwhile, the government's public finances are also under severe strain as Ghanaians call out for more support in view of the hardships they are currently experiencing. Incidentally, government revenues are down sharply, whilst expenditures have risen significantly resulting in a much larger fiscal deficit than originally planned owing to the ramifications of the pandemic. Although government's interventions over the past few years have brought some relief, a lot more needs to be done to help the citizenry.

Although the fight against the COVID-19 pandemic is still ongoing, there is considerable evidence of fruitful private–public partnerships to alleviate the hardships in the country. If Ghana continues to demonstrate this level of shared support, philanthropy and cooperation, it could reap the benefit of a smooth path to recovery from the pandemic. Therefore, one can envision the post-pandemic period as one that can ensure responsible business through the efforts of strong CSR commitment and strategies. There are significant opportunities for companies to help in addressing some of these social issues as their help would be needed in tackling them during this pandemic as well as in the long run.

Ghanaians have, indeed, witnessed the generosity of people, organisations and the government working hand in hand with each other. Community engagements and solidarity exemplified by the relationality of mankind have been commonplace. People checked on each other and also helped in feeding the less privileged ones in society during the pandemic. The amazing initiatives that took place from the initial stages of the pandemic, from the very small acts of kindness that people showed towards each other to big donations from private organisations, churches and hotels supplying empty rooms for quarantined people during this tragedy, are all testimonies of what humanity has brought into people (Sarpong, 2021). Many people ramped up their responses to the pandemic though philanthropic efforts to fulfil a role that should mostly be played by government. The moments of crisis highlighted how important people's priorities are in serving humanity.

13   The Way Forward

The pandemic has had huge impacts on incomes and food access in many parts of the country. It has also magnified existing disparities and inequalities faced by vulnerable groups. One notable thing about this too is that Ghana has been able to demonstrate a strong ability to manage such an emergency thoughtfully. In spite of the pressure from different groups, the government was also able to implement the right decisions for the betterment of the country. This is in spite of the fact that the measures were quite difficult for some groups. The social and economic disruption has laid bare unavoidable truths that can no longer be ignored. First, a significant number of the population live from 'hand to mouth' and in very precarious conditions. This situation obviously needs to be seriously addressed. The need for clean water is also essential for many people especially those in rural areas who still suffer from lack of potable water.

Besides, there is an urgent need to upgrade, build and fund Ghana's health sector to reduce the inequalities within the health delivery system. That said, the government's preparedness to work towards building more hospitals in the country can help in this direction. In addition to emergency relief, strengthening social protection systems and prioritising spending on human capital development would be needed.

Quite significantly, this pandemic is happening against the backdrop of a convergence of global economic and other challenges which, together, are compelling a rethink of policy responses at all levels. Developing countries like Ghana with inadequate health, social and economic systems, therefore, have a dire reality to deal with such issues especially so with many people currently being jobless and facing hard financial times.

14 Conclusion

The primary objective of this chapter was to provide a picture of Ghana's engagement with the COVID-19 pandemic. It was also to explore measures taken by Ghana to stem the spread of the pandemic. The chapter examined the key challenges Ghana faced in the light of the pandemic and also brought to the fore the role played by both government and businesses since the inception of the pandemic. It noted that Ghana has made some considerable gains in the fight against the pandemic in spite of its current challenges.

The chapter attributes this relative success to the prompt response by government, its partnership engagement as well as society's desire to do away with the pandemic. In all these, there has been a marked resilience of the Ghanaian people in adversity. Indeed, COVID-19 has changed the world. It has cost lives, battered health systems, and damaged livelihoods, but, through these challenges, we have seen the best of humanity exemplified through a strong support by people from all walks of life, the business community and government as a whole.

References

Adams, C. N. (2020, May 19). Ghana: President launches Gh¢600 million COVID-19 alleviation programme today. *allAfrica.com*. Retrieved April 20, 2022, from https://allafrica.com/stories/202005190844.html

Akinwotu, E., & Asiedu, K. (2020, May 3). Easing of lockdown a relief to Ghana's poor—Despite fears it is premature. *Guardian*. Retrieved April 3, 2022, from https://www.theguardian.com/global-development/2020/may/03/coronavirus-easing-of-lockdown-a-relief-to-ghanas-poor-despite-fears-it-is-premature

Andrews, L., Higgins, A., Andrews, M. W., & Lalor, J. G. (2012). Classic grounded theory to analyse secondary data: Reality and reflections. *The Grounded Theory Review, 11*(1), 12–26.

Asante, L. A., & Mills, R. O. (2020). Exploring the socio-economic impact of COVID-19 pandemic in marketplaces in urban Ghana. *Africa Spectrum, 55*(2), 170–181.

Ballesteros, L., Useem, M., & Wry, T. (2017). Masters of disasters? An empirical analysis of how societies benefit from corporate disaster aid. *Academy of Management Journal, 60*(5), 1682–1708.

BBC. (2020, April 17). *Coronavirus: Africa could be next epicentre, WHO warns*. Retrieved May 3, 2022, from https://www.bbc.com/news/world-africa-52323375

Bowen, G. A. (2009). Document analysis as a qualitative research method. *Qualitative Research Journal, 9*, 27–40.

CGTN. (2021, December 26). *Ghana president urges citizens to observe COVID-19 protocols during festive period*. Retrieved January 4, 2022, from https://newsaf.cgtn.com/news/2021-12-26/Ghana-president-urges-citizens-to-observe-COVID-19-protocols-16hZXCVUYWQ/index.html

Engzell, P., Frey, A., & Verhagen, M. (2020). *Learning inequality during the COVID-19 pandemic*. Retrieved January 4, 2022, from https://osf.io/preprints/socarxiv/ve4z7/

Human Rights Watch. (2020). *Africa: Covid-19 exposes healthcare shortfalls*. Retrieved May 5, 2022, from https://www.hrw.org/news/2020/06/08/africa-covid-19-exposes-healthcare-shortfalls

IFC. (2020). *Impacts of COVID-19 on the private sector in fragile and conflict-affected situations*. Retrieved May 5, 2022, from https://reliefweb.int/sites/reliefweb.int/files/resources/EMCompass_Note%2093-COVID%20and%20FCS_Nov2020.pdf

ILO. (2020). COVID-19 crisis and the informal economy: Immediate responses and policy challenges. *Briefing note*. Retrieved March 4, 2022, from https://www.ilo.org/global/topics/employment-promotion/informal-economy/publications/WCMS_743623/lang%2D%2Den/index.htm

IMF. (2020, April 13). *IMF executive board approves a US$1 billion disbursement to Ghana to address the COVID19 pandemic*. Retrieved December 21, 2021, from https://www.imf.org/en/News/Articles/2020/04/13/pr20153-ghana-imfexecutiveboard-approves-a-us-1-billion-disbursement-to-ghana-to-address-covid-19

Johnson, B. R., Connolly, E., & Carter, T. S. (2011). Corporate social responsibility: The role of Fortune 100 companies in domestic and international natural disasters. *Corporate Social Responsibility and Environmental Management, 18*(6), 352–369.

Lakemann, T., Lay, J., & Tafese, T. (2020). Africa after the covid-19 lockdowns: Economic impacts and prospects. *GIGA Focus Afrika, 6*. GIGA. https://nbn-resolving.org/urn:nbn:de:0168-ssoar-70106-5

Liu, N., Chee, M. L., Niu, C., Pek, P. P., Siddiqui, F. J., Ansah, J. P., et al. (2020). Coronavirus disease 2019: An epidemic map of medical literature. *BMC Medical Research Methodology, 20*, 177. https://doi.org/10.1186/s12874-020-01059-y

Madsen, P. M., & Rodgers, Z. J. (2015). Looking good by doing good: The antecedents and consequences of stakeholder attention to corporate disaster relief. *Strategic Management Journal, 36*(5), 776–794.

Mahmud, A., Ding, D., & Hasan, M. M. (2021). Corporate social responsibility: Business responses to coronavirus (COVID-19) pandemic. *SAGE Open, 11*(1), 1–17.

Matten, D., & Crane, A. (2005). Corporate citizenship: Toward an extended theoretical conceptualization. *Academy of Management Review, 30*(1), 166–179.

McKnight, B., & Linnenluecke, M. K. (2016). How firm responses to natural disasters strengthen community resilience: A stakeholder-based perspective. *Organization and Environment, 29*(3), 290–307.

Padayachee, N., & du Troit, L. C. (2020, April 13). *Debunking 9 popular myths doing the rounds in Africa about the coronavirus. Conversation*. Retrieved March 9, 2022, from https://theconversation.com/debunking-9-popular-myths-doing-the-rounds-in-africa-about-the-coronavirus-135580

Palayew, A., Norgaard, O., Safreed-Harmon, K., Andersen, T. H., Rasmussen, L. N., & Lazarus, J. V. (2020). Pandemic publishing poses a new COVID-19 challenge. *Nature Human Behaviour, 4*, 666–669.

Qiu, S. C., Jiang, J., Liu, X., Chen, M. H., & Yuan, X. (2021). Can corporate social responsibility protect firm value during the COVID-19 pandemic? *International Journal of Hospitality Management, 93*, 102759.

Reddy, T., Shkedy, Z., van Rensburg, C. J., Mwambi, H., Debba, P., Zuma, K., & Manda, S. (2021). Short-term real-time prediction of total number of reported COVID-19 cases and deaths in South Africa: A data driven approach. *BMC Medical Research Methodology, 21*, 15. https://doi.org/10.1186/s12874-020-01165-x

Reuters. (2022, March 30). *Ghana to start producing own Covid-19 vaccines in January 2024*. Retrieved April 16, 2022, from https://www.reuters.com/world/africa/ghana-start-producing-own-covid-19-vaccines-january-2024-2022-03-30/

Sarpong, S. (2021). Is Covid-19 setting the stage for UN Agenda 2030? In pursuit of the trajectory. In S. Vertigans & S. O. Idowu (Eds.), *Global challenges to CSR and sustainable development, CSR, sustainability, ethics & governance* (pp. 21–38). https://doi.org/10.1007/978-3-030-62501-6_2

Siaw-Frimpong, M., Touray, S., & Sefa, N. (2021). Capacity of intensive care units in Ghana. *Journal of Critical Care, 61*, 76–81.

Silva, D. R. M., & Mont'Alverne, C. (2020). Identifying impacts of COVID-19 pandemic on vulnerable populations. *Survey Research Methods, 14*(2), 141–145. https://doi.org/10.18148/srm/2020.v14i2.7742

Smith, E. (2008). *Using secondary data in educational and social research.* McGraw-Hill Education.

Tabong, P. T., & Segtub, M. (2021). Misconceptions, misinformation and politics of COVID-19 on social media: A multi-level analysis in Ghana. *Frontiers in Communication, 6*(13795). https://doi.org/10.3389/fcomm.2021.613794

Turcotte-Tremblay, A. M., Gali Gali, I. A., & Ridde, V. (2021). The unintended consequences of COVID-19 mitigation measures matter: Practical guidance for investigating them. *BMC Medical Research Methodology, 21*, 28. https://doi.org/10.1186/s12874-020-01200-x

UNESCO. (2020). *Adverse consequences of school closures.* Retrieved March 12, 2022, from https://en.unesco.org/covid19/educationresponse/consequences#:~:text=Increased%20exposure%20to%20violence%20and,common%2C%20and%20child%20labour%20grows

UNICEF. (2022, March 22). *Debunking COVID-19 myths with Ghana's influencers.* Retrieved January 2, 2022, from https://www.unicef.org/ghana/stories/debunking-covid-19-myths-ghanas-influencers

WHO. (2020, May 7). *New WHO estimates: Up to 190 000 people could die of COVID-19 in Africa if not controlled.* Retrieved December 8, 2021, from https://www.afro.who.int/news/new-who-estimates-190-000-people-could-die-covid-19-africa-if-not-controlled

WHO. (2021, July 20). *WHO continues fight against pandemic amid worsening global public health emergency and uneven vaccine rollout.* Retrieved May 5, 2022, from https://www.who.int/news-room/feature-stories/detail/who-continues-fight-against-pandemic-amid-worsening-global-public-health-emergency-and-uneven-vaccine-rollout

Witteveen, D. (2020). Sociodemographic inequality in exposure to COVID-19-induced economic hardship in the United Kingdom. *Research in Social Stratification and Mobility, 69*, 100551. https://doi.org/10.1016/j.rssm.2020.100551

World Bank. (2021). *Ghana COVID-19 emergency preparedness and response project second additional financing.* Retrieved April 6, 2022, from https://documents1.worldbank.org/curated/en/297381619726541950/pdf/Project-Information-Document-Ghana-COVID-19-Emergency-Preparedness-and-Response-Project-Second-Additional-Financing-P176485.pdf

World Bank. (2022a, April 7). *The World Bank in Ghana—The World Bank Group aims to help Ghana towards creating a dynamic and diversified economy, greener job opportunities, for a more resilient and inclusive society.* Retrieved March 12, 2022, from https://www.worldbank.org/en/country/ghana/overview#1

World Bank. (2022b, February 22). *World Bank Group launches new country partnership framework for Ghana.* Retrieved January 10, 2022, from https://www.worldbank.org/en/news/press-release/2022/02/23/world-bank-group-launches-new-country-partnership-framework-for-ghana

Sam Sarpong is an Associate Professor at the School of Economics and Management, Xiamen University, Malaysia (XMU). He obtained his PhD from Cardiff University and his MBA from the University of South Wales. Prior to joining XMU, he taught at Cardiff University, University of London (Birkbeck College), Swansea Metropolitan University, University of Mines and Technology and Narxoz University, respectively. He is also a Visiting Professor at the NIB Doctoral School in Ghana. Sam is currently an Associate Editor of *International Journal of Corporate Social Responsibility* (Springer) and also the Editor of *New Economic Papers—AFR* (NEP-AFR). He serves on the boards of international journals and also acts as a reviewer for some renowned journals.

His research interests lie in the relationship between society, economy, institutions and markets. He tends to explore the nature and ethical implications of social-economic problems and has published in these fields.

Corporate Social Responsibility and the Impact of COVID-19 on Healthcare Institutions in Nigeria

Gloria O. Okafor, Amaka E. Agbata, Innocent C. Nnubia, and Sunday C. Okaro

1 Introduction

Corporate social responsibility (CSR) is the means through which businesses exercise their social obligations to their host communities and society. It is defined as a medium of managing businesses by corporate organisations using economic, environmental and social activities, which will produce a general positive impact on the society (Pohle & Hittner, 2008). The World Business Council for Sustainable Development (2000) defined CSR as the continuing commitment by business to behave ethically and contribute to economic social development while improving the quality of life of the workforce and their families as well as the local community and society at large. It is a show of some worthy behaviours by businesses to the society and government (Adeyanju, 2012) and creation of simultaneous values for businesses and society (Kurucz et al., 2008). Babalola (2012) describes CSR as a notion of 3 Ps, involving people, planet and profit, which includes a wide range of value and means through which success, economic, social and environmental impacts of businesses are measured.

In today's corporate world, CSR has become a debated issue, attracting the attention of owners of business and those who set standards (Nwobu, 2021). Over the last few years, more attention has been placed on social involvement, social responsibility and the ethical behaviour of managers and executives, as a result of public enquiries of different organisations business activities. Socially responsible behaviour is recognised as an attitude that results from internal and external pressure on companies, which is linked to corporate image and reputation (Fernandez-Feijoo et al., 2014). Trustworthy profit-making organisations always support their host

G. O. Okafor (✉) · A. E. Agbata · I. C. Nnubia · S. C. Okaro
Department of Accountancy, Nnamdi Azikiwe University, Awka, Anambra, Nigeria
e-mail: go.okafor@unizik.edu.ng

© The Author(s), under exclusive license to Springer Nature Switzerland AG 2023
S. O. Idowu et al. (eds.), *Corporate Social Responsibility in the Health Sector*, CSR, Sustainability, Ethics & Governance, https://doi.org/10.1007/978-3-031-23261-9_12

communities through the provision of facilities that could enhance the livelihood of the people in their environment and the society at large. The support that reputable companies in Nigeria can give to their host communities is numerous, like educational scholarships, support to health care and healthcare institutions, provision of boreholes for areas without water, good road facilities, security supports and so many other provisions of essential needs to the communities.

The concept of social responsibility emerges with reference to privately owned, profit-making organisations, so it might be assumed that social responsibility is already incorporated into the governance of public healthcare provision (Brandao et al., 2013). The practice of CSR is common among profit-oriented non-health organisations and has not been practised generally by non-profit healthcare organisations. With prominent increase in the service sector globally, there has been a growth in research related to the service sector. Healthcare organisations around the world are under stakeholders' pressure to provide high-quality, cost-effective, accessible and sustainable services (Rodriguez et al., 2020; Ramirez et al., 2013). Like most businesses, the healthcare industry is experiencing an increased focus on connecting business activities to public impact (Collins, 2010). According to Takahashi et al. (2013), CSR is relevant; a possible valuable area of strategic development for healthcare institutions like hospitals has the potential to strengthen the profile of the hospital in the community and to lead a better financial and clinical performance and can be adopted by both profit-oriented and non-profit-oriented hospitals. CSR could enhance the loyalty, reputation and value of healthcare organisations and should be implemented with regard to the core business of the healthcare organisations (Lubis, 2018). It is obvious that CSR is very important to the healthcare institutions, and health in most cases is the main concern for the employer to promote CSR, as good health reduces absenteeism from work; successful CSR is knitted into the fabrics of companies and becomes an essential component of its strategic plan and business model, often with a committed workforce (African Strategy for Health, 2014).

The recent outbreak of COVID-19 has profoundly changed our lives and has affected the foundations of societal well-being. The crisis of COVID-19 pandemic drew attention to the already overburdened public health systems in many countries and to the challenges faced in recruiting, deploying, retaining and protecting sufficient well-trained, supported and motivated healthcare workers. The short-term effects of COVID-19 were particularly severe for most disadvantaged and risk compounding existing socio-economic divides. The COVID-19 did not only bring about economic recession but also caused all countries to take extreme decisions to contain the disease (Vatamanescu et al., 2021). The COVID-19 must have affected the extent of CSR in healthcare sector to and from the healthcare sector. Even though there is no full agreement on the basic way of presenting CSR, discussions on CSR have resulted in a wide-reaching acknowledgement of sustainability reports as the basic tool for its communication and the recognition of the need to develop standards to increase its quality (Fernandez-Feijoo et al., 2014). This study examines the CSR activities and the impact of COVID-19 on the CSR activities in the healthcare sector in Nigeria. To achieve the objective, the CSR from profit-oriented companies in

Nigeria to healthcare institutions before and within the COVID-19 outbreak was reviewed; also, the reports on the website pages of top best hospitals in Nigeria were equally reviewed to see the extent to which the healthcare institutions give account of their socially responsible activities.

2 Review of Related Literature

2.1 Related Theories to Corporate Social Responsibility

The behaviour of corporate organisation towards the support of the healthcare system in Nigeria is in line with the stakeholders' theory, the legitimate theory and the utilitarian theory. The stakeholders' theory propounded by Richard Edward Freeman (1984) advocates that businesses should take into consideration the overall interest of all the corporate stakeholders and not just the shareholders' interests alone. It assumes that the association between a business and its stakeholders is interconnected. Corporate organisations being socially responsible help to create value and reputation in the long run. Majority of the Nigerian companies did not consider only the interest of their shareholders alone but also considered the interest of the entire nation when the outbreak of COVID-19 affected Nigeria. A good number of profit-oriented companies especially those whose economic activities were not adversely affected by the COVID-19 responded to the call to come to the aid of the nation in controlling the COVID-19 through financial and material donations, especially to hospitals. Their social responsiveness helped to reduce the negative impact of COVID-19 on Nigerian healthcare institutions.

The legitimacy theory provides that companies attain legitimate status which is rooted in legitimacy theory when they align their value system to be interconnected with that of the operating environment (Noah, 2017). The theory insists that companies should plan their operations to satisfy societal custom, expectations, demand and worth in order to achieve legitimate status enshrined in CSR (Olateju et al., 2021). According to Guthrie et al. (2006), achieving legitimate status involves obtaining societal support and approval as well as eliminating threat to surviving. Burlea-Schiopoiu and Popa (2016) noted that companies usually relate their corporate image to being visible in their business culture and think that for them to achieve their legitimacy, their business culture has to be enhanced and promoted in the outside environment. COVID-19 affected most parts of the world; laid emphasis on the importance of legitimacy and; and mounted pressure on companies to re-examine their value systems. During the COVID-19 era, many Nigerian companies donated towards the control of the disease, especially to hospitals in equipment, drugs and other facilities to contain the spread of COVID-19, which conferred legitimate status to them.

The utilitarian theory credited to Jeremy Bentham who propounded it in 1781 and John Stuart Mill who popularised it 1861 is a moral theory which propagates acts that will promote happiness. The proponents posit that an act is said to be right if it leads to happiness of the society at large (Driver, 2014). In other words, an act is said to be good if it brings the highest value of good to the greatest number of people. Consequently, corporate organisations have the social responsibility of contributing to happiness of the society through their social responsibility activities. The provisions by some companies to alleviate the effect of COVID-19 pandemic on the masses by donating financially and materially relate to the utilitarian theory. The provisions brought about reductions in both mortality rate and infection rate, thereby making people happy. Being socially responsible promotes the overall happiness of the society especially for the vulnerable ones which is the hallmark of utilitarianism.

2.2 CSR and Healthcare Institutions

Social responsibility is concerned with the means by which an organisation manages its internal operations, as well as the effect of its activities on the social environment. In a healthcare system, social responsibility promotes an ethical obligation that requires hospitals and other organisations to do something valuable in issues such as delivering quality health care to everyone who is entitled to it (Brandao et al., 2013). In the words of Collins (2010), the healthcare institutions are growing from the ones that care for persons only to organisations that function as effective and efficient businesses. In Nigeria, some healthcare organisations like hospitals are set out not only for caring people but also to make profits, while some especially the government-owned health institutions engage in services of caring for human wealth without any commercial or monetary profit in view. These non-profit-oriented healthcare institutions attract donations and supports from other profit-oriented organisations as part of those organisations' corporate social responsibility. On the other hand, healthcare institutions are expected to be socially responsible. The healthcare institutions have both legal and moral obligations to work in the interest of their stakeholders. Companies report their CSR behaviours to be acceptable and transparent in terms of corporate governance as well as environmental, societal and financial sustainability; in line with this, the healthcare organisations should follow a set of values such as equity in access to health care, broad coverage, and effective and efficient services (Siniora, 2017). It has become increasingly apparent that the actions of managers and workers within the healthcare institutions affect more people than just the owners of the organisation (Collins, 2010).

According to Brandao et al. (2013), there are ethical principles that hospitals and healthcare institutions should observe. They include the following:

(i) Supporting flexible labour policies with regards to the breastfeeding period by the women;
(ii) Provisioning the services to assist the handicapped and to satisfy religious need;
(iii) Supporting programmes of social well-being and solidarity;
(iv) Ensuring that all the marketing approaches are ethically based;

(v) Ensuring that all toxic and poisonous wastes are treated carefully with caution;
(vi) Protecting the environment and animal rights; and
(vii) Supporting policies that enrol in the decision-making process of specific groups of patients and non-governmental organisations (NGOs).

Some key areas of CSR according to the reporting guideline of CSR as reported by Moir (2001) are workplace, marketplace, environment, community, ethics and human rights. The workplace describes how the employees are being treated in the workplace, the labour policies and protections available to employees. The marketplace CSR presents reports about consumers and suppliers; in the case of health care, it will involve issues about the management of patients. Environmental aspect describes clearly how toxic and poisonous materials and wastes are handled and disposed to ensure safety of man and animals. The community reports all the activities by an organisation to support its host community development; in the case of health care, it involves how healthcare institutes carry out medical outreaches in their host and other communities, creating awareness of diseases, their controls and ways of preventing them and other supports that are made available to the advantage of the society. Ethics and human rights activities denote having a fair operating environment, ensuring that ethical standards and behaviour prevail in their organisation. Examples of some scopes or opportunities for performing CSR in healthcare organisations according to Adhikari (2017) are health promotion activities; health protection through manufacturing of different health safety goods; complying with occupational safety and health; allocation and distribution of resources for improving the social determinants of health, i.e. the places and conditions where we are born, live, grow, work and age; research and innovation; and support in policy implementation and financial or work effort donations to health-related charities.

2.3 CSR in Health Institutions and COVID-19 Era in Nigeria

Nigerian healthcare institutions like other countries' healthcare institutions were faced with severe challenges due to unexpected outbreak of COVID-19, and this had both positive and negative impacts on the healthcare institutions (Babatunde et al. 2021). In an interview with Conversation Africa, Odubanjo (2020) stated that COVID-19 has positive and negative impacts. The positive impact is that it made people to be aware of the significance of health. In Nigeria, many healthcare institutions were revitalised and re-equipped with personnel and better medical equipment. The negative impact on the Nigerian economy came in form of level of migration to developed countries, some of the developed countries of the world issued special visa and visa exemption to allure healthcare practitioners, and many of the Nigerian health workers grabbed the opportunity and migrated to these developed countries, thus reducing drastically the number of health workers in the Nigerian healthcare system. Meanwhile, Nigerian government does not provide a conducive environment for the health workers, and the health workers still battle with poor remuneration, poor or non-availability of medical/health equipment/facilities, insecurity and so on; these contributed to the massive flight of health workers to other countries for better opportunities. According to Odubanjo (2020), inadequate funds to fight COVID-19 also affected the Nigerian healthcare system as COVID-19 is not only a health pandemic but also a socio-economic pandemic. Consequently, the healthcare system is in dire need of adequate remuneration for the health/medical personnel, availability of well-equipped hospitals in both the urban and rural areas, constant electricity, water supplies and overall improved facilities. The outburst of COVID-19 exposed the inadequacy of medical infrastructure and the pitiable state of the Nigerian healthcare institutions, especially poor medical facilities; it came as a disguised blessing as it contributed to the development of infrastructure in the healthcare sector (Babatunde et al., 2021; Nnubia et al., 2020).

The outbreak of COVID-19 made some corporate entities to be responsive towards their corporate social responsibilty by assisting to contain the effect/spread of the disease. The COVID-19 seems to have led to a sharp increase in governments' and market participants' attention to CSR considerations as social and environmental issues are at the core of the recovery strategy in many countries (Bae et al., 2021). The crisis of the COVID-19 drew attention to the already overburdened public healthcare systems in many countries, especially in Nigeria, where the medical facilities in many health institutions do not function effectively. In the words of Anyanwu et al. (2020), Nigerian healthcare system is dilapidated and weak due to years of neglect and widespread corruption. Most hospitals were not prepared to respond to the COVID-19 outbreak (Ogoina et al., 2021). The healthcare institutions were faced with challenges of protecting and supporting the healthcare professionals, as well as physical facilities. There was increased need for non-healthcare organisations to support the health care system, huge demand for COVID-19-induced CSR to help the healthcare system contain the disease. In response to this outbreak, different organisations in Nigeria reaffirmed their commitments to stakeholders' interests by channelling their CSR towards upgrading and supporting the healthcare institutions.

Nigeria planned that $330 million will be needed in controlling COVID-19 pandemic, but through CSR, corporate organisations provided more than 90% of over $560.52 million actually generated (Aregbeshola & Folayan, 2021). The Central Bank of Nigeria (CBN) released hundreds of billions of Naira to the health sector for the acquisition of health/medical facilities and refurbishing of available ones (Babatunde et al., 2021). Palazzi and Starcher (2006) noted that businesses have social responsibility in addition to fiscal task. In line with this, the CBN through the circular BKS/CSO/DIR/CON/01/089 of 8 April 2020 made a call/request for corporate organisations and individuals to come to the aid of the nation in the fight against COVID-19. They responded to this call and formed a group called "The Nigerian Private Sector Coalition Against COVID-19 (CA COVID)". Through their donations, Nigeria was able to realise more than ₦25 billion. The corporate organisations/individuals that donated and the amount donated by them are tabulated in Table 1.

Table 1 List of companies that donated to Nigeria for COVID-19 pandemic

S/n	Name of organisation/individual	Amount ₦
1	Central Bank of Nigeria (CBN)	2,000,000,000.00
2	Dangote	2,000,000,000.00
3	Flood Relief Fund	1,500,000,000.00
4	BUA Sugar Refinery	1,000,000,000.00
5	Guaranty Trust Bank PLC	1,000,000,000.00
6	United Bank for Africa PLC	1,000,000,000.00
7	First Bank of Nigeria PLC	1,000,000,000.00
8	Zenith Bank PLC	1,000,000,000.00
9	Access Bank PLC	1,000,000,000.00
10	Amperion Power Distribution	1,000,000,000.00
11	African Steel Mills Nigeria Ltd.	1,000,000,000.00
12	Famfa Oil Ltd.	1,000,000,000.00
13	Mike Adenuga Foundation	1,000,000,000.00
14	NDIC	1,000,000,000.00
15	Flour Mills of Nigeria PLC	1,000,000,000.00
16	MTN Nigeria PLC	1,000,000,000.00
17	Pacific Holding Ltd.	500,000,000.00
18	WACOT Rice Ltd.	500,000,000.00
19	Tolaram Africa Enterprise Ltd.	500,000,000.00
20	Bank of Industry	500,000,000.00
21	FrieslandCampina WAMCO	500,000,000.00
22	African Finance Corporation	250,000,000.00
23	APM Terminals Apapa Ltd.	150,000,000.00
24	FSOM Merchant Bank	100,000,000.00
25	FBN Merchant Bank	100,000,000.00
26	Rand Merchant Bank	100,000,000.00
27	Coronation Merchant Bank	100,000,000.00
28	SunTrust Bank	100,000,000.00
29	Providus Bank	100,000,000.00
30	Wema Bank	100,000,000.00
31	Unity Bank	100,000,000.00
32	Heritage Bank	100,000,000.00
33	NOVA Merchant Bank	100,000,000.00
34	Polaris Bank	100,000,000.00
35	Keystone Bank	100,000,000.00
36	KC Gaming Network Ltd.	100,000,000.00
37	Ports & Terminal Multiservice Ltd.	100,000,000.00
38	Ports and Cargo Handling Service	75,000,000.00
39	Five Star Logistics Ltd.	75,000,000.00
40	ENL Consortium	70,000,000.00
41	Josepdam Ports Services Ltd.	60,000,000.00
42	System Specs Ltd.	50,000,000.00

(continued)

Table 1 (continued)

S/n	Name of organisation/individual	Amount ₦
43	Globus Bank	50,000,000.00
44	Titan Trust Bank	50,000,000.00
45	Tak agro & Chemicals Ltd.	50,000,000.00
46	ADAMA Beverages Ltd.	50,000,000.00
47	WA Container Terminal	50,000,000.00
48	Deeper Christian Life Ministry	50,000,000.00
49	Kam Wire Limited	30,000,000.00
50	De damak nig Ltd automobile	25,000,000.00
51	CWAY	20,000,000.00
52	Ahmadu Mahmoud	20,000,000.00
53	Adron Homes & Properties Ltd.	20,000,000.00
54	Other organisations and individuals	2,348,699,791.50
	Total	25,893,699,791.50

Sources: Premium Times (2020); Channels TV (2020) (downloaded on 27th September 2021)

The healthcare sector received much attention from government, international and local organisations due to the COVID-19 pandemic. Nigerian government through the CBN at various times released hundreds of billions of naira as credit intervention to the domestic hospitals, pharmaceutical companies and other health-related stores to increase their scale for health amenities, services, social intervention and other expenses because of COVID-19. International communities provided financial and materials support to Nigeria to help in fighting COVID-19. For example, USA and the UK provided Nigeria with free vaccines against the disease and other supports. The COVID-19 gives Nigeria opportunity of revitalising the healthcare system to avoid having catastrophic experience when it re-strikes another wave (Obi-Ani et al., 2021). Etteh et al. (2020) opined that Nigeria should take extremely important measures in controlling the outburst of contagious deadly diseases now, so that they will be able to stand against unknown future. The outbreak of COVID-19 has placed a clarion call on all African nations to reappraise and strengthen their health policies in other to stand its ground in fighting future pandemic as well as in securing the health of the entire population (Igyuse et al., 2020).

Health is life, and improved health is an essential part of development and sustainability of the economy. Throughout the world, human societies care so much about their health and health maintenance. In developed nations, huge resources are set out to support the healthcare system, and many organisations, both profit-oriented and non-profit-oriented, give huge supports to the maintenance of health. Developing nations like Nigeria need to increase allocation to healthcare system, and policy makers are to encourage both profit-oriented and non-profit-oriented organisations to direct enormous CSR activities towards the healthcare system. Increased resources towards the improvement and maintenance of the healthcare system will surely improve the state of development in the country.

3 Methodology

This study examined the CSR from business organisations to hospitals, pre-COVID-19 and during COVID-19 pandemic era, and also the extent to which the healthcare institutions are seen to be socially responsible. To examine the CSR received by hospitals, the study purposefully reviewed 20 companies out of the total of 177 listed companies on the Nigerian Stock Exchange. The 20 companies were made up of 16 out of the first 20 top-capitalised listed companies in Nigeria that have consistent online published annual reports and four pharmaceutical companies with consistent online annual reports from 2017 to 2020. These companies were from diverse sectors in Nigeria and were reviewed from 2017 to 2020. 2017–2019 reports were reports before the COVID-19 era, while 2020 reports were reports during the COVID-19 pandemic era. The study stopped at 2020 reports because the 2021 reports are yet to be published.

To examine the extent to which the hospitals are socially responsible, the websites of various hospitals were visited. The websites of 46 healthcare institutions were reviewed, 19 government hospitals and 27 private hospitals, to assess the extent of their CSR practices based on the five key areas of CSR as reported by Moir (2001)—workplace, marketplace, environment, community, ethics and human rights. The total existing hospitals (private and government) could not be ascertained with certainty, and the study examined the websites of hospitals listed as the best or top best or notable hospitals in Nigeria by Daily Tipsfinders.com (2021), Nigerian Finders (2021), Lyfboat (2021), Inject sayana press.org (2020) and NigerianInfobusstop.com (2021) all retrieved on 6 November 2021.

3.1 Findings and Discussion

The companies reported differently on their CSR activities. Some companies included the amount spent in their reports, while others gave reports without including the amount spent. Those who gave textual reports (T) explained all the CSR activities carried out to support education, youth and economic empowerment, health, safety and so many others in the community (see Appendix 1). In Appendix 1, we used 'T' to represent years with only detailed textual reports on corporate social responsibility. Table 2 shows the percentage of corporate social responsibilities of different companies attributed to the healthcare institutions.

Before the COVID-19 era (2017–2019), the contributions to healthcare institutions by majority of the sampled companies were below 20% of the entire expenditure on CSR, and only two companies have average contributions of up to 23.67%

Table 2 Percentage of reported CSR value to health institutions

	Company	2017	2018	2019	2017–2019 average	2020
		Health	Health	Health		Health
		%	%	%		%
1	Int Breweries	9.72	2.97	32.71	15.13	30.88
2	Guinness Nigeria	84.92	84.92			
3	Nigeria Breweries PLC	6.62	12.52	21.50	13.55	97.71
4	Nestle	0	0	0	0	92.18
5	BUA Cement PLC	28.82	34.95	7.24	23.67	16.58
6	MTN		0	25.93		
7	Dangote Cement	0	0.39	0	0.13	0.71
8	GTB	6.54	7.18	23.15	12.29	95.57
9	Stanbic IBTC	7.07	5.86	38.74	17.22	46.29
10	Union Bank	0.67	27.15	6.77	11.53	96.61
11	Zenith Bank	11.49	10.28	0	7.26	45.51
12	Access Bank	0	0	0	0	68.68
13	Seplat	2.24	6.65	3.04	3.98	83.97
14	Lafarge Cement	0	0	0	0	0
15	Okomu Oil	0	0	0	0	0
16	Flour Mills	0	0	0	0	95.55
17	Fidson Healthcare	96.10	56.43	64.35	72.29	33.60
18	M&B	0	0	27.11	9.04	75.68
19	Pharma Deko PLC	0	0	39.10	13.03	0
20	GlaxoSmithKline					

Source: Annual reports of sampled companies 2017–2020

and 72.29%. This does not mean that other companies do not support the communities where they exist, but shows that the contributions towards the healthcare institutions are not the priorities of the companies. One would have guessed that more supports will go to hospitals and other healthcare issues than to any other issues, as they help to maintain and improve health. The case changed with the era of COVID-19; Table 2 shows that companies gave more to healthcare institutions during the era than in the previous times. Majority of the companies contributed more than 45% of their total CSR to hospitals and other healthcare-related issues. The COVID-19 pandemic exposed the need to support the hospitals with many equipment and daily hospital supplies, and many companies supported greatly. Some of the CSR reports by the companies whose amounts could not be ascertained (denoted by T) during the COVID-19 era are presented below.

3.2   Extract of CSR Report of Guinness Nigeria PLC in 2020 and 2019

Guinness Nigeria PLC during the COVID-19 era in 2020 partnered with four relevant agencies and 15 state governments tasked with combating the COVID-19 outbreak in Nigeria. The company donated 40,000 hand sanitisers to government agencies—Nigerian Centre for Disease Control (NCDC), Federal Ministry of Humanitarian Disaster Management and Social Development, and Lagos and Edo State Ministries of Health. The hand sanitisers were distributed to the frontline medical personnel and to persons in vulnerable communities across the country. They equally donated non-alcoholic drinks to these organisations. In 2019, Guinness donated 10 million Naira (₦10,000,000) to the Guinness Eye Centre at Lagos University Teaching Hospital (LUTH) and Guinness Eye Centre in Onitsha. They partnered with a non-governmental organisation to provide prepacked maternity packs for 300 expectant mothers in Edo state, and they also partnered with Sightsavers on cataract surgeries for women in the North. 230 women from Kebbi and Sokoto states benefited from the surgeries.

3.3   Extract of CSR Report of GlaxoSmithKline Consumer Nigeria PLC in 2020 and 2019

This company made donations of personal protective equipment (PPE) and surgical masks to the Nigerian COVID-19 presidential task force in 2020. The donations worth 5 million naira only (₦5,000,000). In 2019, the same company provided comfortable and more relaxing environment for infants and their mothers during wait time across 17 hospitals in 12 states in Nigeria. They partnered with a Chief Consultant Paediatrician at National Hospital Abuja to share knowledge, information on the causes, symptoms and prevention of pneumonia. They targeted parents of young children under the ages of 5, and the campaign was centred on 15 states with the highest number of unvaccinated children in Nigeria. The initiative reached an estimated total of 49,293,837 people across the 15 states.

3.4   Extract of CSR Report of MTN Nigeria in 2020

MTN Nigeria Communication PLC donated 1 billion naira (₦1 billion) to the Coalition Against COVID-19 (CACOVID) and 250 million naira (₦250 million) worth of PPE to the Nigeria Centre for Disease Control (NCDC) (Table 3).

Table 3 CSR practices by hospitals

Area of CSR	Private hospitals (27)	Government hospitals (19)	Total (46)
Workplace	18 (66.67%)	10 (52.63%)	28 (60.87%)
Marketplace	5 (18.52%)	2 (10.53%)	7 (15.22%)
Environment	0 (0)	1 (5.26%)	1 (2.17%)
Community	8 (29.63%)	7 (36.84%)	15 (32.61%)
Ethics and human rights	16 (59.26%)	3 (15.79%)	19 (41.30%)
Total	34.81	24.21%	30.43%

Source: Website of 46 hospitals as listed in Appendix 2

Websites of 46 hospitals listed in Appendix 2 were reviewed and presented in Table 3, 27 private hospitals and 19 government hospitals. Findings show that 60.87 % (28 out of 46) of the reviewed hospitals reported employee-related issues in the workplace, and the private hospitals that reported are more than the government hospitals that reported. Workplace is the area of CSR that attracted most attention from all the hospitals. Only 15.2% reported their relationships with patients; majority did not report anything about management of toxic wastes, only 2.17% of which is just one hospital out of 46 hospitals. In the area of community relationship, 32.61% reported their activities with the community, and higher proportion of the public/government hospitals reported more than the private hospitals. 41.30% reported the ethical behaviour of their organisation, and more proportion of the private hospitals reported the ethical behaviour. On average, only 30.43% of the entire sampled hospitals reported their socially responsible activities on the websites. Many other hospitals need to arise to this challenge of making their socially responsible activities public.

4 Conclusion

CSR is a growing area of strategic development in the healthcare institutions. It is very important that every healthcare organisation, both profit-oriented and non-profit-oriented present their social responsibility activities publicly. Data collected and analysed showed that before the era of COVID-19, many hospitals were not being supported appreciably in form of CSR by profit-making companies, but during the era of COVID-19 pandemic, there was a significant increase in the CSR received by the healthcare institutions from the profit-making companies. On the aspect of the healthcare institutions being socially responsible, only about 60.87% of the hospitals reviewed were found to report at least an aspect of their socially responsible activities on their websites. Workplace relationship was the most prominently reported socially responsible activities on the websites of the hospitals, followed by ethical behaviours and human rights, and then community-related activities. There is need to create great awareness towards CSR in general especially, in the areas of marketplace and environmental activities.

The study concludes that CSR has not penetrated the healthcare institutions in Nigeria; therefore, efforts should be made for proper integration of CSR in the healthcare institutions. The CSR to healthcare institutions during the COVID-19 era increased significantly, this may be as a result of quest to save lives, and one may not conclude if the trend will continue after the era of COVID-19. The healthcare institutions in Nigeria are actually underfunded and lack necessary infrastructure; there is need for the healthcare sector policy makers to still call on profit-oriented companies and even non-profit organisations to extend regular supports in terms of provisions of funding, medical equipment and other facilities to healthcare institutions, as they did during the COVID-19 era. The healthcare institutions are encouraged to always present updated reports on how socially responsible they are, in both their annual reports and websites. There is need for more healthcare institutions to report on management of toxic materials and wastes and relationship with patients and the community. There is also need for increased reports on employee-related issues at workplace and on the ethical behaviour of healthcare institutions. Health is life and is an essential part of economic development; therefore, our nation needs to increase resources towards the maintenance and sustainability of the healthcare system. Also, there is need for policy makers to encourage organisations, especially the profit-oriented ones to increase their support towards adequate funding of the healthcare system.

In spite of the contribution of this study, it has some limitations that limited the generalisation of this study and offer possibilities for further research. The study used a convenience study of companies and hospitals; therefore, further research may draw on more robust sampling methods to ensure more generalisable outcome.

Appendices

Appendix 1

Reported Value of Corporate Social Responsibility of Companies and Portions to Healthcare Institute

	Company	2017		2018		2019		2020	
		Total	Health	Total	Health	Total	Health	Total	Health
		₦'000	₦'000	₦'000	₦'000	₦'000	₦'000	₦'000	₦'000
1	Int Breweries	54,548	5300	134,700	4000	92,159	30,149	323,101	99,769
2	Guinnes Nigeria	11,775	10,000	11,775	10,000	T	T	T	T
3	Nigeria Breweries PLC	76,885.99	5092.50	57,700.67	7222.50	259,603.78	55,808.01	634,547.45	620,000
4	Nestle	2088	0	33,965	0	42,905	0	797,710	735,300
5	BUA Cement PLC	17,350	5000	18,598	6500	89,769.25	6500	753,645	124,926
6	MTN			1450	0	21,752,000	5,641,000	T	T
7	Dangote Cement	1,020,098.05	0	1,287,733.42	5000	2,185,000	0	2,851,775.76	20,130.32
8	GTB	867,113.53	58,674.23	928,078.32	66,594.11	505,365.41	117,013.74	1,870,906.23	1,787,933.03
9	Stanbic IBTC	436,629	30,859	233,397	13,678	318,285.76	123,296.75	663,421.15	307,098.15
10	Union Bank	225,522.51	1500	30,200	8200	47,272.53	3200	364,050	351,700
11	Zenith Bank	2,611,000	300,000	3,065,000	315,000	2,729,000	0	3,285,000	1,495,000
12	Access Bank	567,021.16	0	376,753.00	0	353,911.85	0	2,599,664.78	1,785,465.07
13	Seplat	105,361	2363	121,683	8087	72,660.07	2209.63	158,169.83	132,809.77
14	Lafarge Cement	T	T	869,000	0	992,657.63	0	1,260,000	0
15	Okomu Oil	82,343	0	257,204	0	183,632	0	255,352	0
16	Flour Mills	16,000	0	20,700	0	10,850	0	1,202,953.15	1,149,413.15
17	Fidson Healthcare	12,817	12,317	15,928.03	8987.83	26,331.64	16,944.13	58,371.40	19,613.52
18	M&B	5943.73		4687.78	0	4761.06	1290.60	1944.99	1471.90
19	Pharma Deko PLC	100.97	0	396.18	0	67.64	26.45	0	0
20	GlaxoSmithKline	T	T	T	T	T	T	T	T

Source: Annual Reports of Sampled Companies 2017–2020

Appendix 2

Names of Hospitals Whose Websites Were Reviewed

	Private hospitals	Public/govt hospitals
1	St. Catherine Hospital Abuja	Federal Medical Centre Abeokuta
2	Sacred Herat Hospital Abeokuta	Federal Medical Centre Umuahia
3	Bob Hospital	Federal Medical Centre Gombe
4	Limi Hospital Abuja	National Hospital Abuja
5	Cedarcrest Hospitals Ltd. Abuja	University of Maiduguri Teaching Hospital
6	Bingham University Teaching Hospital	Aminu Kano Teaching Hospital Kano
7	Chivar Clinics and Urology Centre Abuja	University of Benin Teaching Hospital
8	Primus Super Specialty Hospital Abuja	Usman Danfodio University Teaching Hospital
9	First Consultants Medical Centre Lagos	Imo State University Teaching Hospital
10	Eko Hospitals Lagos	Nnamdi Azikiwe University Teaching Hospital
11	Lagoon Hospital Lagos	University of Osun Teaching Hospital
12	Sefa Specialist Hospital Kaduna	University of Nigeria Teaching Hospital
13	Duro Soleye Hospital Ikeja	University of Calabar Teaching Hospital
14	Ladkem Eye Hospital Lagos	Jos University Teaching Hospital
15	Divine Medical Centre Ikoyi	Lagos University Teaching Hospital
16	Parklande Specialist Hospital Lagos	University College Hospital Ibadan
17	Ogah Hospital and Urology Centre, Fugar, Edo State	University of Port Harcourt Teaching Hospital
18	Nisa Premier Hospital Abuja	Neuro-Psychiatric Hospital Aro, Ogun
19	Zankli Medical Centre Abuja	Ahmadu Bello University Teaching Hospital Zaria
20	Reddington Hospital Lekki lagos	
21	Noma Hospital Sokoto	
22	Abuja Clinic	
23	St. Nicholas Hospital Lagos	
24	Bridge Clinic	
25	Nordica Fertility Centre Lagos	
26	Meridian Hospital Port Harcourt	
27	Paelon Memorial Hospital Lagos	

References

Adeyanju, O. D. (2012). An assessment of the impact of corporate social responsibility on Nigerian society: The examples of banking and communication industries. *Universal Journal of Marketing and Business Research, 1*(1), 17–43.

Adhikari, S. (2017). *Corporate social responsibility (CSR) in health*. Public Health Entrepreneurship.

African Strategies for Health Project. (2014). *A review of health-related corporate social responsibility in Africa*. A project to the United States Agency for International Development under USAID Contract.

Anyanwu, M. U., Festus, I. J., Nwobi, O. C., Jaja, C. J., & Oguttu, J. W. (2020). A perspective on Nigeria's preparedness, response and challenge of spread of covid-19. *Challenges, 11*(2), 22.

Aregbeshola, B. S., & Folayan, M. O. (2021). *Nigeria's financing of health care during the COVID-19 pandemic: Challenges and recommendations*. Munich Personal Re Pec Archive (MPRA), pp. 1–16.

Babalola, V. A. (2012). The impact of corporate social responsibility on firms' profitability in Nigeria. *European Journal of Economics, Finance and Administration Science, 45*, 39–50.

Babatunde, A. O., Aborode, A. T., & Agboola, P. (2021). Implications of COVID-19 on the healthcare infrastructural development in Nigeria. *Jundishapur Journal of Health Sciences, 12*(4), e112934. https://doi.org/10.5812/jjhs.112934

Bae, K. H., Ghoul, S. E., Gony, Z. J., & Guedhami, O. (2021). Does CSR matter in times of crisis? Evidence from the covid-19 pandemic. *Journal of Corporate Finance, 67*, 101876.

Brandao, C., Rego, G., Duarte, I., & Nunes, R. (2013). Social responsibility: A new paradigm of hospital governance? *HealthCare Analysis, 21*(4), 390–402.

Burlea-Schiopoiu, A., & Popa, I. (2016). Legitimacy theory. In S. O. Idowu, N. Capaldi, L. Zu, A. Gupta, & S. V. B. Heidelberg (Eds.), *Encyclopedia of corporate social responsibility*. https://doi.org/10.1007/978-3-642-28036-8-471. Retrieved from http://www.springerreference.com/docs/html/chapterdbid/33348.html

Channels TV. (2020). *Donations to Nigerian covid-19 relief fund top ₦25bn*. Retrieved from www.channelstv.com/2020/04/18/do...

Collins, S. (2010). Corporate social responsibilities and the future healthcare manager. *The Healthcare Manager, 29*(4), 1–16.

Dailytipsfinders.com. (2021). *Top 10 best hospitals in Nigeria—Latest webometric*. Retrieved November 6, 2011, from https://www.dailytipsfinders.com

Driver, J. (2014). *The history of utilitarianism*. Stanford encyclopedia of philosophy. https://www.Plato.standford.edu/entries/utilitarianisms...

Etteh, C. C., Adoga, M. P., & Ogbaga, C. (2020). COVID-19 response in Nigeria: Health system preparedness and lessons for future epidemics in Africa. *Ethics, Medicine and Public Health*. https://doi.org/10.1016/j.jemep.2020.100580

Fernandez-Feijoo, B., Romero, S., & Ruiz, S. (2014). Commitment to corporate social responsibility measured through global reporting initiative reporting: Factors affecting the behaviour of companies. *Journal of Cleaner Production, 81*, 1–11.

Freeman, R. E. (1984). Strategic management: A stakeholder theory. *Academy of Management Review, 24*(2), 233–236.

Guthrie, J., Cuganesan, S., & Ward, L. (2006). *Legitimacy theory: A Story of reporting social and environmental matters within the Australian food and beverage industry* (pp. 1–35).

Igyuse, S. S., Chia, T., & Oyeniran, O. I. (2020). Implications of corona virus pandemic on public health policy in Africa. *Cumhuriyet Medical Journal, 42*(3), 255–258. https://doi.org/10.7197/cmj.vi:756066

Inject Sayana Press.Org. (2020). *Top 10 best hospitals*. Retrieved November 6, 2021, from www.injectsayanapress.org

Kurucz, E. C., Colbert, B. A., & Wheeler, D. (2008). The business case for corporate social responsibility. In A. Crane, A. McWilliams, D. Matten, J. Moon, & D. Segal (Eds.), *The Oxford handbook on corporate social responsibility* (pp. 83–112). Oxford University Press. isbn:978-0-19-921159-3.

Lubis, A. N. (2018). Corporate social responsibility in health sector: A case study in the government hospitals in Medan, Indonesia. *Business: Theory and Practice, 19*, 15–36.

Lyfboat. (2021). *Best hospitals in Nigeria.* Retrieved November 6, 2021, from https://www.lyfboat.com/hospitals

Moir, L. (2001). What do we mean by corporate social responsibility? *Corporate Governance International Journal of Business Society, 1*(2), 16–22.

Nigerian Finders. (2021). *5 best private hospitals in Nigeria.* Retrieved November 6, 2021, from https://nigerianfinders.com/hospitals

Nigerian Infobusstop. (2021). *Top 10 teaching hospitals in Nigeria 2021.* Retrieved November 6, 2021, from https://nigerianinfobusstop.com

Nnubia, I. C., Egbunike, P. A., Akaegbobi, T., & Okoye, N. J. (2020). Perceived effect of COVID-19 in developing economies: Lessons from Nigeria, South Africa and Kenya. *Journal of Global Accounting, 7*(1), 60–74.

Noah, A. O. (2017). *Accounting for the new environment: The accountability of the Nigerian cement industry.* Unpublished PhD Dissertation, submitted to university of Essex. https://repository.essex.ac.uk/...

Nwobu, O. A. (2021). *Corporate social responsibility and the public health imperative: Accounting and reporting on public health.* Intech Open. https://doi.org/10.5772/intchopen.94356

Obi-Ani, N. A., Ezeaku, O. D., Ikem, O., Isiani, M. C., Obi-Ani, P., & Onu, J. C. (2021). COVID-19 pandemic and the Nigerian primary healthcare system: The Leadership question. *Cogent Arts and Humanities, 8*(1), 1859075.

Odubanjo, D. (2020). *In conversation Africa (2020). Where COVID-19 has left Nigeria's health system.* Retrieved from https://www.theconversation.com/where-covid-19-has-left...

Ogoina, D., Mahmood, D., Oyeyemi, A. S., Okoye, O. C., Kwaghe, V., Habib, Z., et al. (2021). A national survey of hospital readiness during the covid-19 pandemic in Nigeria. *PLoS One, 16*(9). https://doi.org/10.1371/Journal.pone.0257567

Olateju, D. J., Olateju, O. A., Adeoye, S. V., & Ilyas, I. S. (2021). A critical review of the application of the legitimacy theory to corporate social responsibility. *International Journal of Managerial Studies and Research (IJMSR), 9*(3), 01–06. https://doi.org/10.20431/2349-0349.0903001

Palazzi, M., & Starcher, G. (2006). *Corporate social responsibility and business success.* The European Baha'1 Business Forum, pp. 1–41.

Pohle, G., & Hittner, J. (2008). *Attaining sustainable growth through corporate social responsibility.* IBM Global Business Service, Institute for Business Value.

Premium Times. (2020). *COVID-19: Private sector donations hit ₦27.160bn.* Retrieved from www.premiumtimes.ng.com/business...

Ramirez, B., West, D. J., & Costell, M. M. (2013). Development of a culture of sustainability in healthcare organisations. *Journal of Health Organisation and Management, 27*(15), 665–672.

Rodriguez, R., Svensson, G., & Wood, G. (2020). Assessing corporate planning of future sustainability initiatives in private health organisations. *Evaluation and Program Planning, 83*, 101869.

Siniora, D. (2017). *Corporate social responsibility in health care sector.* Duquesne scholarship collections. Graduate Study Research Symposium.

Takahashi, T., Ellen, M., & Brown, A. (2013). Corporate social responsibility: US theory, Japanese experiences and lessons for other countries. *Healthcare Management, Winter,* 176–179.

Vatamanescu, E.-M., Dabija, D. C., Gazzola, P., Gegarro, J. G., & Buzzi, T. (2021). Before and after the outbreaks of covid-19: Linking fashion companies corporate social responsibility approach to consumers' demand for sustainable products. *Journal of Cleaner Production, 321*, 1–13.

World Business Council for Sustainable Development. (2000). *Corporate Social Responsibility: Making good business sense.* World Business Council for Sustainable Development.

Gloria O. Okafor is a Professor in Accountancy Department at Nnamdi Azikiwe University, Awka, Anambra State Nigeria. She has 17 years' experience in accounting lecturing and research. She is a fellow (FCA) of the Institute of Chartered Accountants of Nigeria (ICAN) and a member of the Academy of Management Nigeria. She teaches taxation, environmental management accounting, financial management, financial accounting and accounting standards. She has special research interest in taxation, environmental management and sustainability reporting, corporate social responsibility reporting and accounting ethics. Gloria has published her research in peer-reviewed journals, chapters and conference proceedings.

Amaka E. Agbata (PhD) is a lecturer in the Department of Accountancy, Nnamdi Azikiwe University, Awka, Nigeria. She is an associate member of the Institute of Chartered Accountants of Nigeria. Her academic research interest includes Financial and Corporate Reporting, Public Sector Accounting, Corporate Social Responsibility reporting, Taxation and Auditing. She has a good number of article publications to her credit.

Innocent C. Nnubia, PhD, FIMC, FAMSSRN, CMC, is a lecturer in the Department of Accountancy Nnamdi Azikiwe University Awka, Nigeria. He graduated with First class honors in accountancy in the year 2013 from Chukwuemeka Odumegwu Ojukwu University, Anambra State, Nigeria; he obtained a master's degree in Accountancy from the same University. He obtained a PhD in Accounting in 2019 from the University of Nigeria, Nsukka. Dr Nnubia Innocent has worked at the Chukwuemeka Odumegwu Ojukwu University, Anambra State, for more than 12 years as the staff of the university before securing his current job as a lecturer in Accountancy department at Nnamdi Azikiwe University, Awka. Dr Nnubia Innocent is a member of various local and international bodies. He has also carried out several publications to further enhance his glowing reputation. Dr Nnubia Innocent has served and still serving efficiently in many offices. He is married with children.

Sunday C. Okaro holds a Bachelor's degree, a Master's degree and a Doctorate degree all in Accountancy. He also holds a Master's degree in Banking and Finance. He is a Professional Accountant and holds the Fellowship of the Chartered Association of Certified Accountants of London (FCCA). He is also an Associate member of the Institute of Chartered Accountants of Nigeria (ICAN). Professor Okaro also belongs to the Chartered Institute of Taxation of Nigeria as Associate member. Professor Okaro taught for 40 years on both part-time and full-time basis before retiring in 2021 from full-time teaching career. He however, still doubles as adjunct Professor at UNN Business school and Pro bono Faculty Professor at the prestigious Nnamdi Azikiwe University, Awka, Nigeria. It is not all academics for Professor Okaro as he also had extensive experience in the public sector as a civil servant and a Banker in the private sector. He undertakes consultancy assignments. He is a prolific writer and has over 70 publications to his credit. He is happily married with children and grand children.

The Private Sector's Role in Strengthening Public Hospitals in Zambia During the Coronavirus (COVID-19) Pandemic: A Corporate Social Responsibility (CSR) Perspective

Isaac Kabelenga and Ndangwa Noyoo

1 Introduction

The pandemic that besieged the world in late 2019, referred to as COVID-19, did not spare any country its devastating impacts. On 12 January 2020, the World Health Organization (WHO) confirmed that a hitherto unknown virus known as the novel COVID-19 was the cause of a respiratory illness in a cluster of people in Wuhan City, Hubei Province, China. This was in late 2019, on 31 December. Since then, the transmission of COVID-19 has been significantly greater while affecting many people around the world. This has been signified by a high death toll of patients suffering from the virus. At the time of finalising this chapter on 26 April 2022, the Johns Hopkins University and WHO reported that COVID-19 had rapidly spread across the world, with more than 510 million confirmed cases and more than 6 million deaths reported in 200 countries (Johns Hopkins University, 2022; WHO, 2022). Indeed, since 2022, the WHO has consistently sounded the alarm that COVID-19 is an existential threat to every country in the world, whether it is in the developed or developing countries. Arguably, the pandemic has reshaped the world in unimaginable and profound ways. For instance, hospitals and other health institutions were on the brink of collapse (even in affluent societies of Europe and North America) due to unparalleled pressures on them. Due to this challenge, some public hospitals in certain parts of the world were turning away prospective patients because there were no beds, oxygen or health personnel to look after them (Mohan, 2021). However, full information about how the private sector in Zambia was

I. Kabelenga
School of Humanities and Social Sciences, University of Zambia, Lusaka, Zambia

N. Noyoo (✉)
University of Cape Town, Cape Town, South Africa
e-mail: ndangwa.noyoo@uct.ac.za

© The Author(s), under exclusive license to Springer Nature Switzerland AG 2023
S. O. Idowu et al. (eds.), *Corporate Social Responsibility in the Health Sector*, CSR, Sustainability, Ethics & Governance, https://doi.org/10.1007/978-3-031-23261-9_13

supporting and continues to support public health institutions to cope with the effects of the pandemic is yet to be established. This is because there are few scientific studies that have been undertaken to ascertain the involvement of the private sector, as part of Corporate Social Responsibility (CSR), in strengthening public hospitals to adequately cope with the effects of COVID-19. Even globally, studies are still few that have addressed the said issue. Therefore, the purpose of this chapter is to fill this void while focusing on Zambia.

2 Review of Available Literature

In this section, we review general literature related to CSR and the private sector's involvement in public healthcare as part of the former's CSR initiatives, especially during COVID-19. We also review literature from several countries to situate our chapter so that it can enrich and contribute to a growing body of literature on CSR globally.

2.1 Theoretical Perspectives of CSR

According to Latapí Agudelo et al. (2019, p. 1), there is a long and varied history associated with the evolution of the concept of Corporate Social Responsibility (CSR). The current belief that corporations have a responsibility towards society is not new. In fact, it is possible to trace concern for society from the business sector several centuries back. However, it was not until the 1930s and 1940s when the role of executives and the social performance of corporations began appearing in the literature with authors discussing *what* were the specific social responsibilities of companies. Since the second-half of the twentieth century, many debates related to CSR have unfolded. Arguably, after Howard R. Bowen wrote his seminal book: *Social Responsibilities of the Businessman*, in 1953, the term CSR became more pronounced in academic discourse. In this cited book, Bowen (1953, p. 3) notes that the "decisions and actions of the businessman [and businesswoman] have a direct bearing on the quality of our lives and personalities. His [her]decisions affect not only himself [herself], his[her] stakeholders, his[her] immediate workers, or his[her] customers—they affect all of us." Despite the gender insensitive language, which is perhaps attributable to the period the book was written, Bowen's assertions proffer some insights into the early appraisals of CSR. From that period to the present, there has been a shift in terminology—from the social responsibility of business to CSR. This field has grown significantly and today there is a proliferation of theories, approaches and terminologies pertaining to CSR. Society and business, social issues management, public policy and business, stakeholder management and corporate accountability are just some of the terms used to describe phenomena related to CSR. Recently, renewed interest in CSR and new alternative concepts have been proposed, including corporate citizenship and corporate sustainability (Garriga & Melé, 2004). Some scholars have equated the appraisal of the new concepts of corporate sustainability to the classic notion of CSR (corporate citizenship) (Garriga & Melé, 2004).

Although the debate on the relationship between business and society, and implied responsibilities, has been on-going for decades, there is still no consensus on a commonly accepted definition of CSR (Carroll, 1991; Jones, 1980, 1995, 1999; Mohammed, 2020). To this end, the idea of social responsibility supposes that the corporation has not only economic and legal obligations but also certain responsibilities to society that extend beyond these obligations (McGuire, 1963). Furthermore, Carroll (1979) provides some substance to the argument by noting that CSR involves going beyond the law. Thus, a definition of social responsibility, if it is to fully address the whole range of obligations business has to society, must embody the economic, legal, ethical and discretionary categories of business performance (Mohammed, 2020). Echoing the former standpoint, Votaw (1972, p. 25) argues that:

> corporate social responsibility means something, but not always the same thing to everybody. To some, it conveys the idea of legal responsibility or liability; to others, it means socially responsible behaviour in the ethical sense; to still others, the meaning transmitted is that of 'responsible for' in a causal mode; many simply equate it with a charitable contribution; some take it to mean socially conscious; many of those who embrace it most fervently see it as a mere synonym for legitimacy in the context of belonging or being proper or valid; a few see a sort of fiduciary duty imposing higher standards of behaviour on business men [and women] than on citizens at large.

In the light of the aforementioned, see also Carroll (1991, 1999) and Garriga and Melé (2004), who touch on the same topic. In this chapter, we will be guided by a definition of CSR proffered by the World Business Council for Sustainable Development (WBCSD) (2000). The WBCSD (2000) defines CSR as the continuing commitment by business to behave ethically and contribute to economic and social development, while improving the quality of life of the workforce and their families, as well as the local community and society at large. Taken further, it can be noted that there are various theories of CSR. These are inter alia: (1) *instrumental theories*, where the corporation is seen as only an instrument for wealth creation, and its social activities are only a means to achieve economic results; (2) *political theories*, which concern themselves with the power of corporations in society and a responsible use of such power in the political arena; (3) *integrative theories*, in which the corporation is focused on the satisfaction of social demands; (4) *ethical theories*, based on ethical responsibilities of corporations to society: in practice, each CSR theory presents four dimensions related to profits, political performance, social demands and ethical values (Garriga & Melé, 2004; Melé, 2008; Mohammed, 2020); and (5) *Corporate Social Performance*, which is based on the theoretical perspective that power requires responsibility. Society allows companies to operate and in return they must serve society not only by creating wealth but also by contributing to social needs and satisfying social expectations towards business (Garriga & Melé, 2004; Melé, 2008; Mohammed, 2020). The sixth element relates to *Corporate Citizenship*.

Here, the key concept is participation in society and going beyond fulfilling legal duties (as in the case of state citizenship) to actively contributing to the wellness of society or the world as a whole. This theory recovers the position of the company in society and suggests that the company stands shoulder-to-shoulder with citizens, who together form a community. It expands the functionalist vision that would reduce business to an economic purpose. Moreover, it has a global scope (Garriga & Melé, 2004; Melé, 2008; Mohammed, 2020).

From the above-presented literature, it can be inferred that CSR is a concept that has many different meanings and definitions. The way it is understood and implemented differs greatly for each company and country as well. However, during times of community and societal crises, the common denominator in all of the above-mentioned theoretical perspectives is an understanding that CSR actually means firms recognising that they do have ethical obligations to respond pragmatically to social pressures. Thus, with each passing day, companies are expected to play a more active role in building and improving society (Siegel & McWilliams, 2001; Melé, 2008; Mohammed, 2020).

2.2 CSR in Public Healthcare During COVID-19

This chapter focusses on CSR not in a normal situation but in a time of uncertainty due to a global health pandemic. Mahmud et al. (2021) refer to challenging times that include recessions or other crises/disasters and point out that these have adverse effects on social stability and economic growth. In such an environment, some companies may withdraw from their usual strategic CSR activities due to severe resource shortages and increasing uncertainty in the macroeconomic environment. Still, many businesses may make voluntary donations to society, for example, by giving money, in-kind and time to social activities. For Nwobu (2021), the 'new normal' in corporate social responsibility can be created by focusing extensively on public health issues. Public health is crucial because no corporation operates in a vacuum. Hence, corporations depend on profit to continue to be in business. On the other hand, the population must remain healthy because the business customers are a subset of the population. However, research on CSR and public health provisions is yet to assess how corporations can account for the impact of public health on business operations and vice versa (Nwobu, 2021). Nevertheless, Dănescu and Popa (2020) are able to show the increasingly important role played by corporate governance and corporate social responsibility in pharmaceutical companies in improving public health in countries with emerging economies. This issue is of critical importance especially when we juxtapose it against COVID-19. Indeed, some scholars have argued that the private sector has a key role to play in the COVID-19 battle. This is because governments cannot be expected to fight the pandemic alone. They need help in regard to testing, treatment, provision of PPEs and a lot more (Bhushan, 2020). In his global analysis on impacts of COVID-19, Bhushan (2020) established that countries across the development spectrum were grappling with an extraordinary situation. Despite this, in India, the authorities had responded decisively with a strong whole-of-government approach. Perhaps, this can be attributed to India's history of community work done by the corporate sector, which was decreed by India's government under the Companies Act. India's new Companies Act 2013 introduced a provision for CSR, which rests upon the ideology that companies need to give back to the society in the same way as they also take from it in the form of raw materials, human resources and so on (Fernandes & Aquinas, 2020).

Kenya is another country where the private sector helped to strengthen the public health system during COVID-19. Many private companies had donated money to help expand the screening capabilities of public healthcare facilities (Business Daily, 2020). Another interesting case is Greece. According to Kritas et al. (2020), while COVID-19 was spreading, the Greek government attempted to increase the number of intensive care units (ICU) and their accessibility to the public. The number of such facilities was far below the European average, and thus the government also tried to provide all the necessary means to help healthcare personnel perform their duties effectively. From the onset of the COVID-19 pandemic, the Greek government asked for voluntary assistance from the private sector in order to support the national healthcare system and provide all the necessary materials to healthcare personnel, as a way to strengthen CSR during this period. That being said, Navickas et al. (2021) remind us that this difficult time has shown that socially responsible companies act responsibly in social, environmental and economic terms, regardless of the challenges of the period. Stakeholders, both internal and external, are at the centre of their attention regardless of changes in the business environment. Socially responsible companies work in intensive interaction with all stakeholders, because the importance of such interaction becomes apparent precisely in times of crisis. This snapshot of countries where private bodies were engaged in helping the public health sector to combat COVID-19 is by no means exhaustive. It is meant to merely highlight the positive roles played by the private sector in the light of CSR's contribution to public hospitals or the healthcare sector in different parts of the world.

2.3   CSR in Zambia

Even though there is burgeoning literature on CSR in Zambia, especially its links to the mining sector, small and medium enterprises and social policy (see Lungu & Mulenga, 2005; Noyoo, 2010, 2016, 2021; Choongo, 2017), there is still not much work that shows its role in uplifting the standards of living of ordinary people. In addition, there is little information related to the efforts made by the private sector in strengthening public hospitals in Zambia to cope with the effects of COVID-19. Furthermore, more research is still needed to establish the impacts of COVID-19 on the Zambian society.

3   Background Information: COVID-19 in Zambia

The Republic of Zambia is a landlocked country neighbouring the Democratic Republic of the Congo (DRC) to the north, Tanzania to the north-east, Malawi to the east, Mozambique, Zimbabwe, Botswana and Namibia to the south and Angola to the west. Zambia's total population in 2021 was just over 19 million, with a life expectancy at birth of 52 years. Zambia recorded its first confirmed two cases of COVID-19 on 18 March 2020, involving a family returning from a vacation in France and a local businessman who had travelled to Pakistan (Ministry of Health [MOH], 2020). While measures were put in place to slow the spread of the virus, such as social distancing, limiting operations of some businesses considered potential spreaders of the COVID-19 pandemic such as cross-border trading, clearing and forwarding agencies, bars, night clubs, restaurants, places of worship and markets among others, Zambia shied away from a complete lockdown. Thus, a full-scale lockdown was not instituted. As a land-locked country relying on imported goods, fuel, some food and most of its medicines, Zambia decided that the negative impact of a total shutdown would outweigh any benefits. This decision has resulted in a continuous increase in cases of COVID-19 in Zambia. For instance, from having two cases of COVID-19 in March 2020, on 6 June 2020, the number rose to 1089 cases, with 912 recoveries and 7 deaths. On 18 August 2020, COVID-19 cases increased to 9981 with 8776 recoveries and 260 deaths. By 26 April 2022, Zambia had a cumulative total number of 319,431 cases of COVID-19 with 314,577 recoveries and 3976 deaths (Kabelenga & Chola, 2021; MOH, 2022; WHO, 2022; Zambia National Public Health Institute [ZNPHI], 2022).

Zambia's economy was negatively impacted as a result of the closure of almost all of the country's boarders, except with Tanzania, as a mitigation measure to prevent the spread of COVID-19 during the first, second and third waves. As a result, imports and exports of goods and services experienced a negative shock. This resulted in a reduction in domestic and international revenue for the country to provide basic services such as primary healthcare in public hospitals. The immediate outcome from the loss of revenue was an increase in poverty and inequality. According to Kalikeka et al. (2021), during the pandemic, the headcount poverty had worsened by 2.2%, the poverty gap by 2% and inequality as measured by the Gini increased by 1%. Across industries, the majority experienced negative growth, with the arts, entertainment and recreation sectors (-74%) experienced the worst shock. If it were not for government measures such as the COVID-19 Emergency Cash Transfer, which was introduced in July 2020, to cushion the harsh effects of the pandemic, the situation might have been extremely dire. Given the foregoing, it was imperative to undertake a research study that could establish some facts about CSR's role in public hospitals in Zambia during COVID-19.

4   Methodology

This chapter is based on a desk review study that was undertaken in Zambia to ascertain the effects of the COVID-19 pandemic on public hospitals as well as to determine the role of the private sector in strengthening the former. The secondary data that was assessed covered the period from March 2020 to April 2022 (i.e. during the first, second, third and fourth waves of the pandemic). This involved reviewing daily and weekly reports provided by the Zambian government through the Ministry of Health (MOH) and the Zambia National Public Health Institute (ZNPHI), United Nations (UN), World Health Organizations (WHO), Johns Hopkins University, public and private media and individual researchers focusing on COVID-19. Other issues related to the effects of COVID-19 on public hospitals were the donations made by the private sector to the government as part of CSR to help hospitals cope with the pandemic. Both published and grey literature and verbatim quotes of various actors in this arena, as captured by the public and private media, were reviewed.

4.1   Data Analysis

Data was analysed using content, thematic and discourse analyses. That is, by reviewing literature, major contents were drawn. The contents were summarised into themes, and then supported by discourses that were provided by government officials and representatives of the private sector. The analysis involved several steps:

(a) First, through reading and listening to literature several times. This enabled the researchers to get a sense of the issues raised by the literature.
(b) Second, by means of coding the accounts in reviewed literature. This involved marking different effects of COVID-19 on the Zambian society and on public hospitals and how the private sector stepped in to strengthen public hospitals to cope with the effects of COVID-19.
(c) Third, to categorise the codes under main themes which revolved around effects of COVID-19 in Zambia and how the private sector responded to the effects of COVID-19 as part of their CSR mandate.
(d) Fourth, in order to enhance credibility of this chapter, the researchers searched for representative statements from government officials and those from the private sector in terms of recorded interviews as reported by public and private electronic media to support the themes.
(e) Fifth, the examining of the interplay of themes reflectively. The researchers' reflections upon the data in relation to available literature on CSR, COVID-19 and public hospitals served as a background that helped to elicit more detailed accounts of the data.

(f) Sixth, the search for concepts from the literature to use in interpreting the data. The concepts of corporate citizenship, community social relationships and power were established as analytical tools to interpret the data. These concepts were chosen after attentive reflections upon the data. That is, all the issues raised in the literature seemed to revolve around the above three scientific concepts. Furthermore, in the quotations which were recorded by electronic media, the participants' names were anonymised to protect their identity. However, the actual verbatim accounts were maintained.

Furthermore, the literature was triangulated with various sources of information such as those provided by the government, Zambia's MOH, ZNPHI, UN, WHO, Southern African Development Community (SADC), Civil Society Organizations (CSOs), public and private media houses and studies from individual scientific researchers. All these sources of information enriched the findings.

4.2 Limitations of the Study

A key limitation of this study is that it did not collect primary data from individual donors in order to have nuanced information about their COVID-19 Emergency Corporate Social Responsibility (ECSR). Thus, this chapter relied on secondary data.

5 Findings

The overall findings of the study show that the COVID-19 pandemic had negatively impacted on public hospitals in Zambia, and that the private sector had played and continues to play a part in strengthening operations of public hospitals to cope with the ramifications of COVID-19. We present the findings of the study according to various themes listed in the following sections.

5.1 Impacts of COVID-19 on Public Hospitals in Zambia

During the period when there was a sharp rise in COVID-19 cases (during the first wave, March–July 2020; second wave, December 2020–March 2021; third wave, May–July 2021; and fourth wave, December 2021–February 2022), there was a high demand for COVID-19 testing kits, personal protective equipment (PPE) for medical staff and patients, hospital beds, oxygen for COVID-19 patients and medicine, among other things. However, public hospitals were under strain during the pandemic and struggled to meet the increased demand for hospital care, for example,

Levy Mwanawasa Hospital in Lusaka was designated to hospitalise COVID-19 patients. Other cases were diagnosed at other government and private hospitals where COVID-19 patients were often hospitalised (MOH, 2020). Similar impacts were recorded in other public hospitals in the rural districts of Zambia. For instance, on 20 August 2020, Zambia's Ministry of Health (MOH) and WHO reported the following extract on the impacts of COVID-19 on hospitals:

> Shortly after Zambia reported its first cases of COVID-19 in March 2020, the country's National Public Health Institute launched intensive, targeted screenings for cases in high-risk populations and locations. In Nakonde, a rural town on the border with Tanzania, the screenings revealed a huge number of infections. We recorded 400 cases in three days and knew hospital capacity in the area would not be able to handle so many patients at a time. The northern Nakonde town is a bustling commercial hub with cross-border movement of people and trucks.

In addition, the analysis of data indicates that as a result of home-based care, brought-in-dead (BIDs) cases also increased:

> In June and July, 2021, the number of BIDs for certification began to markedly increase followed by increases in the number of adult in-patients with respiratory conditions requiring admissions. Many healthcare workers in Zambia's public sector, especially junior doctors and nurses, have been providing heroic care to large numbers of COVID-19 patients while rotating through busy inpatient services that require quarantine periods away from their homes and families. During the same period, decreased numbers of patients have been presenting for other conditions, presumably due to COVID fears. This has been especially notable among pediatric cases. (MOH, 2022; WHO, 2022).

The above-mentioned issues connote that the onset of COVID-19 in Zambia had resulted in high morbidity and mortality rates with many people being hospitalised for treatment. In addition, there was insufficient supplies of drugs, and severe shortages of healthcare professionals. Therefore, COVID-19 weakened the capacity of public hospitals in Zambia to adequately provide healthcare to those who needed it. Thus, COVID-19 had several negative multiplier effects which extended from infected individuals to public hospitals and the whole Zambian society. Given the weak state of Zambia's public healthcare system and its relative inability to cope with patients with severe or complex needs during the COVID-19 pandemic (SADC, 2020; Bwalya, 2021; MOH, 2022), there was a growing realisation from the citizens that the healthcare system in Zambia could not meet the unprecedented demands placed on it by the COVID-19 health crisis. Moreover, it may not be able to adequately cope with future epidemics and pandemics. Due to the foregoing con-clusion, it can be unambiguously stated that Zambia's public health system needs to be urgently overhauled because the majority of Zambians depend on it. Thus, the government needs to invest more financial and other resources into the country's public healthcare system (SADC, 2020; Bwalya, 2021).

5.2 Involvement of the Private Sector in Strengthening Public Hospitals to Cope with the Effects of COVID-19 Pandemic

From the data that was collected and analysed, the authors were able to decipher that during COVID-19, a new approach known as Emergency Corporate Social Responsibility (ECSR) emerged in Zambia. ESCR is anchored in two main areas, namely monetary ECSR and non-monetary ECSR. Monetary ECSR was mainly in the form of financial donations. Notable private companies that made COVID-19 donations are listed in Table 1.

It can be noted from Table 1 that monetary ECSR involved donations of money by the private sector to the government to buy essential COVID-19 medicines to reduce the negative effects of COVID-19 on patients. Other essential items were PPEs, facemasks, hospital beds, oxygen ventilators and hand sanitisers. The Alliance for Community Action (2020) listed COVID-19 cash donations received by the government between March and 30 April 2020 (Table 2).

From the above tables, it can be argued that some companies from the private sector made financial contributions to strengthen public hospitals in Zambia to withstand the negative impacts of COVID-19. The above statistics are supported by some monetary ECSR discourses that dominated the Zambia print and electronic media in the same period. For instance, on 15 April 2020, the following news headline stood out in the Zambia media:

Trade Kings Donates 28 million to COVID

Trade Kings Foundation has given K28 million (about USD 1.51 Million) to the Ministry of Health towards the fight against the Coronavirus pandemic. Trade Kings Group Manager presented the cheque on behalf of Trade Kings Foundation to Minister of Health today. The Manager says the K28 million will cover three aspects, which include: medical supplies, prevention and information dissemination on COVID-19. The manager said bus stations, markets, prisons, COVID-19 quarantine centres, hospitals, frontline medical personnel and shopping malls will be covered under prevention (Zambia Daily Mail, 2020).

Arguably, some companies demonstrated that they cared for the people of Zambia by responding positively to the public outcry for donations during the pandemic. Such actions are likely to enhance good community relations with the local communities where they operate. Globally, this is one way through which the private sector is expected to provide CSR, during societal crises (see Garriga & Melé, 2004; Melé, 2008; Mohammed, 2020). However, a critique is necessary for the foregoing approach, which is akin to charity. Again, Zambian corporates seem to easily gravitate towards philanthropic actions than broad-based CSR as argued by Noyoo (2021). Due to this, CSR is still not well defined in Zambia. On the other hand, non-monetary ESCR, which can also be referred to as in-kind contributions involved making any donation, other than money, deemed best by an individual private company to strengthen operations of public hospitals to adequately cope with the effects of COVID-19. The donations included inter alia, facemasks, food,

Table 1 Donations from private companies

Name of company	Value of contribution in Zambian Kwacha (ZMW)	Value of contribution in US dollar (USD)
Nemchem	4,033,700	221,853.5
World Vision Zambia	2,160,000	118,800
Industrial Development Corporation	1,735,484	95,451.62
Zambia National Commercial Bank (ZANACO)	1,477,250	81,248.75
Association of Indian Community	1,436,076	78,984.18
Zhong Mei Engineering Group Limited	870,000	47,850
Lafarge Zambia PLC	801,780	44,097.9
University of Lusaka (UNILUS)	775,000	42,625
Emerald International	758,100	41,695.5
Cekan General Dealers	691,458	38,030.19
Fritech Networks	511,000	28,105
Water Aid	405,970	22,328.35
Barrick Lumwana Co., Limited	365,609	20,108.5
Puma Energy (Z)	349,996	19,249.78
Surya Bio-Fuels	307,500	16,912.5
Juijang Chambers of Commerce in Southern Africa	300,976.5	16,553.71
China Civil Engineering Cooperation	300,000	16,500
Trading Investments Limited	298,500	16,417.5
Tradany Investments Limited	280,000	15,400
Marcopolo Group of Companies	212,600	11,693
Bullion Bureau De Change	201,933	11,106.32
Longrich	201,600	11,088
Quadruple Innovations	182,160	10,018.8
Hindu Swayam Sevak Sanga	80,424	4423.32
Trade Kings	52,360	2879.8
Zambeef Products Plc	30,000	1650
Superior Milling	29,988	1649.34
Savenda Group of Companies	22,680	1247.4

Source: Alliance for Community Action (2020) and Office of the Auditor General (2020)

oxygen ventilators and hand sanitisers. Table 3 summarises non-monetary contributions made by different private companies.

The above social contributions are supported by verbal remarks made by some donors and the government at the time of donations. This demonstrates that the contributions were part of CSR. For instance, the following electronic newspaper headline and story attest to this:

Table 2 Cash donors and amount donated

Name of donor	Amount in Zambian Kwacha (ZMW)	Amount in US dollars (USD)
Airtel	5,000,000	275,000
United Bank of Africa (UBA)	2,650,725	145,790
Health Professionals Council of Zambia	200,000	11,000
Savenda	200,000	11,000
Southern SDA	160,000	8800
Patriotic Front (PF) Party	150,000	8250
Poweng Zambia Ltd.	150,000	8250
National Health Research Authority	144,000	7920
Pet And Me	100,644	5535
Concord Construction	100,000	5500
EMN Minerals	100,000	5500
Petroleum Association of Zambia	100,000	5500
Anonymous Individual	53,786	2958
Anonymous Individual	24,600	1353
Hubert Ghislain Marie Van De Ghinste	20,000	1100
Eliyshama Ministries Chalala	10,000	550
Kukula Seed For Sdho	10,000	550
Total	9,780,755	537,942

Table 3 In-kind donations and their monetary value

Category	Value in Zambian Kwacha (ZMW)	Value in US dollar (USD)
Value of donations in-kind per category		
Healthcare and laboratory supplies and services	34,572,738.5	1,901,500.62
PPE	29,635,505	1,629,952.78
Communications and Professional Services	1,802,650	99,145.75
Food	984,719	54,159.55
Logistics equipment, supplies and services	447,032	24,586.76
Educational services and supplies	8000	440
IT, Telecoms and Digital Solutions	1500	82.5
Grand total	68,317,659.5	3,757,471.27

Kalumbila Mine Donates Towards Fighting COVID-19

Kalumbila Minerals Limited (KML), a subsidiary of First Quantum Minerals, has pledged to donate in-kind over K600,000 towards combating the coronavirus in Zambia. KML Manager, has also pledged to provide the necessary health support that shall fall within the means of the mining firm. (Lusaka Times, 2020)

5.3   Discussions of the Findings

When the findings of this study are linked to the CSR theories, it can be ascertained that they are in tandem with some of them. For instance, with regard to the integrative theories, corporate citizenship and corporate social performance, the key concept when talking about CSR is participation of the private sector in solving societal problems by actively contributing to the wellness of society or the world as a whole (Siegel & McWilliams, 2001; Garriga & Melé, 2004; Melé, 2008; Mohammed, 2020). From the data, it can further be determined that some of the private-sector organisations contributed to the wellness of the Zambian society during the COVID-19 pandemic by providing financial donations, PPE, oxygen concentrators, medicine, hand sanitisers, fuel, transport, communication and food, among other things, to public hospitals. This is likely to make those organisations to have a positive corporate image both in Zambia and around the world. This is because they demonstrated that they were in Zambia not only to do business and make profits from their businesses but also to participate in solving problems faced by the Zambian society. This is, in part, what CSR entails (see Bowen, 1953; Davis, 1960; McGuire, 1963; Votaw, 1972; Carroll, 1979; Siegel & McWilliams, 2001; Mohammed, 2020).

Interestingly, the findings of this study are similar to CSR initiated by the private sector in other countries in the Global South, especially in India and Kenya. From the literature review, it was evident that in these countries, which were also severely impacted by the COVID-19 pandemic, such as Zambia, the private sector had shored up the public healthcare system by supporting public hospitals though donations of both monetary and non-monetary resources (Bhushan, 2020; Business Daily, 2020). This means that despite ESCR being context specific, it also has some commonalities in the way it is being provided in different countries in the Global South. It is also interesting to note that, unlike in India, where the private sector was compelled by the law to contribute part of their profits to public hospitals in accordance with Section 135 of Companies (CSR) Rules 2014 and Schedule VII of the Companies Act 2013 (Fernandes & Aquinas, 2020), there is no evidence in Zambia that the private sector was compelled by any law to embark on ESCR. The result of this study suggests that the state of CSR in Zambia during the COVID-19 pandemic is in line with what Votaw (1972) referred to as voluntary CSR, where the private sector decides to address societal needs during crises out of free will and moral obligation.

Furthermore, it can be speculated that ESCR occurred spontaneously and was not coordinated by either the private sector or the government. Thus, there were no specified amounts of money or resources that businesses were required to donate to public hospitals. Each situation was determined by the discretion of the donor. Whether the contributions were significant or minuscule, it can be assumed that they could have helped the health facilities in their daily operations to combat COVID-19. This is because public hospitals did not receive a lot of support from the government, and therefore the private sector's contributions were timely. If the resources were used according to the intended purpose as provided by the donors' verbatim quotes above, it can be speculated that the support could have likely strengthened the capacity of public hospitals to cope with the effects of COVID-19. At the very least, it can be argued that the private sector had more resources to support its responses to COVID-19. Overall, the study was able to ascertain that the government was and is still not fighting the COVID-19 pandemic alone. Indeed, some private-sector organisations have also played vital roles in the fight against COVID-19.

6 Conclusions and Policy Implications

Given some private-sector organisations' willingness to embark on ECSR and considering that the COVID-19 pandemic is still unfolding in Zambia and the whole world, this study highlights the policy implications of the ECSR and some considerations for the future. First, if the ECSR was spontaneous, it means that the private sector in Zambia has an appetite for such endeavours and should be encouraged to formalise them. This can be done by the government through the development of CSR-friendly policies and legislation. Simubali (2021) alerts us to the fact that there is no policy in place governing CSR in Zambia and there is no clear legal framework to regulate the same. In 2006, the government made proposals for the development of a policy on public–private partnerships (PPPs) but did not go any further. This area needs urgent attention so that the CSR environment is anchored in a progressive and forward-looking policy and legislative framework. The emergence of ECSR during COVID-19 is a clear sign that public hospitals and the whole healthcare sector could be emboldened by private-sector investments. Second, in the light of foregoing assertion, CSR should be legislated in Zambia by including a specific clause in the Companies Act, 2017. The clause should cover the essential prerequisites pertaining to definition, execution, fund allotment and reporting for successful CSR projects that are implemented. This will result in CSR activities undertaken by the private sector to significantly benefit communities (Simubali, 2021). Third, there is a need to link CSR initiatives to social policy imperatives (Noyoo, 2010). Such a focus should hinge on the government's broad development agenda and specifically its social development goals.

The purpose of this chapter was to discuss the private sector's role in strengthening public hospitals in Zambia during the Coronavirus (COVID-19) pandemic. It did this while using a Corporate Social Responsibility (CSR) analytical lens. In addition, the chapter was based on a desktop research study that endeavoured to ascertain the role that the private sector had played in bolstering the capacity of public hospitals while COVID-19 ravaged Zambia. The study discovered that a new phenomenon referred to as COVID-19 Emergency Corporate Social Responsibility (ECSR) emerged in Zambia. The study was able to determine that ECSR helped public hospitals by donating money, PPE, oxygen concentrators, medicine, food and so on. However, ECSR was ad hoc and could dissipate anytime. Hence, in this concluding section, the chapter has put forth some policy, legislative and other recommendations that could leverage ECSR and strengthen Zambia's CSR environment.

References

Alliance for Community Action. (2020). *Break-down of COVID-19 donations received by the Government of Zambia between March and 30th April 2020*. Alliance for Community Action.

Bhushan, I. (2020). *COVID-19: Private sector has a key role in the battle*. Retrieved from https://www.hindustantimes.com/analysis/covid-19-private-sector-has-a-key-role-in-the-battle/story-UBAvajhkYvYox82KbUB1lK.html

Bowen, H. R. (1953). *Social responsibilities of the businessman*. Federal Council of the Churches of Christ in America.

Business Daily. (2020). *Covid-19kitty hits Sh1.29bn*. Retrieved from https://www.businessdailyafrica.com/economy/Covid-19-kitty-hits-Sh1-29bn/3946234-5530646-107vrir/index.html

Bwalya, J. (2021). *SADC and the Abuja declaration: Honouring the pledge*. Retrieved from https://www.file:///C:/Users/kh64bgk/AppData/Local/Temp/Policy-Briefing-230-bwalya.pdf

Carroll, A. B. (1979). A three-dimensional concept of corporate performance. *The Academy of Management Review, 4*(4), 497–505.

Carroll, A. B. (1991). The pyramid of corporate social responsibility: Toward the moral management of organizational stakeholders. *Business Horizons, 34*(4), 39–48.

Carroll, A. B. (1999). Corporate social responsibility: Evolution of a definitional construct. *Business and Society, 38*(3), 268–295.

Choongo, P. A. (2017). Longitudinal study of the impact of corporate social responsibility on firm performance in SMEs in Zambia. *Sustainability, 9*(8), 1300. https://doi.org/10.3390/su9081300

Dănescu, T., & Popa, M.-A. (2020). Public health and corporate social responsibility: Exploratory study on pharmaceutical companies in an emerging market. *Global Health, 16*, 117. https://doi.org/10.1186/s12992-020-00646-4

Davis, K. (1960). Can business afford to ignore social responsibilities? *California Management Review, 2*(3), 70–76.

Fernandes, M. S., & Aquinas, P. G. (2020). CSR in health care—An overview. *Indian Journal of Applied Research, 10*(7), 25–26.

Garriga, E., & Melé, D. (2004). Corporate social responsibility theories: Mapping the territory. *Journal of Business Ethics, 53*, 51–71.

Johns Hopkins University. (2022). *COVID-19 behaviors dashboard—Saturday 5 February, 2022*. Retrieved from https://covidbehaviors.org/

Jones, T. M. (1980). Corporate social responsibility revisited, redefined. *California Management Review, 22*(3), 59–67.

Jones, T. M. (1995). Instrumental stakeholder theory: A synthesis of ethics. *Academy of Management Review, 20*(2), 404–437.

Jones, T. M. (1999). The institutional determinants of social responsibility. *Journal of Business Ethics, 20*, 163–179.

Kabelenga, I., & Chola, J. (2021). *Social protection and COVID-19: Impacts on informal economy workers in rural and urban Zambia*. Fredrich Ebert Stiftung (FES).

Kalikeka, M., Bwalya, M., Adu-Ababio, K., Gasior, K., McLennan, D., & Rattenhuber, P. (2021). *Distributional effects of the COVID-19 pandemic in Zambia*. Retrieved from https://www.wider.unu.edu/sites/default/files/Publications/Policy-brief/PDF/PB2021-2-distributional-effects-covid-19-pandemic-zambia.pdf

Kritas, D., Tzagkarakis, S. I., Atsipoulianaki, Z., & Sidiropoulos, S. (2020). The contribution of CSR during the covid-19 period in Greece: A step forward. *HAPSc Policy Briefs Series, 1*(1), 238–243. https://doi.org/10.12681/hapscpbs.24971

Latapí Agudelo, M. A., Jóhannsdóttir, L., & Davídsdóttir, B. A. (2019). literature review of the history and evolution of corporate social responsibility. *International Journal of Corporate Social Responsibility, 4*(1), 1–23. https://doi.org/10.1186/s40991-018-0039-y

Lungu, J., & Mulenga, C. (2005). *Corporate social responsibility practices in the extractive industry in Zambia* (2nd ed.). Mission Press.

Lusaka Times. (2020). *Kalumbila mine donates towards fighting COVID-19.* Retrieved from https://www.lusakatimes.com/2020/04/01/kalumbila-mine-donates-towards-fighting-covid-19/

Mahmud, A., Ding, D., & Hasan, M. (2021). Corporate social responsibility: Business responses to coronavirus (COVID-19) pandemic. *SAGE Open, 11*(1), 1–17. https://doi.org/10.1177/2158244020988710

McGuire, J. (1963). *Business and society.* McGraw Hill.

Melé, D. (2008). Corporate social responsibility theories. In A. Crane, D. Matten, A. McWilliams, J. Moon, & D. S. Siegel (Eds.), *The Oxford handbook of corporate social responsibility* (pp. 47–82). Oxford University Press.

Ministry of Health (MOH). (2020). *COVID-19 updates.* Ministry of Health.

Ministry of Health (MOH). (2022). *COVID-19 updates.* Ministry of Health.

Mohammed, S. (2020). Components, theories and the business case for Corporate Social Responsibility. *International Journal of Business and Management Review, 8*(2), 37–65.

Mohan, R. (2021). *Shortage of hospital beds and oxygen as India battles second Covid-19 wave.* Retrieved from https://www.straitstimes.com/asia/south-asia/shortage-of-hospital-beds-and-oxygen-as-india-battles-second-wave

Navickas, V., Kontautienc, R., Stravinkiene, J., & Balin, Y. (2021). Paradigm shift in the concept of corporate social responsibility: COVID-19. *Green Finance, 3*(2), 138–152. https://doi.org/10.3934/GF.2021008

Noyoo, N. (2010). Linking corporate social responsibility and social policy in Zambia. In P. Utting & J. C. Marques (Eds.), *Corporate social responsibility and regulatory governance: Towards inclusive development?* (pp. 105–123). Palgrave Macmillan.

Noyoo, N. (2016). Corporate Social Responsibility forays in Southern Africa: Perspectives from South Africa and Zambia. In S. Vertigans, S. O. Samuel, & R. Schmidpeter (Eds.), *Corporate social responsibility in Sub-Saharan Africa: Sustainable development in its embryonic form* (pp. 69–83). Springer International.

Noyoo, N. (2021). Beyond the mining sector: Broad-based corporate social responsibility in Zambia. In S. O. Idowu (Ed.), *Global practices of corporate social responsibility* (pp. 579–593). Springer International.

Nwobu, O. A. (2021). *Corporate social responsibility and the public health imperative: Accounting and reporting on public health.* Retrieved from https://www.intechopen.com/chapters/74802

Office of the Auditor General of Zambia. (2020). *Interim report of the auditor general on the audit of utilisation of COVID-19 resources as at 31st July 2020.* Government of the Republic of Zambia (GRZ).

Siegel, D., & McWilliams, A. (2001). Corporate social responsibility: A theory of the firm perspective. *Academy of Management Review, 26*(1), 117–127.

Simubali, M. (2021). *A critical examination of the enactment of specific law on corporate social responsibility in Zambia* (Unpublished master's thesis). Cavendish University.

Southern African Development Community (SADC). (2020). *SADC region: COVID-19 status update.* Retrieved from https://www.sadc.int/issues/COVID-19/

Votaw, D. I. (1972). Genius became rare: A comment on the doctrine of social responsibility Pt. 1. *California Management Review, 15*(2), 25–31.

World Business Council for Sustainable Development (WBCSD). (2000). *Corporate social responsibility (CSR) in brief*. Retrieved from https://growthorientedsustainableentrepreneurship.files. wordpress.com/2016/07/csr-wbcsd-csr-primer.pdf
World Health Organisation (WHO). (2022). *COVID-19 updates*. WHO.
Zambia Daily Mail. (2020). *Trade Kings donates 28 million to COVID*. Retrieved from https:// www.facebook.com/ZAMBIADAILYMAIL/photos/trade-kings-donates-28-million-to-covid-19-fight-in-zambiathe-k28000000-donation/2975676842526309/
Zambia National Public Health Institute (ZNPHI). (2022). *COVID-19 updates*. Ministry of Health.

Isaac Kabelenga, PhD is a Social/Public Policy and Development Analyst; Social Worker; Planning, Monitoring & Evaluation (P, M & E) and Social Protection expert by training. Isaac is an academic product of the Lapin University—Finland (Ph.D, Doctor of Social Sciences), London School of Economics and Political Science (LSE—UK, MSc. Social Policy and Development), University of Zambia (Zambia, Bachelor's Degree—BSW) and Economic Policy Research Institute (EPRI)—South Africa, Kenya and the Netherlands (Designing and Implementing Social Protection Programmes in Africa specializing in Social Protection Policy). He works at the University of Zambia as a full-time lecturer for postgraduate (PhD and Masters Degree) and undergraduate students in the Department of Social Development Studies—School of Humanities and Social Sciences.
   Other major contributions to the Zambian Society: Isaac is the Founder of Zambian Center for Poverty Reduction and Research Limited (ZCPRR). ZCPRR is a Zambian-based Organization formed to help in reducing poverty across various categories of the most poverty stricken people in both rural and urban Zambia. This is being done by consistently searching for new innovative approaches to poverty reduction through scientific research informed by local ways of knowing and doing, service delivery, and providing relevant education to the poor people, and sharing relevant information with the Government of Zambia, Political Parties, Academic Institutions, Civil Society Organizations, International Donor Community and other development actors on how to reduce poverty in rural and urban Zambia. *See details on this website*: https://zcprr.com/
   Isaac is also the founder of Zambian Think Tank for Social Protection Foundation Limited. This foundation is formed at the request of the Southern African Social Protection Experts Network (SASPEN) with financial support from Open Societies Initiative in Southern Africa (OSISA) for all the ten (10) Southern Africa Development Community (SADC) countries that participated in the study titled: *Impact Assessment on the state of Social Protection in the context of COVID-19 in Southern Africa*, namely Botswana, Mozambique, Malawi, Zimbabwe, Eswatini, Tanzania, Namibia, Mauritius, South Africa and Zambia to establish social protection organizations in their countries to sustain the social protection conversations in their countries and at regional level and to use the evidence of the research project to influence policy beyond the life of the research. Being an expert of Social Protection and being country researcher for Zambia, Isaac formed the above organization.

Ndangwa Noyoo, PhD is a Professor at the University of Cape Town (UCT). He is currently a Visiting Professor at the Catholic University in Munich Germany. He previously worked at the University of Johannesburg in the Department of Social Work as an Associate Professor. He was also employed by the South African Government in the Department of Social Development as a Chief Director/Social Policy Specialist, and the University of the Witwatersrand's Department of Social Work as a Senior Lecturer and Deputy Head of Department. He holds a Doctor of Philosophy (Ph.D) from the University of the Witwatersrand, Master of Philosophy (MPhil) in Development Studies from the University of Cambridge and Bachelor of Social Work (BSW) from the University of Zambia. He was also a postdoctoral fellow at the Fondation Maison des Sciences de l'Homme (FMSH) Paris, France. He has published widely in the areas of social policy, social development, social work, human rights, corporate social responsibility and indigenous knowledge systems. His recent publications include: *The Coronavirus Crisis and Challenges to Social Development: Global*

Perspectives, Cham: Springer (2022) (Co-editor); *Social Welfare and Social Work in Southern Africa*, Stellenbosch: Sun Media Press (2021) (Editor); *Promoting Healthy Human Relationships in Post-Apartheid South Africa: Social Work and Social Development Perspectives*, Cham: Springer (2020) (Editor); *Social Policy in Post-Apartheid South Africa: Social Re-engineering for Inclusive Development*, Oxon: Routledge (2019).

Part III
CSR and COVID-19 Pandemic in Asia

India
    Malaysia

Business Responses to COVID-19 Through CSR: A Study of Selected Companies in India

Sumona Ghosh

1 Introduction

Currently, the world is passing through very difficult and alarming times. We have faced three coronaviruses during the twenty-first century—SARS in 2002 in China and MERS in 2012 in Saudi Arabia resulting in severe respiratory disease outbreaks and finally death and now COVID-19 in 2019. We observe that fatality rate in case of SARS and MERS were higher than COVID-19; however, dispersion was much easier in case of COVID-19, which resulted in an increase in the number of positive cases and deaths. On 11 March 2020, World Health Organization (WHO) declared 'COVID-19 as a pandemic' (Hewings-Martin, 2020).

The COVID-19 pandemic has brought about far-reaching changes globally. It is being regarded as the most harmful health emergency having a deleterious effect not only on the health of people but also on the global economy. This pandemic has resulted in millions suffering due to stringent social isolation and death of near and dear ones, and simultaneously it has given severe shocks to the global economy where it was in the verge of stagnation. The Director-General of the WHO, Dr Tedros Adhanom Ghebreyesus, expressing his concerns about the multiple impacts of this pandemic had observed "This is not just a public health crisis; it is a crisis that will touch every sector. So, every sector and every individual must be involved in the fights" (Ducharme, 2020). This pandemic saw people struggling with health disaster and financial crisis simultaneously. The pain the people suffered during this time was "personal, emotional, psychological, societal, economic, and cultural; and it will leave scars. In many regards, we view Covid-19 as analogous to that which Taleb (2008) calls a 'Black Swan Event'—a shocking event that changes the world" (He & Harris, 2020). During this time of uncertainty, companies as a part of their Corporate

S. Ghosh (✉)
Department of Commerce, St. Xavier's College (Autonomous), Kolkata, West Bengal, India

© The Author(s), under exclusive license to Springer Nature Switzerland AG 2023
S. O. Idowu et al. (eds.), *Corporate Social Responsibility in the Health Sector*, CSR,
Sustainability, Ethics & Governance, https://doi.org/10.1007/978-3-031-23261-9_14

Social Responsibility (CSR) can play a pivotal role in supporting and reviving communities and society in general, rather than just employees and customers, by undertaking various social initiatives through diverse CSR programs, as they did during the Asian Tsunami in 2004, Hurricane Katrina in the United States in 2005, the Central Java Earthquake in Indonesia in 2006, Hurricane Ike in the United States in 2008, the Sichuan Earthquake in China in 2008, the Haitian Earthquake in 2010, the Great Eastern Japan Earthquake and Tsunami in 2011, to name a few.

The Director-General of the WHO, Dr Tedros Adhanom Ghebreyesus, urged the companies to come forward and contribute in his opening speech on 6 March 2020. He said, "We look forward to businesses stepping up to play their part. We need you. WHO is working with the World Economic Forum to engage companies around the world, and earlier this week I spoke to more than 200 CEOs about how they can protect staff and customers, ensure business continuity and contribute to the response? We're all in this together, and we all have a role to play." (Ducharme, 2020).

The physically or economically vulnerable people unfortunately face more uncertainties and risks with respect to their health, income and safety. Therefore, the priority during the time of pandemic is health, safety and supporting the vulnerable people. In this regard, corporate social responsibilities undertaken by the companies can be treated as an excellent mechanism to accomplish sustainable development (Mahmud et al., 2020). On the one hand, companies can enhance their reputation and financial performance while also deliver social benefits that can help people at large to sustain the COVID-19 pandemic period and overcome the crises (Bapuji et al., 2020; Guan et al., 2020; Guerriero et al., 2020; Kucharska & Kowalczyk, 2019).

However, detailed understanding and research about the corporates' response to the pandemic situation is still little explored, especially with respect to health. There is limited knowledge about the ways in which the corporates are protecting the society and, in particular, the most vulnerable individuals and those most affected by the COVID-19 pandemic. Besides, "the studies are limited to the understanding of the phenomenon for developed economies" (McKibbin & Fernando, 2020). Given this background, the main aim of this present chapter is twofold. First, we would like to study the responses of the business to COVID-19 with respect to health through CSR. Second, we would also like to construct a Corporate Health Disclosure Index (CHDI) for the time period 2019–2021 across all companies and to differentiate the companies on that basis in order to find out the extent to which the companies in India are being responsive to such a disclosure during the pandemic years.

The remaining article is organised as follows. Section 'Literature Review' gives an overview of the literature review. Section 'Methodology' details the methodological aspects as implemented for the study in the chapter. In the 'Results and Discussion' section, the results along with a discussion is presented followed by concluding remarks and limitations in the 'Conclusion and Discussion' section.

2   Literature Review

COVID-19 pandemic has affected the world severely in two ways—first, the world is finding it difficult to provide proper treatment to the millions of people affected by this disease and, second, it is finding difficult to support a large number of people to survive since many of them are without wages (Bapuji et al., 2020). The COVID-19 pandemic has widened social inequalities. Millions of part-time or full-time job-holders lost their jobs. Lockdowns called by most of the cities resulted in millions becoming unemployed, demanding security benefits, especially in the developed part of the world. This would have a major impact on the future social stability and economic growth affecting life, health, education, in other words, overall human development (Bapuji et al., 2020). According to Peeri et al. (2020), "It seems that we did not learn from the two prior epidemics of Coronavirus and were ill-prepared to deal with the challenges the COVID-19 epidemic has posed."

Since countries across the world are trying to curb the rapid spread of COVID-19 either by imposing quarantine nationwide, or closure of educational institutions and specific business operations, or bans on public gatherings, the companies are struggling to keep their global supply chains active. Most of the world's population were under some quarantine or the other and most were following 'Stay Home, Stay Safe' policy (Kaplan et al., 2020). Strict nationwide lockdown forced many small businesses to shut down creating pressure on the other businesses to fill the demand–supply gap in the country (Sibley et al., 2020). Restrictions imposed on the transport services affected the supply of many critical raw materials for production, resulting in shortages of many essential commodities (Beck & Hensher, 2020; Gostin & Wiley, 2020). Disrupted working environment resulted in increased corporate failures (Willcocks, 2020). The pandemic also resulted in disruptions in country's geopolitical condition, enhanced financial shocks and oil prices which decreased the firm's value of the profit (Zhang et al., 2020; Sharif et al., 2020).

Thus, COVID-19 pandemic forced the companies to shift their business model to contingent models to take care of the health measures of the various stakeholders (He & Harris, 2020; Anser et al., 2020a, 2020b). Thus, according to Euro News (2020), "The impact of Covid-19 on the global economy is likely to be unprecedented since the 1930s Great Depression."

During any crisis moment, a company is responsible and accountable to all its stakeholders be it consumers, employees or society at large (Georage, 1981). According to Boadi et al. (2019), "Community trust is the crucial rope to tie the mutual interests of business and community, and empathy is an important cue for establishing trust". The problems faced by society during such times, especially job loss, can be solved by helping each other (Aknin & Whillans, 2020). Exemplary performances of the firms could be seen during the pandemic, especially with respect to CSR policies related to health, economic and social needs of the community (Aguinis et al., 2020). It was observed that major corporations across the world extended their support either in cash or in-kind to prevent the spread of this virus (Gokarna & Krishnamoorthy, 2021). Most recently, Mahmud et al. (2020) observed that US firms in collaboration with government authorities and NGOs responded quickly to any kind of help demanded by the affected people.

From the Indian perspective, Chaudhary et al. (2020) assessed the impact of COVID-19 on affected sectors, such as aviation, tourism, retail, capital markets, MSMEs and oil. Duvendack and Sonne (2021) presented a case study based on 'systems thinking and complexity theory' on how the city of Mumbai has responded to COVID-19 where non-profit organizations, businesses and citizen volunteers along with the government had got together to fight against the pandemic. Singh et al. (2020) used 'principal component analysis' to identify the latent dimensions regarding people's preventive measures during the pandemic and found that they were following lockdown and social distancing rules and using naturopathy for immunity. Ladha (2020) presented a framework so as to decide between 'national and local lockdown' in response to the coronavirus pandemic and find the "optimum level of lockdown that balances the benefits (avoided ailments) of lockdown against its cost (arising from reduced economic output)." Rao et al. (2022) in their paper analysed the 'psychological impact of the COVID-19 pandemic' on engineering undergraduates in South India, who are in the age group of 19–22, and found that there was severe inclination towards depression and stress amongst the students.

It can therefore be concluded that firms should design effective CSR policies and programmes, which would efficiently deal with environmental and social challenges imposed by COVID-19 pandemic (Donthu & Gustafsson, 2020; Aguinis et al., 2020; García-Sánchez & García-Sánchez, 2020).

3 Methodology

3.1 Data Source and Selection of Companies

An empirical and analytical study of Corporate Health Disclosure during the pandemic was made for the financial years 2019–2020 to 2020–2021. Secondary sources were used for the study whereby the company's annual reports, websites and sustainability reports were analysed. 'Longitudinal Qualitative Document Analysis' was used to generate data for the specific period. Qualitative Document Analysis, according to Glaser and Strauss (1967), described "the meanings, prominence and the theme of messages and emphasized the understanding of the organization as well as how it was presented." For the data coding procedure used in this study, the author read between the lines of the annual reports, sustainability reports and the documents uploaded on the websites. Identification and classification of Corporate Health Disclosure was made by the author from the data generated through the document analysis since there was absence of any standardized guidelines regarding such disclosures. Data analysis was done through Microsoft Office excel sheets in the research line-up of Voegtlin and Greenwood (2016) and Xiao et al. (2020). From the ET 500 list published by the *Economic Times*, top 100 companies were selected for the years 2019–2020 and 2020–2021. From the list of

Table 1 Sample data set

Non-service sector	BFSI sector
Reliance Industries	General Insurance Corporation of India
Indian Oil Corporation Ltd.	Bajaj Finserv Ltd.
Rajesh Exports	Indian Bank
ONGC	Bank of India
Vodafone Idea Ltd.	Kotak Mahindra Bank
Power Grid Corporation of India Ltd.	HDFC Life Insurance Company Ltd.
Redington (India) Ltd.	SBI Life Insurance Company Ltd.
Motherson Sumi Systems Ltd.	Axis Bank
Maruti Suzuki India Ltd.	Union Bank of India
UltraTech Cement Ltd.	ICICI Prudential Life Insurance Company
Hindustan Unilever Ltd.	Bank of Baroda
Jindal Steel and Power Ltd.	Canara Bank
HCL Technologies Ltd.	Punjab National Bank
JSW Steel Ltd.	HDFC Bank Ltd.
Grasim Industries Ltd.	ICICI Bank Ltd.
GAIL India Ltd.	State Bank of India
Mahindra and Mahindra Ltd.	Housing Development Finance Corporation of India
Steel Authority of India	Power Finance Corporation Ltd.
Wipro Ltd.	
ITC Ltd.	
Coal India Ltd.	
NTPC	
Bharti Airtel Ltd.	
Hindalco Industries Ltd.	
Vedanta Ltd.	
Infosys Ltd.	
Tata Motors	
Bharat Petroleum Corporation Ltd.	
Hindustan Petroleum Corporation Ltd.	
Tata Consultancy Services Ltd.	
Larsen and Toubro	
Tata Steel	

100 most valuable companies, the first 50 companies were chosen for the study on the basis of market capitalization. These companies were ranked on the basis of market capitalization since this parameter gives us an indication of not only the present but future prospects of the company as well. The companies were analysed for the time period 2019–2021. Out of the selected data set of 50 companies, shown in Table 1, 18 companies belonging to the Banking, Financial Services and Insurance (BFSI) sector (4 insurance companies, 3 finance companies, 11 banking companies) and the remaining 32 belonged to the non-service sector. We have included

both service and non-service sectors for our study because the disruptions caused by COVID-19 affected both the sectors.

3.2 Research Design and Methodology

We begin with the construction of the CHDI (Corporate Health Disclosure Index). We begin with 41 indicators of business responses to health during COVID-19, grouped into six main categories: (a) Health Care Services, (b) Health system strengthening, (c) Financial Support, (d) Wellness and Health programmes, (e) Innovative responses to combat COVID-19 and (f) Others. Each indicator was assigned a value of 1 if the company disclosed information about this and 0 otherwise in order to get the score.

In Table 2, the basis for assignment of binary to the different parameters relating to Health Care Services (HCS1–HCS14), Health system strengthening (HSS1–HSS12), Financial Support (FS1–FS6), Wellness and Health programmes (WHP1–WHP4), Innovative responses to combat COVID-19 (IRCC1) and Others (OTH1–OTH4) for developing CHDI is given below.

To arrive at the CHDI value, we added the scores of all the 41 individual parameters and divided it by the maximum possible value (see Equation 1). The maximum value for CHDI is 41.

Equation 1 Corporate Health Disclosure Index
The CHDI can be calculated as follows:

$$\text{CHDI} = \left(\sum di/n\right) \times 100 = (\text{TS}/n) \times 100$$

where CHDI is Corporate Health Disclosure Index (CHDI), $di = 1$ if indicator i is disclosed and 0 if indicator i is not disclosed, n is no. of indicators (i.e. maximum score) and TS is total score.

3.3 Research Framework

A structured process was followed for defining the data, selection of content, data coding and data analysis. This structured framework is shown in Fig. 1. The concept of this framework has been developed taking insights from Hasan et al. (2019, 2020) and Xiao et al. (2020).

Table 2 Parameters used for construction of CHDI (Corporate Health Disclosure Index)

The basis for assignment of binary values for each parameter is explained as follows:
(a) Health care services score
HCS1. Daily COVID-19 symptom checker—*This variable is assigned a value of 1 if the company discloses information about this and 0 otherwise*
HCS2. Digital health consultations provided—*This variable is assigned a value of 1 if the company discloses information about this and 0 otherwise*
HCS3. Free video consultation with doctors and receive virtual healthcare, counselling, diagnosis and prescription—*This variable is assigned a value of 1 if the company discloses information about this and 0 otherwise*
HCS4. Daily monitoring of health and reinforcing safety and hygiene practices—*This variable is assigned a value of 1 if the company discloses information about this and 0 otherwise*
HCS5. Assistance in vaccination programme—*This variable is assigned a value of 1 if the company discloses information about this and 0 otherwise*
HCS6. Implementing aggressive RT-PCR and antigen testing for employees—*This variable is assigned a value of 1 if the company discloses information about this and 0 otherwise*
HCS7. Provided access to medical care facilities—*This variable is assigned a value of 1 if the company discloses information about this and 0 otherwise*
HCS8. Assistance in doctor consultations—*This variable is assigned a value of 1 if the company discloses information about this and 0 otherwise*
HCS9. Support towards hospitalisation—*This variable is assigned a value of 1 if the company discloses information about this and 0 otherwise*
HCS10. Assistance with medical amenities—*This variable is assigned a value of 1 if the company discloses information about this and 0 otherwise*
HCS11. Assisting in home testing—*This variable is assigned a value of 1 if the company discloses information about this and 0 otherwise*
HCS12. Community-based testing for the vulnerable—*This variable is assigned a value of 1 if the company discloses information about this and 0 otherwise*
HCS13. Providing experienced and well-trained medical professionals to take care of employees—*This variable is assigned a value of 1 if the company discloses information about this and 0 otherwise*
HCS14. Telehealth consultation/video conferencing—*This variable is assigned a value of 1 if the company discloses information about this and 0 otherwise*
(b) Health system strengthening
HSS1. Developing COVID-19 testing capabilities—*This variable is assigned a value of 1 if the company discloses information about this and 0 otherwise*
HSS2. Supported healthcare infrastructure—*This variable is assigned a value of 1 if the company discloses information about this and 0 otherwise*
HSS3. *Providing* efficient ambulance services—*This variable is assigned a value of 1 if the company discloses information about this and 0 otherwise*
HSS4. Providing COVID-19 protective gears—*This variable is assigned a value of 1 if the company discloses information about this and 0 otherwise*
HSS5. Setting up a 24x7 emergency helpline number for employees and families for any emergency situation—*This variable is assigned a value of 1 if the company discloses information about this and 0 otherwise*
HSS6. Providing Mobile Medical Units/clinic—*This variable is assigned a value of 1 if the company discloses information about this and 0 otherwise*

(continued)

Table 2 (continued)

HSS7. Providing mobile health vans—*This variable is assigned a value of 1 if the company discloses information about this and 0 otherwise*

HSS8. Tie-ups with leading hospitals for providing treatment—*This variable is assigned a value of 1 if the company discloses information about this and 0 otherwise*

HSS9. Providing tele-medicine centres—*This variable is assigned a value of 1 if the company discloses information about this and 0 otherwise*

HSS10. Providing mobile testing vans, swab collection booths—*This variable is assigned a value of 1 if the company discloses information about this and 0 otherwise*

HSS11. Providing dedicated covid testing laboratory—*This variable is assigned a value of 1 if the company discloses information about this and 0 otherwise*

HSS12. Establishment of hubs to train healthcare professionals on various aspects related to COVID-19—*This variable is assigned a value of 1 if the company discloses information about this and 0 otherwise*

(c) Financial support

FS1. Contribution to the PM CARES Fund/Covid Relief measures—*This variable is assigned a value of 1 if the company discloses information about this and 0 otherwise*

FS2. Medical benefits/medical reimbursements/treatment support/financial assistance/financial support for covid research—*This variable is assigned a value of 1 if the company discloses information about this and 0 otherwise*

FS3. Medical insurance scheme/COVID Indemnity Health Cover—*This variable is assigned a value of 1 if the company discloses information about this and 0 otherwise*

FS4. Contribution to Chief Ministers' Relief Fund/State Disaster Management Authorities—*This variable is assigned a value of 1 if the company discloses information about this and 0 otherwise*

FS5. Funding for upgrading health infrastructure/procurement of medical equipment—*This variable is assigned a value of 1 if the company discloses information about this and 0 otherwise*

FS6. Financial support to access to COVID-19 care packages—*This variable is assigned a value of 1 if the company discloses information about this and 0 otherwise*

(d) Wellness and health programmes

WHP1. Health talks/mental health talks—*This variable is assigned a value of 1 if the company discloses information about this and 0 otherwise*

WHP2. Emotional and wellness support—*This variable is assigned a value of 1 if the company discloses information about this and 0 otherwise*

WHP3. Online counselling—*This variable is assigned a value of 1 if the company discloses information about this and 0 otherwise*

WHP4. Tele-counselling services for frontline health and childcare workers—*This variable is assigned a value of 1 if the company discloses information about this and 0 otherwise*

(e) Innovative responses to combat COVID-19

IRCC1. Innovative responses undertaken by the company to combat COVID-19—*This variable is assigned a value of 1 if the company discloses information about this and 0 otherwise*

(f) Others

OTH1. Online learning program, on the fundamentals of infection prevention and control of COVID-19—*This variable is assigned a value of 1 if the company discloses information about this and 0 otherwise*

OTH2. Training and mentoring healthcare providers on prevention and management of COVID-19 and social behaviour—*This variable is assigned a value of 1 if the company discloses information about this and 0 otherwise*

(continued)

Table 2 (continued)

OTH3. Conducting periodic COVID-19 health risk assessment of employees through online surveys: calling, and contact tracing—*This variable is assigned a value of 1 if the company discloses information about this and 0 otherwise*

OTH4. Supporting the wards of deserving paramedical staff and housekeeping staff (providing groceries)—*This variable is assigned a value of 1 if the company discloses information about this and 0 otherwise*

Source: Author's own development

4 Results and Discussion

Out of the 41 parameters listed in Table 2, we observed that four parameters have turned out to be highly significant and they are HSS2, HSS4, HCS5 and FS1. HSS2 highlights companies' assistance in providing ventilators, oxygen concentrators, beds in hospitals, isolation centres, Covid care centres, critical medical equipment, saline stand, locker sets and trolleys, and 86% of the companies have provided such assistance to strengthen health system. HSS4 highlights companies providing COVID-19 protective gears such as PPE kits, sanitisers, masks, gloves, soaps, immunity booster kits and medical protective clothing, and 84% of the companies have distributed such protective gears to the society. HCS5 highlights companies' augmenting the country's vaccination programme including rolling out own vaccination programme across locations for all employees of the firm and eligible family members, organising vaccination camps and creating vaccination tracking portal, and 68% of the companies have indulged in healthcare services and have provided assistance with respect to vaccination programme for the employees and the society at large.

Financial Services 1 (FSI) signifies companies' contribution to the PM CARES Fund/Covid Relief measures, and 76% of the companies have been responsive to this parameter. It is noteworthy to specify here that the Ministry of Corporate Affairs (vide its General Circular No. 10/2020 dated 23 March 2020) has clarified that funds spent for COVID-19 will be considered as an eligible CSR activity. According to General Circular No. 15/2020, F. No. CSR-01/4/2020-CSR-MCA Government of India, Ministry of Corporate Affairs, the following will constitute CSR expenditure:

- 'Contribution made to the PM CARES Fund': The *Prime Minister's Citizen Assistance and Relief in Emergency Situations Fund (PM CARES Fund)* was created on 27 March 2020 for 'combating, containment and relief efforts' against the coronavirus outbreak and similar situations in the future.
- 'Contribution made to State Disaster Management Authority to combat COVID-19 shall qualify as CSR expenditure'.
- 'Spending CSR funds for COVID-19-related activities will qualify as CSR expenditure'
- 'If any ex-gratia payment is made to temporary/casual workers/daily wage workers over and above the disbursement of wages, specifically for the purpose of fighting COVID-19, the same will be admissible towards CSR expenditure.'

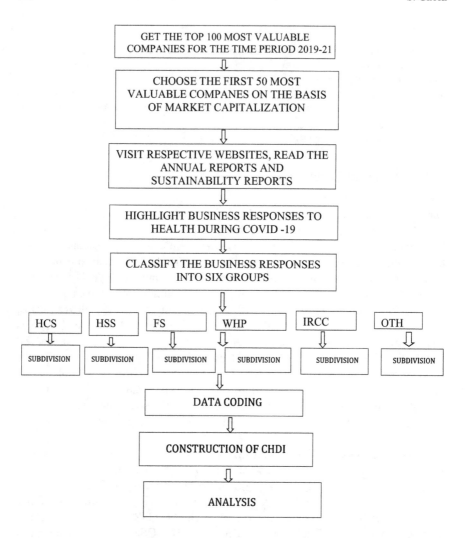

Source. Illustration made by author.

Note: HCS- Health Care System, HSS-Health System Strengthening, FS-Financial Services, WHP-Wellness and Health Programmes, IRCC-Innovative Response to Combat Covid -19, OTH- Other Programmes.

Fig. 1 Research framework. Source. Illustration made by author. Note: *HCS* Health Care System, *HSS* Health system strengthening, *FS* Financial Services, *WHP* Wellness and Health programmes, *IRCC* Innovative response to combat COVID-19, *OTH* Other programmes

The overall results are provided in Table 3.

The 41 indicators of business responses to health during COVID-19 were grouped into six main categories: (a) Health Care Services (HCS), (b) Health system strengthening (HSS), (c) Financial Support (FS), (d) Wellness and Health

Table 3 Responsiveness of the companies towards individual CHDI parameters (%)

CHDI parameters	Responsiveness of the companies towards CHDI parameters (%)
HCS1	2.00
HCS2	8.00
HCS3	4.00
HCS4	4.00
HCS5	**68.00**
HCS6	34.00
HCS7	18.00
HCS8	10.00
HCS9	4.00
HCS10	4.00
HCS11	0.00
HCS12	4.00
HCS13	8.00
HCS14	10.00
HSS1	4.00
HSS2	**86.00**
HSS3	24.00
HSS4	**84.00**
HSS5	28.00
HSS6	6.00
HSS7	2.00
HSS8	10.00
HSS9	8.00
HSS10	2.00
HSS11	4.00
HSS12	2.00
FS1	**76.00**
FS2	32.00
FS3	14.00
FS4	18.00
FS5	6.00
FS6	18.00
WHP1	20.00
WHP2	24.00
WHP3	8.00
WHP4	4.00
IRCC1	16.00
OTH1	2.00
OTH2	4.00
OTH3	2.00
OTH4	2.00

Source: Author's own computation

Table 4 Responsiveness of the companies towards CHDI parameters categorized into groups

CHDI parameters categorized into groups	Responsiveness of the companies (%)
HCS	13.00
HSS	**22.00**
FS	**27.00**
WHP	14.00
OTH	3.00
IRCC	16.00

Source: Author's own computation

programmes (WHP), (e) Innovative responses to combat COVID-19 (IRCC) and (f) Others (OTH). Table 4 presents companies responsiveness towards these six groups. Out of the six groups, HSS and FS turned out to be significant where in 22% of the companies invested in strengthening the health system of our country to fight against the pandemic and 27% of the companies made financial contribution to support the country to fight against COVID-19.

Further analysis of the group scores of individual companies as presented in Table 5 highlights the fact that *Reliance Industries*, an Indian multinational conglomerate company with a diverse range of businesses including energy, petrochemicals, natural gas, retail, telecommunications, mass media and textiles, made the highest contribution in *Health Care Services group (HCS)* securing a score of 0.43. They contributed through efforts such as (1) organising for daily COVID-19 symptom checker for all employees; (2) health consultations provided through Jio Health Hub app; (3) employees and family members could book a free video consultation with RIL doctors and receive virtual healthcare, counselling, diagnosis and prescription; (4) encouraging daily monitoring of health and reinforcing safety and hygiene practices; (5) 'rolling out own vaccination programme, R-Surakshaa, across locations for all the employees and eligible family members' and (6) implemented aggressive RT-PCR and antigen testing for employees entering the premises.

Jindal Steel and Power Ltd., an Indian steel and energy company, made the highest contribution towards *Health Service Strengthening (HSS)* group, securing a score of 0.58. It helped strengthening the healthcare system by (1) supporting the healthcare infrastructure by providing ventilators, oxygen concentrators and 30-bed isolation centres; (2) providing COVID-19 protective gears such as PPE kits, sanitisers, masks, gloves, soaps, etc. to the employees and society at large; (3) providing mobile clinic and ambulance; (4) providing dedicated Covid testing laboratory and (5) providing tele-medicine centres.

Bajaj Finserv Ltd., a core investment company, *Kotak Mahindra Bank*, a banking company and *Redington (India) Ltd.*, a supply chain solution provider, had made the highest contribution towards *Financial Services group (FS)*, and their score being 0.67. Their efforts included (1) reimbursing expenses towards procurement of oxygen cylinders; (2) reimbursement of purchase of medicines for the employees and their family members; (3) reimbursement towards servicing of generators for the healthcare centres; (4) reimbursement towards repairing charges of medical

Table 5 Individual company group score

Name of the companies	HCS	HSS	FS	WHP	IRCC	OTH
NTPC	0.36	0.42	0.33	**0.50**	0.00	0.00
Reliance Industries	**0.43**	0.33	0.17	**0.50**	0.00	0.00
Mahindra and Mahindra Ltd.	0.21	0.33	0.50	0.25	**1.00**	0.00
Bajaj Finserv Ltd.	0.21	0.25	**0.67**	0.25	0.00	**0.25**
Jindal Steel and Power Ltd.	0.14	**0.58**	0.33	0.00	0.00	0.00
HCL Technologies Ltd.	0.21	0.33	0.33	0.25	**1.00**	0.00
Bharti Airtel Ltd.	0.29	0.25	0.33	0.25	0.00	0.00
Hindalco Industries Ltd.	0.21	0.33	0.33	0.25	0.00	0.00
Larsen and Toubro	0.14	0.33	0.33	**0.50**	0.00	0.00
Wipro Ltd.	0.29	0.25	0.33	0.00	0.00	0.00
Vedanta Ltd.	0.07	0.42	0.33	0.25	0.00	0.00
Kotak Mahindra Bank	0.07	0.17	**0.67**	**0.50**	0.00	0.00
ICICI Bank Ltd.	0.14	0.33	0.50	0.00	0.00	0.00
Tata Consultancy Services Ltd.	0.14	0.25	0.17	0.25	**1.00**	**0.25**
ICICI Prudential Life Insurance Company	0.21	0.17	0.33	0.25	**1.00**	0.00
Redington (India) Ltd.	0.07	0.08	**0.67**	0.25	0.00	**0.25**
Grasim Industries Ltd.	0.14	0.33	0.33	0.00	0.00	0.00
Infosys Ltd.	0.14	0.25	0.17	0.25	**1.00**	0.00
Axis Bank	0.21	0.17	0.17	0.50	0.00	0.00
HDFC Bank Ltd.	0.14	0.25	0.50	0.00	0.00	0.00
State Bank of India	0.07	0.25	0.33	0.00	**1.00**	**0.25**
GAIL India Ltd.	0.14	0.33	0.17	0.00	0.00	0.00
ITC Ltd.	0.21	0.25	0.17	0.00	0.00	0.00
Bharat Petroleum Corporation Ltd.	0.21	0.25	0.17	0.00	0.00	0.00
Tata Steel	0.14	0.25	0.17	0.25	0.00	0.00
Housing Development Finance Corporation of India	0.07	0.25	0.33	0.25	0.00	0.00
Maruti Suzuki India Ltd.	0.14	0.25	0.00	0.25	**1.00**	0.00
Hindustan Unilever Ltd.	0.14	0.17	0.17	0.25	**1.00**	0.00
ONGC	0.07	0.25	0.17	0.00	0.00	**0.25**
SBI Life Insurance Company Ltd.	0.07	0.17	0.50	0.00	0.00	0.00
Power Finance Corporation Ltd.	0.07	0.25	0.17	0.25	0.00	0.00
Indian Oil Corporation Ltd.	0.07	0.17	0.33	0.00	0.00	0.00
UltraTech Cement Ltd.	0.07	0.25	0.17	0.00	0.00	0.00
JSW Steel Ltd.	0.14	0.17	0.17	0.00	0.00	0.00
Steel Authority of India	0.14	0.17	0.17	0.00	0.00	0.00
Coal India Ltd.	0.07	0.25	0.17	0.00	0.00	0.00
Tata Motors	0.14	0.17	0.00	0.25	0.00	0.00
Hindustan Petroleum Corporation Ltd.	0.07	0.17	0.33	0.00	0.00	0.00
HDFC Life Insurance Company Ltd.	0.07	0.08	0.17	**0.50**	0.00	0.00
Canara Bank	0.07	0.08	0.50	0.00	0.00	0.00
Power Grid Corporation of India Ltd.	0.07	0.08	0.33	0.00	0.00	0.00
Motherson Sumi Systems Ltd.	0.07	0.17	0.17	0.00	0.00	0.00

(continued)

Table 5 (continued)

Name of the companies	HCS	HSS	FS	WHP	IRCC	OTH
Indian Bank	0.00	0.17	0.33	0.00	0.00	0.00
Union Bank of India	0.07	0.08	0.33	0.00	0.00	0.00
Punjab National Bank	0.00	0.17	0.33	0.00	0.00	0.00
Bank of Baroda	0.07	0.08	0.00	0.00	0.00	0.00
Vodafone Idea Ltd.	0.00	0.08	0.00	0.00	0.00	0.00
General Insurance Corporation of India	0.00	0.00	0.17	0.00	0.00	0.00
Bank of India	0.00	0.00	0.17	0.00	0.00	0.00
Rajesh Exports	0.00	0.00	0.00	0.00	0.00	0.00

Source: Author's own computation

equipment of the healthcare institutions; (5) Corona Kavach insurance policy/medical insurance cover provided to the employees; (6) contribution to the PM CARES Fund/Chief Minister's Public Relief Fund; (7) Covid Care Allowance for the employees to support Covid-related care such as tests, scans, medicines and buying other items like oximeter, masks and sanitisers; (8) financial assistance to employees for medical expenditure; (9) financial assistance to upgrade healthcare infrastructure and (10) 'Protecting the Protectors', where the employees raised money to support COVID-19 relief operations undertaken by the doctors.

NTPC, India's largest energy conglomerate; *Reliance Industries* and *Larsen and Toubro*, Indian multinational conglomerates, with a diverse range of business including engineering, construction, manufacturing, technology and financial services; *Kotak Mahindra Bank* and *HDFC Life Insurance Company Ltd.*, life insurance companies—all have contributed immensely in *Wellness and Health programmes group (WHP)*, with a score of 0.50. They took the following initiatives: (1) multiple health initiatives for employees such as virtual medical consultations, yoga sessions and guidance on mental health/various health issues, emotional and wellness-related initiatives were organised; (2) psychological guidance sessions were arranged on well-being and shared with employees; (3) wellness initiatives, popularly recognised as 'Health to the Power Infinity', and various health and wellness-related initiatives through online and onsite interactions such as Emotional Assistance program, Employee Outreach program were organised; (4) 'Mindful Morning Movement' series for mindfulness, meditation and yoga; 'November Movement', with focus on men's health; 'Walkathon', with focus on walking and contributing to a cause, were organised; (5) SNEHAL-24 × 7 online counselling program was introduced and (6) 'iCall' online counselling was introduced.

Interestingly, *Innovative Responses to Combat COVID-19 (IRCC)* was observed in few companies. *Mahindra and Mahindra Ltd.*, an Indian multinational conglomerate (score 1), launched Mhealthy, a comprehensive solution powered by new-generation technologies to enable data-driven digital diagnostics to enable workforce and community safety against COVID-19 (antibody and screening tests). The company also contributed to UNICEF's 'Flush the Virus'—a Community

Toilet Sanitisation initiative. *HCL Technologies Ltd.*, an Indian multinational information technology services and consulting company (score 1), launched the #TakeCareHCL program to create COVID-19 awareness. *Tata Consultancy Services Ltd.*, an Indian multinational information technology services and consulting company (score 1), provided "TCS Data Marketplace solution, to support Indigenisation of Diagnostics, an ambitious new project launched by the Government of India to scale up indigenous Covid-19 diagnostic test-kit production capacity to a million test kits a day". *ICICI Prudential Life Insurance Company*, a life insurance company (score 1), created an app—'iWorkSafe' for employees to submit their health status daily. *Infosys Ltd.*, an Indian multinational information technology company (score 1), "created mobile applications like 'Crush Covid RI' and 'Apthamitra' to help local governments in their fight against COVID-19". *State Bank of India*, an Indian multinational public sector bank and financial services (score 1), came up with Project Praana, which aimed to design an electro-mechanical ventilator with locally available components so that the production of the ventilators can be rapidly scaled up. They also launched the India Health Alliance (IHA), a collaborative healthcare programme and HelloSwasti™, a comprehensive tele-care solution for the vulnerable. *Maruti Suzuki India Ltd.*, an Indian automobile manufacturer (score 1), developed a health application named 'Kushal Mangal' and provided to Tier-1 supplier partners at free of cost towards their COVID-19 management. *Hindustan Unilever Ltd.*, a consumer goods company (score 1), launched 'Project U CARES' a Health Improvement Program. The Rural Telehealth program 'Jio Mobile Doctornia' was launched as a revolutionary telehealth ecosystem. 'Mum's Magic Hands' (MMH) handwashing behaviour change programme was initiated in order to reduce health and hygiene risks related to diarrhoea and COVID-19.

Apart from the above, some companies (score 0.25) have also undertaken certain other measures (OTH) such as conducting periodic COVID-19 health risk assessment of employees through online surveys: calling and contact tracing by *Bajaj Finserv Ltd.* "TCSiON CoronaWarriors, an online learning program, was specifically created for paramedical and professional healthcare workers on the fundamentals of infection prevention and control of COVID-19" by *Tata Consultancy Services Ltd. Redington (India) Ltd.* supported the wards of deserving paramedical staff and housekeeping staff by providing groceries. *State Bank of India*, partnered with ECHO India for training and mentoring healthcare providers on prevention and management of COVID-19. *ONGC*, an Indian government-owned crude oil and natural gas corporation, initiated training by in-house doctor/medical staff on Covid protocols and sensitized about Covid-appropriate social behaviour.

Using the 41 parameters listed in Table 2 above, we have computed the CHDI score using the method explained above. The CHDI computed in the above manner will help us to compare the health disclosure standards across our sample firms during COVID-19. The CHDI scores are in the ranges (≤ 0.05), (>0.05–0.10), (>0.10–0.15), (>0.15–0.20), (>0.20–0.25), (>0.25–0.30) and (>0.30–0.34). The minimum CHDI score is 0.00 and the maximum CHDI score is 0.34. For instance, NTPC has disclosed 14 parameters out of the 41 parameters specified in Table 2, hence its CHDI score is 0.34 (14/41). The rest of the companies CHDI have been calculated similarly. This implies that while all the companies, except one, have disclosed their responses to COVID-19, they have failed to be responsive to many crucial actions (Table 2) that had to be taken to curb such a pandemic. The companies have been classified according to these ranges wherein the range (>0.30–0.35) denotes Group 7 and (≤0.05) Group 1. The remaining ranges denote Groups 6, 5, 4, 3 and 2, respectively. Group 7 denotes companies whose CHDI fall within the highest range (>0.30–0.35) and Group 1 denotes companies whose CHDI fall within the lowest range (≤0.05). This has been highlighted in Table 6.

Out of a sample set of 50 companies, we observe about 28% (14) of the companies fall in Group 3 where the CHDI range is (>0.10–0.15) and only 4% (2) companies fall in the highest tier Group 7 where the CHDI range is (>0.30–0.35) as presented in Table 7.

5   Conclusion and Discussion

Out of the 41 parameters, we observed that four parameters turned out to be highly significant and they are supporting healthcare infrastructure (HSS2), providing COVID-19 protective gears (HSS4), assisting in vaccination programme (HCS5) and contributing to the PM CARES Fund/Covid Relief measures (FS1). Out of the six groups, HSS and FS turned out to be significant wherein 22% of the companies invested in strengthening the healthcare system of our country to fight against the pandemic and 27% of the companies made financial contribution to support the country to fight against COVID-19. Reliance Industries, an Indian multinational conglomerate company with a diverse range of businesses including energy, petrochemicals, natural gas, retail, telecommunications, mass media and textiles, made the highest contribution in Health Care Services group (HCS), securing a score of 0.43. Jindal Steel and Power Ltd., an Indian steel and energy company, made the highest contribution towards Health Service Strengthening (HSS) group, securing a score of 0.58. Bajaj Finserv Ltd., a core investment company, Kotak Mahindra Bank, a banking company and Redington (India) Ltd., a supply chain solution provider, had made the highest contribution towards Financial Services group (FS), their score being 0. 67. NTPC, India's largest energy conglomerate; Reliance Industries and Larsen and Toubro, Indian multinational conglomerates; Kotak Mahindra Bank and HDFC Life Insurance Company Ltd., life insurance companies—all have contributed immensely in Wellness and Health programmes group (WHP), with a score of 0.50. Interestingly, 'Innovative responses to combat COVID-19 (IRCC)' was observed in few companies. Apart from the above, some companies such as Bajaj Finserv Ltd., Tata Consultancy Services Ltd., Redington (India) Ltd., State Bank of India and ONGC have also undertaken certain other measures (OTH) such as conducting periodic COVID-19 health risk assessment of employees through online surveys, online learning program. From our study we also

Table 6 Individual company CHDI score

Name of the companies	≤0.05	>0.05–0.10	>0.10–0.15	>0.15–0.20	>0.20–0.25	>0.25–0.30	>0.30–0.34	Group nos.
NTPC							0.34	7
Reliance Industries							0.33	7
Mahindra and Mahindra Ltd.						0.28		6
Bajaj Finserv Ltd.						0.30		6
Jindal Steel and Power Ltd.						0.28		6
HCL Technologies Ltd.					0.25			5
Bharti Airtel Ltd.					0.25			5
Hindalco Industries Ltd.					0.25			5
Larsen and Toubro					0.25			5
Wipro Ltd.					0.23			5
Vedanta Ltd.					0.23			5
Kotak Mahindra Bank					0.23			5
ICICI Bank Ltd.					0.23			5
Tata Consultancy Services Ltd.				0.20				4
ICICI Prudential Life Insurance Company				0.20				4
Redington (India) Ltd.				0.20				4
Grasim Industries Ltd.				0.20				4
Infosys Ltd.				0.18				4
Axis Bank				0.20				4
HDFC Bank Ltd.				0.20				4
State Bank of India				0.18				4
GAIL India Ltd.				0.18				4
ITC Ltd.				0.18				4
Bharat Petroleum Corporation Ltd.				0.18				4
Tata Steel				0.18				4

(continued)

Table 6 (continued)

Name of the companies	≤0.05	>0.05-0.10	>0.10-0.15	>0.15-0.20	>0.20-0.25	>0.25-0.30	>0.30-0.34	Group nos.
Housing Development Finance Corporation of India				0.18				4
Maruti Suzuki India Ltd.			0.15					3
Hindustan Unilever Ltd.			0.15					3
ONGC			0.15					3
SBI Life Insurance Company Ltd.			0.15					3
Power Finance Corporation Ltd.			0.15					3
Indian Oil Corporation Ltd.			0.13					3
UltraTech Cement Ltd.			0.13					3
JSW Steel Ltd.			0.13					3
Steel Authority of India			0.13					3
Coal India Ltd.			0.13					3
Tata Motors			0.13					3
Hindustan Petroleum Corporation Ltd.			0.13					3
HDFC Life Insurance Company Ltd.			0.13					3
Canara Bank			0.13					3
Power Grid Corporation of India Ltd.		0.10						2
Motherson Sumi Systems Ltd.		0.10						2
Indian Bank		0.10						2
Union Bank of India		0.10						2
Punjab National Bank		0.10						2
Bank of Baroda	0.05							1
Vodafone Idea Ltd.	0.03							1
General Insurance Corporation of India	0.03							1
Bank of India	0.03							1
Rajesh Exports	0.00							1

Source: Author's own computation

Table 7 Frequency of companies and CHDI score range

CDHI score	Frequency of companies	Percent	Cumulative percent
≤0.05	5	10.0	10.0
>0.05–0.10	5	10.0	20.0
>0.10–0.15	**14**	**28.0**	**44.0**
>0.15–0.20	13	26.0	70.0
>0.20–0.25	8	16.0	88.0
>0.25–0.30	3	6.0	96.0
>0.30–0.35	2	4.0	100.0
Total	50	100.0	

Source: Author's own computation

observe that out of a sample set of 50 companies, about 28% (14) of the companies fall in Group 3 where the CHDI range is (>0.10–0.15) and only 4% (2) of the companies fall in the highest tier Group 7 where the CHDI range is (>0.30–0.35).

We developed a CHDI score by using 41 parameters developed through Longitudinal Qualitative Document Analysis of the company's annual reports, websites and sustainability reports for the financial years 2019–2021. Our results showed that business response towards health during COVID-19 was average. Businesses have mostly concentrated on short-term plans, primarily supporting healthcare infrastructure, assisting in vaccination programmes and contributing to the PM CARES Fund. Long-term plans for reintegrating people, especially the vulnerable one, into normalcy were not so much emphasized by companies in general. This study attempted to develop a Corporate Health Disclosure Index during the pandemic era, and we believe that the results will be useful to the companies and administrators in evaluating their performance in terms of health-related steps taken during COVID-19 and what necessary plans can be prepared to improve the situation. Since literature shows that study in this regard is very limited, we have tried to be as innovative as possible in our approach, taking the first 50 large listed companies. Therefore, the sample was restricted.

References

Aguinis, H., Villamor, I., & Gabriel, K. P. (2020). Understanding employee responses to COVID-19: A behavioral corporate social responsibility perspective. *Management Research, 18*(4), 421–438.

Aknin, L. B., & Whillans, A. V. (2020). Helping and happiness: A review and guide for public policy. *Social Issues and Policy Review, 15*, 1–32.

Anser, M. K., Islam, T., Khan, M. A., Zaman, K., Nassani, A. A., Askar, S. E., Abro, M. M. Q., & Kabbani, A. (2020a). Identifying the potential causes, consequences, and prevention of communicable diseases (including COVID-19). *BioMed Research International, 2020*, 1–13. Article ID: 8894006.

Anser, M. K., Yousaf, Z., Khan, M. A., Sheikh, A. Z., Nassani, A. A., Abro, M. M. Q., & Zaman, K. (2020b). Communicable diseases (including COVID-19)—Induced global depression: Caused by inadequate healthcare expenditures, population density, and mass panic. *Front Public Health, 8*, 398.

Bapuji, H., Patel, C., Ertug, G., & Allen, D. G. (2020). Corona crisis and inequality: Why management research needs a societal turn. *Journal of Management, 46*(7), 1205–1222.

Beck, M. J., & Hensher, D. A. (2020). *Insights into the impact of Covid-19 on household travel, working, activities and shopping in Australia–the early days under restrictions*. Retrieved from https://ses.library.usyd.edu.au/handle/2123/22247

Boadi, E. A., He, Z., Bosompem, J., Say, J., & Boadi, E. K. (2019). Let the talk count: Attributes of stakeholder engagement, trust, perceive environmental protection and CSR. *SAGE Open, 9*(1), 1–15.

Chaudhary, M., Sodani, P. R., & Das, S. (2020). Effect of COVID-19 on Economy in India: Some Reflections for Policy and Programme. *Journal of Health Management, 22*(2), 169–180.

Donthu, N., & Gustafsson, A. (2020). Effects of COVID-19 on business and research. *Journal of Business Research, 117*, 284–289.

Ducharme, J. (2020, March 11). World Health Organization declares COVID-19 a "Pandemic." Here's what that means. *The Time*. https://time.com/5791661/who-coronavirus-pandemic-declaration/

Duvendack, M., & Sonne, L. (2021). Responding to the multifaceted COVID-19 crisis: The case of Mumbai, India. *Progress in Development Studies., 21*(4), 361–379.

Euronews. (2020). *COVID-19: World economy in 2020 to suffer worst year since 1930s Great Depression, says IMF*. Retrieved https://www.euronews.com/2020/04/14/watch-live-interna tional-monetary-fund-gives-world-economic-outlook-briefing-oncovid-19

García-Sánchez, I. M., & García-Sánchez, A. (2020). Corporate social responsibility during COVID-19 pandemic. *Journal of Open Innovation: Technology, Market, and Complexity, 6*(4), 126.

Georage, R. D. (1981). Moral responsibility and the corporation. *Philosophic Exchange, 12*(1), Article 3.

Glaser, B. G., & Strauss, A. (1967). *The discovery of grounded theory: Strategies for qualitative research*. Aldine Publishing.

Gokarna, P., & Krishnamoorthy, B. (2021). Corporate social responsibility in the time of COVID-19 pandemic: An exploratory study of developing country corporates. *Corporate Governance and Sustainability Review, 5*(3), 73–80.

Gostin, L. O., & Wiley, L. F. (2020). Governmental public health powers during the COVID-19 pandemic: Stay-at-home orders, business closures, and travel restrictions. *Jama, 323*(21), 2137–2138.

Guan, D., Wang, D., Hallegatte, S., Davis, S. J., Huo, J., Li, S., et al. (2020). Global supply-chain effects of COVID19 control measures. *Nature Human Behaviour, 4*, 577–575.

Guerriero, C., Haines, A., & Pagano, M. (2020). Health and sustainability in post-pandemic economic policies. *Nature Sustainability, 3*, 494–496.

Hasan, M. M., Nekmahmud, M., Yajuan, L., & Patwary, M. A. (2019). Green business value chain: A systematic review. *Sustainable Production and Consumption, 20*, 326–339.

Hasan, M. M., Popp, J., & Oláh, J. (2020). Current landscape and influence of big data on finance. *Journal of Big Data, 7*, 21.

He, H., & Harris, L. (2020). The impact of Covid-19 pandemic on corporate social responsibility and marketing philosophy. *Journal of Business Research, 116*, 176–182.

Hewings-Martin, Y. (2020, April 10). How do SARS and MERS compare with COVID-19? *Medical News Today*. https://www.medicalnewstoday.com/articles/how-do-sars-and-merscompare-with-covid-19

Kaplan, J., Frias, L., & McFall-Johnsen, M. (2020, July 11). A third of the global population is on coronavirus lockdown—Here's our constantly updated list of countries and restrictions. *Business Insider*. https://www.businessinsider.com/countrieson-lockdown-coronavirus-italy-2020-3

Kucharska, W., & Kowalczyk, R. (2019). How to achieve sustainability? Employee's point of view on company's culture and CSR practice. *Corporate Social Responsibility and Environmental Management, 26*(2), 453–446.

Ladha, R. S. (2020). Coronavirus: A framework to Decide between national and local lockdown. *Journal of Health Management., 22*(2), 215–223.

Mahmud, A., Ding, D., Kiani, A., & Hasan, M. (2020). Corporate social responsibility programs and community perceptions of societal progress in Bangladesh: A multimethod approach. *SAGE Open, 10*(2), 1–17.

McKibbin, W. J., & Fernando, R. (2020, March 2). *The global macroeconomic impacts of COVID-19: Seven scenarios.* CAMA Working Paper No. 19/2020, Available SSRN: https://ssrn.com/abstract=3547729 or https://doi.org/10.2139/ssrn.3547729

Peeri, N. C., Shrestha, N., Rahman, M. S., Zaki, R., Tan, Z., Bibi, S., Baghbanzadeh, M., Aghamohammadi, N., Zhang, W., & Haque, U. (2020). The SARS, MERS and novel corona-virus (COVID-19) epidemics, the newest and biggest global health threats: What lessons have we learned? *International Journal of Epidemiology, 49*, dyaa033.

Rao, S. S., Pushpalatha, K., Sapna, R., & Monika Rani, H. G. (2022). An analysis of the psychological implications of COVID-19 pandemic on undergraduate students and efforts on mitigation. In D. Garg, S. Jagannathan, A. Gupta, L. Garg, & S. Gupta (Eds.), *Advanced computing. IACC 2021. Communications in computer and information science* (Vol. 1528). Springer.

Sharif, A., Aloui, C., & Yarovaya, L. (2020). COVID-19 pandemic, oil prices, stock market, geopolitical risk and policy uncertainty nexus in the US economy: Fresh evidence from the wavelet-based approach. *International Review of Financial Analysis, 70*, 101496.

Sibley, C. G., Greaves, L., Satherley, N., Wilson, M. S., Lee, C., Milojev, P., et al. (2020). Short-term effects of the COVID-19 pandemic and a nationwide lockdown on institutional trust, attitudes to government, health and wellbeing. *The American Psychologist, 75*, 618–630.

Singh, A. K., Agrawal, B., Sharma, A., & Sharma, P. (2020). COVID-19: Assessment of knowl-edge and awareness in Indian society. *Journal of Public Affairs, 20*(4), e2354.

Taleb, N. N. (2008). Infinite variance and the problems of practice. *Complexity, 14*(2).

The Ministry of Corporate Affairs. (2020). *Vide General Circular No. 10/2020 dated 23rd March, 2020.* Government of India.

Voegtlin, C., & Greenwood, M. (2016). Corporate social responsibility and human resource management: A systematic review and conceptual analysis. *Human Resource Management Review, 26*(3), 181–197.

Willcocks, L. P. (2020). The Covid-19 pandemic shines a spotlight on the systemic risk to global business. *LSE Business Review.* Retrieved from http://eprints.lse.ac.uk/104378/1/businessreview_2020_04_28_the_covid_19_pandemic_shines_a_spotlight_on.pdf

Xiao, M., Cooke, F. L., Xu, J., & Bian, H. (2020). To what extent is corporate social responsibility part of human resource management in the Chinese context? A review of literature and future research directions. *Human Resource Management Review, 30*(4), 100726.

Zhang, D., Hu, M., & Ji, Q. (2020). Financial markets under the global pandemic of COVID-19. *Finance Research Letters, 36*, 101528.

Sumona Ghosh has been associated with St. Xavier's College Kolkata since 2002. She was the head of the Department of Law from 2003 to 2018. Presently she is the Joint Coordinator of the Foundation Course. After completing her post-graduation in Commerce with rare distinction, Prof. Ghosh has been conferred with the Degree of Philosophy in Business Management by the University of Calcutta on 31st of July 2014. Her area of research was on Corporate Social Responsibility (CSR). The title of her doctoral dissertation was "Pattern of participation of Public and Private sector companies in Corporate Social Responsibility Activities".

She has published in journals of national and international repute. Dr Ghosh has been highly acclaimed for her guest lectures on CSR in premier institutes of higher learning including the Indian

Institute of Management (Calcutta), Indian Institute of Management (Shillong). She has taken sessions in Management Development Programmes conducted by premier institutes on CSR. She has presented papers on CSR at various national and international conferences. Her research interest lies in Corporate Social Responsibility, Sustainable Development, Integrated Reporting and Philosophies of management. Dr Ghosh is also a Certified Assessor for Sustainable Organizations (CASO), certification conferred upon her by UBB GmBH Germany. Dr Ghosh has been appointed as the Council member of the Sustainable Businesses Council of the very first independent National Business Chamber for Women that has been established in India—Women's Chamber of Commerce and Industry (WICCI) (www.wicci.in).

The Rippling Effect of COVID-19 in Malaysia: Now and Then

Sam Sarpong and Ali Saleh Alarussi

1 Introduction

The spread of the COVID-19 pandemic has had a significant socio-economic impact on the lives of people over the past years. Since its onset, the COVID-19 pandemic has caused much havoc to many countries. Social and physical distancing measures and lockdowns of businesses, schools and overall social life became commonplace to curtail the spread of the disease. It also put considerable pressure on healthcare systems worldwide. Malaysians also experienced the same trend, as the pandemic ravaged its economy, affected its healthcare systems and laid bare the vulnerability of many families in the face of the unprecedented crisis.

Malaysia has undergone major waves of COVID-19 outbreaks and months of lockdowns since the announcement of the pandemic in March 2020. Emotional stories of families with depleted savings and surviving on one meal a day also made the news since the pandemic emerged. Police statistics have also showed that suicides went up during the pandemic (Idrus, 2021). Women, alongside the poor, elderly, disabled and migrant populations, have particularly borne the brunt of the fallout from the pandemic. Other vulnerable groups, economically at risk prior to the crisis, include those with lower levels of education and self-employed. Malaysia is currently on the path of challenging recovery, with the impact of COVID-19 remaining a major concern for the country.

The chapter provides an insight into how the pandemic ravaged Malaysia and the interventions the government made to redress the situation. The analysis is made from secondary materials sourced from government statements, policy briefs, media

S. Sarpong (✉) · A. S. Alarussi
School of Economics and Management, Xiamen University Malaysia, Sepang, Selangor, Malaysia
e-mail: samsarpong@xmu.edu.my; alisaleh@xmu.edu.my

© The Author(s), under exclusive license to Springer Nature Switzerland AG 2023
S. O. Idowu et al. (eds.), *Corporate Social Responsibility in the Health Sector*, CSR, Sustainability, Ethics & Governance, https://doi.org/10.1007/978-3-031-23261-9_15

reports and other COVID-19 materials. These were complemented with personal observations and discussions with noted authorities.

It begins with a background on the COVID-19 situation in Malaysia. It then explores recent studies done on Covid. The methodology used by the chapter is then described in detail. The chapter analyses the impact of the pandemic in the following areas: health, economy, socio-economic and humanitarian aspects that are important to the country.

2 The Emergence of COVID-19 in Malaysia

Malaysia was quite swift in terms of its response after it recorded its first case of COVID-19 on 25 January 2020. It subsequently introduced social and public health measures, including a nationwide Movement Control Order (MCO) on 18 March 2020. A special session of the National Security Council was convened daily to monitor the COVID-19 situation, and briefings by the Director-General of Health kept Malaysians informed on infection clusters, recoveries and updates on the pandemic response.

The Malaysian government implemented different levels of MCO to effectively control the COVID-19 situation, ranging from high-risk to low-risk areas.

- Phase 1—MCO from 18 March 2020 till 3 May 2020.
- Phase 2—Conditional MCO (CMCO) from 4 May 2020 till 9 June 2020.
- Phase 3—Recovery MCO (RMCO) from 10 June 2020 till 31 December 2020.
- Phase 4—CMCO in the areas with high COVID-19 cases from 14 December 2020 till 31 December 2020.
- Phase 5—RMCO nationwide from 1 January 2021 till 31 March 2021; certain states with high COVID-19 cases were placed under MCO 2.0 from 13 January 2021 till 4 March 2021.
- Recent measures—currently, the borders have been opened and major restrictions have been lifted.

3 General Situation

Malaysia reopened its borders to international tourists on 1 April 2022. It has also lifted the mandatory use of masks, effective 1 May 2022. These are part of measures by the government to ease the restrictions imposed since the emergence of the pandemic. In easing the restrictions, the government explained that the reported cases have gone down considerably. Besides, 81.3% of Malaysia's total population had been fully vaccinated, whilst 68% of the adult population has taken a booster dose. In addition, 42% of the children population (aged 5–12) has completed at least one dose and nearly 20% has completed two doses.

The impact of COVID-19 on the world economy as a whole has been devastating. According to the Organisation for Economic Co-operation and Development (OECD), the COVID-19 pandemic has led to social distress around the world, as well as huge economic disruption (OECD, 2020). The massive spread of the virus in Malaysia significantly disrupted social and business activities. It affected people's incomes and caused economic chaos in the country. To minimise the economic impact of this pandemic, Malaysia took several actions to recover the economy. At the end of February 2020, the government announced a RM20 billion financial stimulus package intended to mitigate the impact of COVID-19. This was based on three major strategies, namely (1) lessen the effect of COVID-19, (2) people-based economic growth and (3) encourage quality investments (The Star, 2020a).

The swift measures the government put in place in response to the pandemic also attracted domestic and international praise. The country's anti-Covid blueprint revolved around three pivotal tenets: encompassing the enhancement of screening capabilities and hospitalisation capacity in the public health sector; the adoption of contact tracing technologies aimed at bringing disease clusters under control; as well as the launch of an effective communication strategy to present new Standard Operating Procedures (SOP) and win the support of the population. Other than the loss of hundreds of lives, this pandemic paralysed the country in many aspects. On top of that, the closure of Malaysia's borders also dealt a huge blow to the country's tourism industry, which is the third-largest contributor to Malaysia's gross domestic product (GDP), after the manufacturing and commodities sectors.

4 Literature Review

Numerous studies have been conducted on COVID-19 globally since its inception. Cases involving Malaysia have also been covered extensively too. For instance, Amaran et al. (2021) looked at the early response by the Malaysian authorities to the pandemic. They chronicled the series of events from the time the pandemic got to Malaysia and described the experiences of the Malaysian healthcare system in combatting the pandemic. Meanwhile, Vicknasingam et al. (2021) evaluated how service providers and recipients were adapting and coping during the initial periods of the COVID-19 response. Nga et al. (2021), on the other hand, also examined how policies enacted during the COVID-19 pandemic affected unemployment in Malaysia by focussing on the situation in Sabah, a state in Malaysia. Others such as Cheng (2022) sought to capture the socio-economic impacts of COVID in Malaysia. In effect, numerous studies have been conducted on the pandemic and its impact on Malaysia within particular timeframes and using different methodologies.

5 Methodology

This chapter aims to contribute to the body of knowledge by bringing us up to date on how the pandemic has panned out and its impact on Malaysian society. It uses secondary data in the light of the many studies, which have been done on the subject matter. With a lot of studies done on Covid, the authors felt such information cannot be ignored and therefore ensured these outcomes were also examined in undertaking this project.

A document review was performed to identify publicly available information on the measures employed in Malaysia to address the COVID-19 pandemic. Government and the health ministry websites were also searched for relevant information/documents containing COVID-19 measures/strategies and case occurrence in Malaysia. Besides, WHO global research on the coronavirus disease database (WHO COVID-19 database) and PubMed (https://pubmed.ncbi.nlm.nih.gov/) were searched for relevant literature.

6 Malaysia's Healthcare System

One critical area that has borne the brunt of the pandemic has been Malaysia's healthcare system (KPMG, 2020). The Malaysian government took the responsibility of ensuring that its healthcare system can manage a public health response to the pandemic right from the time that COVID-19 emerged. The National Crisis Preparedness and Response Centre (CPRC) under the Ministry of Health (MOH), Malaysia, was activated on 5 January 2020 in response to that. The centre was established with the aim of minimising the potential impact of the epidemic on healthcare systems, social services and economic activity. These levels of engagements showed the level of preparedness that was put in Malaysia.

Several reasons may contribute to the overall level of preparedness of Malaysia to mitigate and manage the outbreak. One main reason is the structure of centralised commanding authorities at different levels that helped in the implementation of coherent strategies involving many stakeholders. Furthermore, effective communications on decisions undertaken by the government and awareness campaigns made public compliance possible.

Malaysia experienced intense emotions at the beginning of the pandemic, with great uncertainty regarding the pandemic's outcome, as the world saw a frighteningly high COVID-19 mortality (Amaran et al. 2021). However, in view of the fact that Malaysia had learnt from its previous experiences of managing infectious diseases such as Severe Acute Respiratory Syndrome (SARS), Pandemic Influenza 2009, Middle-East Respiratory Syndrome (MERS-CoV), Avian Influenza, Zika and Nipah virus, it was able to have a speedy primary healthcare response. The learning experiences from these outbreaks have strengthened the core public health functions as well as many key healthcare systems areas such as the health workforce, service delivery, information and technology system, as well as leadership and governance to support a more resilient healthcare system.

In addition, historically, the foundation of the Malaysian healthcare system has created a widespread and integrated public healthcare delivery system, distributed across the country including remote and rural areas that is universal and low cost. Malaysia has a dual public–private healthcare system in which the provision of public health services allows for the mobilisation of appropriate resources from national to state levels to support response at the facilities. This coordination highlights the importance of having a public healthcare system supported by government during a public health crisis, at least in Malaysia. In order to manage the increasing number of COVID-19 cases in Malaysia, rigorous and routine screening efforts and disinfection process in public or overcrowded places were important to control the spread of the infection too. The well-planned vaccination exercises were also quite instrumental in ensuring an effective way to reduce the effect of the pandemic on the people.

7 Pandemic Misinformation

In spite of the major efforts made in the vaccination exercise, there were some people who felt at odds with or were quite hesitant to undergo the vaccination. A key element in the rise of vaccine hesitancy and refusal, therefore, was the widespread misinformation and disinformation that went with that. This led to a widening trust deficit on vaccination, resulting in a decrease in vaccine confidence. As a result, the need for more active rebuttals to anti-vaccine messaging so as to reduce vaccine hesitancy and refusal in order to maintain high vaccination rates became necessary. The MOH and other relevant agencies in Malaysia such as National Security Council and Malaysian Communications and Multimedia Commission, therefore, utilised both conventional and digital media to communicate with the public to allay the fears of the public.

Despite the measures that the allied agencies took to dispel the wrong impressions of vaccinations, rumours and fake news were still circulated privately on social media, such as WhatsApp or Facebook posts, that required clarification from MOH and these agencies. The need, therefore, arose for a consistent and standardised presentation of information to the public to avoid any misinformation and also to ensure public confidence in the government's handling of the pandemic.

8 The Impact of Covid on Health Workers

The COVID-19 pandemic has tested the medical fraternity in Malaysia as it posed quite a serious challenge to those involved in handling it. The uncertainty of COVID-19, high risk exposure and other emerging issues made the situation quite a challenging one for the healthcare team. Many health workers suffered tremendously at the beginning of the pandemic. A total cumulative number of 224 health workers were diagnosed with COVID-19 as of 11 April 2020 (Nienhaus & Hod, 2020). Incidentally, 80% of these cases were community acquired. They got infected whilst attending religious and social gatherings. They, in turn, infected their colleagues when they started working at their various health facilities. Globally, surveillance data reported to World Health Organization (WHO) between January 2020 and May 2021 showed 6643 health and care workers died as a result of the pandemic (reliefweb, 2021). This is the officially reported figure; however, indications are that this could even be higher since the reporting processes are somewhat different in some countries.

Apart from the infection rate, which was quite noticeable amongst Malaysian health workers, many were also deemed to have experienced burnout. Roslan et al. (2021) attribute this to the high demands that came from their jobs. In order to provide care for the growing number of COVID-19 cases, healthcare professionals had to work with a great deal of pressure due to their increasing workload. They were also exposed to risk of getting infected by COVID-19 patients. Aside from that, there were inadequate personal protection equipment (PPE) for them and, more particularly, they also had less time to rest and spend with their families. Exhaustion appeared to be the major issue, whilst many others also reported of physical, occupational, psychological, and socially related negative impacts emanating from burnout.

Healthcare workers also experienced discomforting situations in view of their risk attached to their role as COVID-19 frontline health workers. The risky nature of their work compounded their stress levels and also led to severe depression among some of them. In order to reduce the high level of burnout of healthcare workers at their workplace, strategies such as increasing the availability of PPE and giving psychological support to them were made. Some also had their workload reduced. In addition, individual and organisational wellness strategies were implemented for healthcare providers throughout the country.

9 The Impact of Covid on Health Tourism

Given Malaysia's position as a fast-growing competitor in the global health and medical tourism sector, the advent of COVID-19 was indeed quite devastating for the country. In the years before COVID-19, this sector had shown a great potential as a major foreign exchange earner for the country. As a result, the government had taken a series of proactive measures to enhance Malaysia as a preferred health tourism destination. Health tourism in Malaysia consists of two main categories, which are medical tourism and wellness programme. Patients often opt for medical treatments in one of the country's internationally recognised hospitals and stay on during the convalescence or recovery period. Others come for a holiday by exploring the various forms of wellness programmes that are available in Malaysia.

Competitive medical fees and modern medical facilities remain two vital factors that make Malaysia a popular destination among health tourists. It offers a wide choice of state-of-the-art private medical centres boasting an impressive array of sophisticated diagnosis, therapeutic and in-patient facilities. These establishments are well-equipped and staffed to ensure the highest level of professionalism, safety and care to patients. Most private medical centres have certifications for internationally recognised quality standards or have been accredited by the Malaysian Society for Quality of Health.

Such is the state in which the government holds the health and medical tourism sector that when COVID-19 emerged, it clearly derailed the ambitions and set targets made in connection with the potential of the sector. Before COVID-19, the MOH set up a Corporate Policy and Health Industry Division to promote medical tourism and related healthcare products including traditional medicine, etc. The Ministry's promotional efforts are more focused with the support from relevant government agencies and the private sector.

In 2019, health tourism in Malaysia achieved more than MYR1.7 billion in hospital receipts, resulting in total economic impact of MYR7 billion. The pandemic disrupted several of the nation's key economic sectors including healthcare travel, which saw an impact of over 50% in revenue reduction. However, the sector is expected to pick up shortly by playing to its strengths and ensuring industry resilience through digital services and platforms to address the needs of healthcare travellers. It is predicted and expected that the numbers might return to at least RM1 billion by 2022.

10 The Impact of Covid on the Economy

Malaysia remains an attractive investment destination, with the availability of well-educated labour, investor-friendly policies and incentives, well-developed infrastructure and a preferred gateway to the Southeast Asia market. In addition, Malaysia's technologically inclined economy has been quite robust until the pandemic set in. The pandemic has derailed the huge achievements Malaysia was set to achieve. The economy was suffered greatly during the implementation of the first MCO. For the first quarter of 2021 when the second MCO was implemented, the country's GDP contracted by 0.5%, which is significantly lower than that of 2020 (The Malaysianreserve.com, 2021).

The movement restrictions caused disruptions to workflow and service delivery, especially to vulnerable groups. Adapting to new ways of remote working also proved quite challenging for some staff, and many reported working longer hours to accommodate the needs of clients. Furthermore, workers were also expected to frequently participate in virtual meetings and discussions. The lockdown also disrupted family life and daily routines, with severe consequences for children's welfare and well-being. Measures to prevent transmission of COVID-19 clearly exacerbated existing vulnerabilities.

The sudden enforcement of the MCO put various sectors of the economy in jeopardy. The direct damage caused by the virus can be seen in the tourism and travel industries, manufacturing, construction, mining and agriculture, with many workers being laid off and others being placed on unpaid leave (Murugiah, 2020). The tourism industry suffered an estimated loss of RM3.37 billion in the first 2 months of the year (Dzulkifly, 2020). The forced closure of small businesses, mainly the small- and medium-sized enterprises (SMEs) and services, also led to shutdowns and many losing their jobs, as well as individuals going bankrupt (Cheng, 2020).

Malaysia's government responded by promptly introducing the Prihatin Rakyat Economic Stimulus Package or PRIHATIN. This initiative was meant to ensure a level of support for businesses and strengthening the economy. In addition to enhancing the support for households and businesses, various programmes were also initiated, such as the Wage Subsidy Programme, moratorium on loan repayments, waivers or discounts to Small and Medium Scale Enterprises (SMEs) and a discount scheme on monthly electricity bills for 6 months, commencing 1 April 2020.

11   The Impact of Covid on the Lives of Malaysians

The real impact on the economy and people's lives, however, has been quite immense. The length of 'stay-at-home' policy and the duration of the MCO created a lot of difficulties for many people. Jobs have since been lost, workers have been retrenched or had their pay cut and aside that, many businesses have had to close. COVID-19 has also shown deep societal differences. Existing inequalities and inequities deepened, impacting hard on vulnerable groups. The pandemic impacted seriously on disadvantaged communities and different income groups quite differently. Following the first reported cases, the situation was compounded by some groups such as migrants or homeless people with poor living conditions who were living in overcrowding environments (Kluge et al., 2020).

There is every indication that the health and economic impacts of the virus are being borne disproportionately. People without access to running water, as well as refugees, migrants and displaced people, reportedly suffered disproportionately both from the pandemic and its aftermath, due to limited movement, fewer employment opportunities, and other Covid-induced problems. Signs of economic distress started appearing in neighbourhoods across Kuala Lumpur, the capital, and other Malaysian cities in the course of the pandemic, with people putting up white flags outside their houses to call for assistance either for food or for other pressing issues. The flags, sometimes little more than T-shirts or strips of cloth, were a cry for help from mostly low-income families who were financially affected by the second lengthy coronavirus lockdown. The campaign, shared on social media as *#benderaputih* (white flag), was a way for families to appeal for food, work or other essentials as many businesses closed and joblessness rose.

Non-governmental organisations (NGOs) actively helped those who were affected by this pandemic. They provided food, shelter for the homeless, and even gave out money to help those in need. Some NGOs also helped by providing protective masks, disinfection chambers (The Star, 2020b), whilst others helped to educate citizens on COVID-19 (The Star, 2020c). Several NGOs and public figures have helped to prepare PPE for medical frontliners. For example, several Malaysian fashion designers associated with the Malaysian Official Designers Association (MODA) have produced PPE for local medical staff (Cheong, 2020). Volunteer tailors also helped to prepare PPE for frontline staff.

Malaysia's repeated lockdowns, meanwhile, lowered demand for labour, with the number of registered jobs dropping by 130,000 in just the first quarter of the year, according to government data from the Department of Statistics Malaysia (2022). Suicide rates also increased significantly during the first 5 months of 2021, which was partly attributed to the pandemic (Idrus, 2021). According to Idrus, the phenomenon of suicide showed that people were in critical conditions following the pandemic.

Like elsewhere, the lockdown in Malaysia disrupted family life and daily routines, with severe consequences for children's welfare and well-being. Measures to prevent the transmission of COVID-19 exacerbated existing vulnerabilities and also took a toll on families. Other vulnerable groups such as migrants or homeless people with poor living conditions became increasingly susceptible towards the spread of the disease (Kluge et al., 2020). Meanwhile, the execution of institutional quarantining of people who have been in contact with confirmed or probable cases overwhelmed the system until the authorities realised that self-quarantine could be a more realistic measure (Ebrahim et al., 2020).

12   The Impact of Covid on Education

School closures affected pre-schooling, lower primary schools, secondary schools and universities. Online schooling became the norm, which proved to be a new experience for many families and their children. Malaysia, like elsewhere, turned to technology as an alternative to in-school instruction. Although there was relative success in the transition to remote learning in general in urban areas, it did not turn out to be the panacea for children in rural areas and remote indigenous areas during the time of the COVID-19 crisis. Some parents reportedly found online schooling challenging or tedious and were unable to support their children to learn from home. Parents with limited education, especially, struggled to supervise and assist students. There were issues too in some rural areas where people lacked the use of smartphones, laptops, tablets or access to the internet.

The pandemic's impact on higher education was also quite paramount. Although higher education institutions were quick to replace face-to-face lectures with online learning, these closures affected learning and examinations as well as the safety and legal status of international students in the country.

13   The Supportive Role of Business During the Pandemic

Local companies stepped up their efforts to assist communities following the outbreak of the pandemic. They contributed to the Food Aid Foundation to help feed people who were suffering from severe hunger. Others provided gloves and sanitisers to both the public and healthcare workers. Alongside that, the companies also contributed to the government's COVID-19 fund, which was aimed at supporting government efforts in easing the burden faced by the pandemic.

Support from companies covered many areas. For instance, the Axiata Group launched a RM150 million COVID-19 programme to give immediate assistance to micro-SMEs in financial difficulties. The Sunway Group also provided more than RM34 million to help Malaysians cope with the pandemic. As part of its Corporate Social Responsibility (CSR) initiatives, UDA Holdings Berhad contributed goods, equipment and electronics worth more than RM100,000 to Sungai Buloh Hospital to care for COVID-19 patients under its watch. Petronas, an oil and gas company in Malaysia, contributed RM20 million worth of medical equipment and supplies, including ventilators, hospital beds and mattresses, thermal imaging cameras, digital thermometers, virus test kits and PPE, to healthcare frontline workers in Malaysia, as well as hand sanitisers to senior citizens at selected homes.

Many companies have proactively engaged in various CSR activities, particularly those that can offer immediate help and assistance to the fight against the virus. Undoubtedly, the current pandemic offers a wide range of significant opportunities to those with a more mindful and acumen approach to CSR. Some of these companies have maintained their commitment to CSR throughout the pandemic as they saw this as an opportunity to display their citizenry. The businesses invested in CSR to achieve a meaningful engagement with society at a crucial period of the pandemic. As often the case, moral obligation and inter-industry collaboration are key solutions driving supportive organisational level responses to global health crises. Besides, since businesses are an integral part of society, it is also their responsibility to ensure that they take care of society in difficult times. These philanthropic CSR actions have further confirmed that business and society are intertwined (McLennan & Banks, 2019).

During crises, such as the pandemic for instance, society and local communities have emergent needs and require prompt assistance for a full recovery. Thus, companies with the capacity and resources to mobilise promptly and systemically can be especially effective in executing disaster relief and recovery management (Ballesteros et al., 2017).

COVID-19 pandemic has, indeed, exposed and exacerbated some ingrained social issues, such as poverty and inequality. CSR contributions have been quite remarkable and have helped such vulnerable people to survive in the face of adversity. Corporate philanthropy has been visible everywhere in Malaysia. Although companies also face their own financial challenges, they have shown their concerns and continue to fulfil their corporate citizenship. Through the implementation of CSR initiatives, companies have highlighted their significance in society and also helped to ease the government's burdens.

14 The New Norm for Malaysians

The recent pandemic has changed the way people live in Malaysia. It led to social distancing measures, the crave for personalised space, small group-based activities as well as an increase in the use of digital platforms in the country. People's lives remained on hold for a quite a long time as the disease held out. This disease has altered and even devastated many facets of people's lives. Now, Malaysians would need to turn the current situation into real opportunities to do things that can usher them into the future with renewed hope.

15 Overcoming the Economic Downturn due to COVID-19

Malaysia's Budget 2022 is about bolstering economic recovery by building resilience and driving reform after the strict lockdown measures that were used to curb COVID-19. Themed as 'Malaysian Family, Prosperous and Peaceful', it has a focus on job creation initiatives to support many Malaysians to transition back into the workforce and also to assist businesses to recruit the people they need to support their recovery.

To minimise the impact of the COVID-19 pandemic on businesses, several actions to restore business capabilities including access to financing, driving strategic investments and reviving targeted sectors are also expected to be rolled out with a 40-billion-ringgit package. This includes direct loans, financing guarantees and equity-based schemes. The aim is to benefit all businesses regardless of their size. Meanwhile, a total of 1.6 billion ringgits has been provided to drive economic recovery in the tourism sector, which has been severely affected by the pandemic. Initiatives in the budget include wage subsidy, financing, special assistance and maintenance of tourism infrastructure.

Despite these extraordinary challenges that the pandemic brought along, there are also opportunities to drive positive change through these difficult times (KPMG, 2020). New ways of doing business and diversifying the economy are highly required by embracing technological advances and digitalisation of business. This pandemic has created opportunities too for Malaysian researchers to play their part by developing different technologies to help Malaysians facing the pandemic. Examples of these developments include COVID-19 rapid test kits (Gomes, 2020), creation of face shields using 3D printing, laser cutting or DIY builds (Tariq, 2020) and manufacture of sanitizing tunnels (Mohamad, 2020). All of these creative ideas show that Malaysians are acting together to battle against COVID-19.

The pandemic has precipitated both comprehensive strategic reviews and the use of every investment lever by companies to position themselves for a reshaped competitive landscape. The movement restriction orders and the aggressive screening exercises have given the much needed hope to Malaysia in fighting the COVID-19 pandemic.

In the immediate term, protecting jobs and workers has to be at the core of the crisis response. Immediate action is also needed to ensure that people can effectively access healthcare whilst social protection should be a central element of the immediate stimulus package. An employment strategy for the medium- to long-term recovery of jobs and incomes would also be needed. Addressing the needs of the most vulnerable groups should remain a major priority too, with particular focus on informal and non-standard workers. Pre-existing inequalities and discrimination have exacerbated the vulnerabilities of these groups, which has been compounded by the impact of COVID-19.

Malaysia's post-pandemic recovery phase is also expected to work the accomplishment of productive development strategies, that is, to bring the potential of green economy solutions, e-commerce, the digital economy and the paid care economy into sharper focus. These potentials should be identified and developed in the longer term. Interestingly, Malaysia has young people who are technologically savvy and innovative.

16   Conclusion

Malaysia has come a long way since the inception of COVID-19. It has endured a lot since. It is now on a path to recovery. As such, it has shown a keenness to support its businesses and people to have a renewed hope. To be able to better prepare for any future health and economic shocks of this nature, Malaysia needs to continue with enhancing its healthcare system and an economy that is stronger, greener and more resilient and a society that is inclusive. Support for new strategic sectors and business models would require attention to addressing underlying structural and institutional weaknesses, including improving the data monitoring system, simplifying bureaucratic processes, promoting social dialogue and addressing inequality and discrimination.

References

Amaran, S., Kamaruzaman, A. Z. M., Esa, N. Y. M., & Sulaiman, Z. (2021). Malaysia healthcare early response in combatting COVID-19 pandemic in 2020. *Korean Journal of Family Medicine, 42*(6), 425–437.

Ballesteros, L., Useem, M., & Wry, T. (2017). Masters of disasters? An empirical analysis of how societies benefit from corporate disaster aid. *Academy of Management Journal, 60*(5), 1682–1708.

Cheng, C. (2020). *COVID-19 in Malaysia: Economic impacts & fiscal responses*. Institute of Strategic and International Studies (ISIS) Malaysia. Retrieved April 20, 2022, from https://www.isis.org.my/2020/03/26/covid-19-in-malaysia-economicimpacts-fiscal-responses/

Cheng, C. (2022). The socioeconomic impacts of Covid-19 in Malaysia. *Asia Policy-National Bureau of Asian Research, 17*(1), 35–44.

Cheong, B. (2020). Malaysian designers sew PPE gowns for medical frontliners: 'Togetherness is the key'. *The Star*. Retrieved April 21, 2022, from https://www.thestar.com.my/lifestyle/style/2020/04/02/039togetherness-is-the-key039-malaysian-designers-sew-ppe-gowns-for-medical-frontliners

Department of Statistics Malaysia. (2022, April 27). *Labour force survey report, 2021*. Retrieved May 10, 2022, from https://www.dosm.gov.my/v1/index.php?r=column/cthemeByCat&cat=126&bul_id=L1kxcjNmdDduMXBHUll2VGlweCsxQT09&menu_id=Tm8zcnRjdVRNWWlpWjRlbmtlaDk1UT09

Dzulkifly, D. (2020, March 13). Muhyiddin: Tourism industry hit hardest by Covid-19, faces RM3.37b loss. *Malay Mail*. Retrieved May 12, 2022, from https://www.malaymail.com/news/malaysia/2020/03/13/muhyiddin-tourism-industry-hit-hard-by-covid-19-to-lose-rm3.37b-while-gdp-s/1846323

Ebrahim, S. H., Ahmed, Q. A., Gozzer, E., Schlagenhauf, P., & Memish, Z. A. (2020). COVID-19 and community mitigation strategies in a pandemic. *British Medical Journal, 368*, m1066. Retrieved April 13, 2022, from https://doi.org/10.1136/bmj.m1066

Gomes, V. (2020, March 27). Malaysian companies develop Covid-19 rapid test kits. *The Edge Markets*. Retrieved April 10, 2022, from https://www.theedgemarkets.com/article/malaysiancompanies-develop-covid19-rapid-test-kits

Idrus, P. G. (2021). Suicide rising in Malaysia due to hardships amid coronavirus pandemic. *Anadolu Agency*. Retrieved May 11, 2022, from https://www.aa.com.tr/en/asia-pacific/suicide-rising-in-malaysia-due-to-hardships-amid-coronavirus-pandemic/2293079

Kluge, H. H. P., Jakab, Z., Bartovic, J., D'Anna, V., & Severoni, S. (2020). Refugee and migrant health in the COVID-19 response. *Lancet, 395*, 1237–1239.

KPMG. (2020). *COVID-19 and healthcare -Navigating reaction, resilience, recovery and the new reality*. Retrieved from https://home.kpmg/my/en/home/campaigns/2020/07/covid-19-and-healthcare.html

McLennan, S., & Banks, G. (2019). Reversing the lens: Why corporate social responsibility is not community development. *Corporate Social Responsibility and Environmental Management, 26*(1), 117–126.

Mohamad, N. I. (2020, April 6). USM researchers prove automatic sanitising tunnel can be built for RM1,500. *New Straits Time*. Retrieved April 21, 2022, from https://www.nst.com.my/news/nation/2020/04/581781/usm-researchers-prove-automatic-sanitising-tunnel-can-be-built-rm1500

Murugiah, S. (2020, March 26). Covid-19 to place Malaysian economy into technical recession, says AmBank. *The Edge Markets*. Reteieved April 20, 2022, from https://www.theedgemarkets.com/article/covid19-place-malaysian-economy-technical-recession-says-ambank

Nga, J. L. H., Ramlan, W. K., & Naim, S. (2021). Covid-19 pandemic and unemployment in Malaysia: A case study from Sabah. *Civil Societies: An Interdisciplinary Journal, 13*-(2) Retrieved May 9, 2022, from https://doi.org/10.5130/ccs.v13.i2.7591

Nienhaus, A., & Hod, R. (2020). COVID-19 among Health Workers in Germany and Malaysia. *International Journal of Environmental Research and Public Health*. Retrieved May 17, 2022, from https://www.mdpi.com/1660-4601/17/13/4881/htm

OECD. (2020, June 24). *The impact of the coronavirus (COVID-19) crisis on development finance*. Retrieved May 12, 2022, from https://www.oecd.org/coronavirus/policy-responses/the-impact-of-the-coronavirus-covid-19-crisis-on-development-finance-9de00b3b/

Reliefweb. (2021). *The impact of COVID-19 on health and care workers: A closer look at deaths*. Retrieved April 17, 2022, from https://reliefweb.int/report/world/impact-covid-19-health-and-

care-workers-closer-look-deaths?gclid=Cj0KCQjw-JyUBhCuARIsANUqQ_LJutX6V7GPs0
mXFts7qvQtwz1gltY0DDYZM8MNaMYwLUjjybb2lVcaAvrCEALw_wcB

Roslan, N. S., Yusoff, M. S. B., Razak, A. A., & Morgan, K. (2021). Burnout prevalence and its associated factors among Malaysian healthcare workers during COVID-19 pandemic: An embedded mixed-methods study. *Healthcare, 9*(1), 90. Retrieved April 12, 2022, from https://doi.org/10.3390/healthcare9010090

Tariq, Q. (2020). Covid-19: Malaysian 3D printing enthusiasts produce face shields to aid frontliners in fight against pandemic. *The Star.* Retrieved May 12, 2022, from https://www.thestar.com.my/tech/tech-news/2020/03/23/malaysian-3d-printers-produce-face-shields-to-aid-frontliners-in-the-fight-against-covid-19

The Malaysianreserve.com. (2021, May 12). *Malaya's GDP continues to decline.* Retrieved April 12, 2022, from https://themalaysianreserve.com/2021/05/12/malaysias-gdp-continues-to-decline/

The Star. (2020a, February 28). *Lots of incentives in stimulus package.* Retrieved May 12, 2022, from https://www.thestar.com.my/news/nation/2020/02/28/lots-of-incentives-in-stimulus-package

The Star. (2020b, March 16). *NGO educates public on disease.* Retrieved April 11, 2022, from https://www.thestar.com.my/metro/metro-news/2020/03/16/ngo-educates-publicon-disease

The Star. (2020c, March 14). *COVID-19: Cops urge those with info on tabligh to contact Health Ministry.* Retrieved April 12, 2022, from https://www.thestar.com.my/news/nation/2020/03/14/covid-19-cops-urge-those-with-info-on-tabligh-to-contact-health-ministry

Vicknasingam, B., Salleh, N. A. M., Chooi, W.-T., Singh, D., Zaharim, N. M., Kamarulzaman, A., & Chawarski, M. C. (2021). COVID-19 impact on healthcare and supportive services for people who use drugs (PWUDs) in Malaysia. *Frontiers in Psychiatry.* https://doi.org/10.3389/fpsyt.2021.630730

Sam Sarpong is an Associate Professor at the School of Economics and Management, Xiamen University, Malaysia (XMU). He obtained his PhD from Cardiff University and his MBA from the University of South Wales. Prior to joining XMU, he taught at Cardiff University, University of London (Birkbeck College), Swansea Metropolitan University, University of Mines and Technology and Narxoz University, respectively. He is also a Visiting Professor at the NIB Doctoral School in Ghana. Sam is currently an Associate Editor of *International Journal of Corporate Social Responsibility* (Springer) and also the Editor of *New Economic Papers—AFR* (NEP-AFR). He serves on the boards of international journals and also acts as a reviewer for some renowned journals.

His research interests lie in the relationship between society, economy, institutions and markets. He tends to explore the nature and ethical implications of social-economic problems and has published in these fields.

Ali Saleh Alarussi is an Assistant Professor at the School of Economics and Management, Xiamen University. He has experience in accounting scholarship. He had his MSc and PhD in International Accounting from University Utara Malaysia (UUM), Malaysia.

Part IV
CSR and COVID-19 Pandemic in South America

Bolivia

Corporate Social Responsibility in Bolivia: Hospital Responses to the COVID-19 Pandemic

Boris Christian Herbas-Torrico, Carlos Alejandro Arandia-Tavera, and Alessandra Villarroel-Vargas

1 Introduction

Nowadays, the COVID-19 pandemic has caused significant changes globally, some of the most notorious in economics and health (Garrett & McNolty, 2020). While commercial firms adapt to the market, hospitals must adjust to the COVID-19 infection waves and changes in WHO recommendations (Rosenbaum, 2020). Specifically, regarding hospitals, Latin America has been one of the regions most affected by the pandemic due to its lack of infrastructure, medical equipment, resources, and trained personnel (Hummel et al., 2021). The general population's high demand for medical care and the medical leaves of health staff generated much pressure on hospitals. Hence, Latin American societies faced difficult situations resulting from the COVID-19 pandemic.

In Bolivia, the effects of the COVID-19 pandemic worsened due to economic informality, inequality, and health disparities. Deaths in the poorest departments were more than seven times higher in July 2020 than in July 2019. Compared to the same period in 2019, the peak of fatalities in the wealthiest departments was only two times higher (Hummel et al., 2021), because more than 70% of the working population lacks employment contracts or employer-based social security, and tight control of health measures is challenging (Baker et al., 2020). Government agencies and municipalities imposed mask and glove mandates, but low-wage employees were expected to obtain and pay for their own personal protective equipment (PPE) and testing. Hence, in Bolivia, numerous firms experienced widespread infections among their staff. For example, many Bolivian hospitals shut down their operations

B. C. Herbas-Torrico (✉) · C. A. Arandia-Tavera · A. Villarroel-Vargas
Exact Sciences and Engineering Research Center (CICEI), Bolivian Catholic University, Cochabamba, Bolivia
e-mail: bherbas@ucb.edu.bo

© The Author(s), under exclusive license to Springer Nature Switzerland AG 2023
S. O. Idowu et al. (eds.), *Corporate Social Responsibility in the Health Sector*, CSR, Sustainability, Ethics & Governance, https://doi.org/10.1007/978-3-031-23261-9_16

due to the high number of infections among patients and staff (France24, 2021). Consequently, the preventive measures taken by the Bolivian government to stop the spread of the SARS-CoV-2 resulted in a deepening of poverty, which manifested in precariousness or lack of access to resources (monetary and non-monetary), job opportunities, education, and health care (CEDLA, 2021).

In Bolivia, precarious work and the COVID-19 pandemic impacted many people due to job insecurity, low pay, and lack of contributing social safety coverage. The precariousness of working conditions has had a significant influence on low-income workers. According to CEDLA (2021), nine out of ten people in extremely precarious jobs are poor, and three out of ten have seen their status worsen throughout the COVID-19 pandemic. Thus, private and public firms, especially hospitals, must show how they live the social purposes and values that guide and represent them. Because it is at times like the one we are currently going through, they can demonstrate with facts that social responsibility is not just a discourse but that their commitment is real. Available literature suggests that in other countries, the COVID-19 pandemic forced hospitals to adopt new practices, such as Corporate Social Responsibility (CSR) practices, by adopting new measures to improve their response (Manuel & Herron, 2020). Hence, based on the previous arguments, our study examines the perceptions of CSR practices implemented by Bolivian hospitals during the COVID-19 pandemic from the point of view of three stakeholders: hospital management, hospital staff, and patients. In particular, we surveyed 5 hospital managers, 66 hospital staff, and 230 patients from Bolivian hospitals. Our findings imply that Bolivian CSR practices are primarily philanthropic and related to financing or assistance plans for low-income patients and offering different COVID-19 screening tests at an affordable price or even free of charge. Moreover, our results show that hospital staff are satisfied with their jobs. However, patients are somewhat satisfied with the quality of hospital services. In the next section, we present the background and related literature followed by methodology, results, conclusion, and discussion.

2   Background and Related Literature

2.1   COVID-19 and Corporate Social Responsibility

At the end of 2019, scientists found a new virus known as SARS-CoV-2, which can cause COVID-19 disease (WHO, 2021). The COVID-19 pandemic brought challenges to human functioning, including the global economy and health systems (Chakraborty & Maity, 2020). This pandemic has infected over 498 million individuals as of April 18, 2022 (Statista, 2022), with over 6 million people dying from the infectious disease (Statista, 2022), making it one of the century's worst pandemics (Chakraborty & Maity, 2020; Su et al., 2021). Due to this fact, the world has changed dramatically and forced people to adapt to new lifestyles. Likewise, it has forced organizations to adapt to new realities or face losing their markets.

As history constantly shows, difficulties also create opportunities. Hence, firms saw the COVID-19 pandemic as an opportunity to improve their CSR practices by making their organizations more genuine and oriented toward contributing solutions to pressing issues (He & Harris, 2020; Manuel & Herron, 2020). For example, Vodafone implemented free access to unlimited mobile data for vulnerable UK customers during the COVID-19 quarantine (BBC, 2020a). These genuine and authentic CSR practices show firms' commitment to society's welfare.

Unfortunately, the difficult economic conditions brought by the COVID-19 pandemic caused some firms to abandon ethical practices and try to take advantage of their customers or employees (He & Harris, 2020). Compared to the United States and Europe, in Latin America, stakeholders do not trust the transparency and reliability of firms. This mistrust in firms has increased because some firms were detected doing business irregularly. Moreover, public officials were not complying with their own restraint rules (Balog-Way & McComas, 2020; Estabrooks et al., 2020; Bargain & Aminjonov, 2020). Furthermore, Sharma and Yogi (2020) point out that the COVID-19 pandemic exposed governments and private firms to prioritize their self-interests while neglecting their employees, suppliers, and other stakeholders. For example, in Bolivia, public officials and some private firms did illegal deals by selling overpriced health-care equipment, PCR tests, and biosafety materials (BBC, 2020b). Hence, this contributes to the decreasing credibility of governments and private firms in developing countries, such as Bolivia.

Despite these issues, Latin American firms continue to prioritize environmental and philanthropic CSR practices (Diaz & Sánchez, 2021). Particularly, Coffie and Hinson (2021) point out that even before the COVID-19 pandemic, charitable practices were one of the most widely used CSR practices in developing countries, second only to environmental practices. Loureiro et al. (2012) and Zeler and Capriotti (2019) suggest that this is due to the common practice among firms from developing countries of not sharing adequate CSR information with their consumers and the wider public. As a result, in developing countries, CSR practices have become a marketing tool to share the firm's philanthropic and environmental practices with the public (De los Salmones & Del Bosque, 2005) to increase its market share instead of being socially responsible.

2.2 CSR Practices in Bolivia

Regarding Bolivia, we found that CSR literature is quite limited. Available research suggests that most CSR practices are focused on protecting the environment (Cameron, 2011; Herbas-Torrico et al., 2018, 2021). Mainly, Cameron (2011) indicates that the mining industry is the industry that most heavily uses CSR practices in Bolivia due to its economic importance and its environmental impact. For example, their CSR practices are related to industry compliance with soil studies in drilling regions and reporting monthly emission measurements.

Nowadays, environmental CSR practices positively influence stakeholders, thanks to the new generation of managers educated on the latest eco-friendly trend. According to Herbas-Torrico et al. (2021), the CSR practices that Bolivians approve of the most are those related to the environment and community support. For example, the Bolivian firm CBN implemented efficient water resources management. Banco Bisa implemented fund-raising campaigns for education centers for people with disabilities following this trend. In addition, given the impact of this trend today, the Bolivian government has implemented guidelines to promote the implementation of CSR practices. Among some of these guidelines, we can mention that the CSR implementation guideline launched by the Ministry of Economy in 2019 provides systematic guidelines and evaluation lists for CSR practices (Ministerio de Desarrollo Productivo y Economía Plural, 2021).

There are also private firms that encourage sustainability and promote CSR monitoring and success stories. For example, the Observatory of Corporate Social Responsibility of the Federation of Private Entities regularly evaluates the sustainability reports of the federation's member firms. Similarly, the Bolivian Institute of Foreign Trade (IBCE) regularly gathers and publishes examples of successful CSR practices to encourage good initiatives. For example, the IBCE (2019) showed a successful case from Embol, a subsidiary of Coca-Cola Company in Bolivia, indicating that the firm has implemented CSR practices for water reclamation, women empowerment, and glass and plastic bottle recycling. Similarly, the IBCE (2019) reported that UNAGRO, a company dedicated to producing ethanol and its derivatives, had certified most of its processes as green due to the use of solar panels and methods of water reclamation.

Finally, similar to other developing countries, the COVID-19 pandemic revealed that Bolivia's economic context was too complex. Specifically, Bolivia is the poorest country in the Americas (Pagina Siete, 2021a; La Republica, 2019); hence during the pandemic, unemployment has risen to 70%, economic informality was at an all-time high, and people were working in precarious conditions (e.g., without contracts, health insurance, social safety nets) (Hummel et al., 2021). These findings suggest that the government efforts to mitigate COVID-19 pandemic consequences have been ineffective. Specifically, unemployment has increased by 1.4% between January 2020 and January 2021, closing the year with a rate of 8.4% (Pagina Siete, 2021b), inflation increased by 2.95% in comparison to 2020 (Statista, 2021), and a staggering 14.9% of firms have filed for bankruptcy between January 2020 and January 2021 (Fundempresa, 2021). Due to these dire economic conditions, CSR practices have become less important in a firm's operations and fight for survival. However, some private firms have flourished during the pandemic and kept working on implementing CSR practices. Consequently, private-sector donations became the most significant contributor to hospitals, improving health services for low-income patients (CBN, 2021). Hence, this example suggests that philanthropic and charitable activities are among the most common CSR practices, even during the COVID-19 pandemic.

2.3   CSR and Hospitals

Due to the COVID-19 pandemic, social and economic structures have been tested worldwide. Consequently, society demands more from firms, such as hospitals. Before the pandemic, stakeholders expected firms to take more social responsibility because they had experienced an unprecedented economic expansion due to globalization (Puaschunder, 2018). Now, this health crisis has brought a diversity of new expectations. Hence hospitals should identify, assess, and prioritize stakeholder demands to build a realistic and pragmatic corporate responsibility in this new period (Duan et al., 2020). COVID-19 social, health, and environmental concerns imply that the health-care industry, especially hospitals, has been given additional and varied responsibilities. Specifically, the number of stakeholders' responsibilities to consider in hospitals has increased and now includes more than simply patients: physicians, administrative employees, nurses, managers, and service and supplier providers are just a few examples.

Complications in organizing stakeholders' care and the optimal use of resources to meet their needs give many improvement opportunities. Therefore, the relationship between CSR and hospitals has evolved in the context of health-care delivery as a new paradigm for improving their shortcomings (Brown, 2000). Moreover, in a health-care context, CSR means that there is also an ethical responsibility that requires hospitals and other organizations to do something beneficial for the quality of their health-care service (Gharaee et al., 2013). Duan et al. (2020) suggest that CSR becomes a tool to facilitate the relationship between stakeholders and defend the patient's rights against possible abuses by the hospital staff. Moreover, it protects hospital staff from inappropriate orders or mistreatment at work and protects the organization from immoral profit-oriented managers.

Furthermore, the COVID-19 pandemic has highlighted the existing hospital problems even more. According to Creixans-Tenas et al. (2020), ambiguous social mission policies impeded patient treatment. Moreover, Garrett and McNolty (2020) indicate that hospitals have functioned unorganized due to continual changes in WHO COVID-19 safety guidelines, lack of control over infections, and medical leaves. As a result, Virani et al. (2020) indicate that hospitals' CSR practices must focus on patient care: for example, allow separate rooms for COVID-19 patients to be cared for and receive better treatments. Other consequences of the COVID-19 pandemic were anxiety and depression problems experienced by people when hospitals experienced high demand due to COVID-19 infections (Xie et al., 2021; Alsharif, 2020). To avoid these problems, hospitals experimented with different possible solutions, such as offering rest rounds to the medical and administrative staff so that, in addition to resting, they had medical check-ups due to their high virus exposure (Virani et al., 2020; Garrett & McNolty, 2020).

Because of their limited health-care capacity, inadequate emergency response, and deficient hospital infrastructure, Latin American nations have been worst impacted by the COVID-19 pandemic (Hummel et al., 2021). Many individuals could not visit a hospital due to lack of income, poverty, and informal employment

(Callejas et al., 2020; Martinez, 2021). As a result, many individuals used untested alternative medicine or just ignored their symptomatology, resulting in severe sickness (Callejas et al., 2020). Despite the COVID-19 pandemic, Latin American firms copied CSR practices from industrialized nations (Sharma & Swati, 2020). For example, in countries such as Perú and Brazil, firms moved big factories from urban zones to places far from cities to improve air quality and environmental conditions for neighboring areas (Sharma & Swati, 2020). Furthermore, they were attempting to enhance their methods in support of the local community to improve community members' lifestyles (Agrawal & Sharma, 2022). Some of the CSR practices used by Latin American firms included the use of positive messaging to promote prevention behaviors such as "*stay at home*" or "*get vaccinated*" (Argote et al., 2021) and the donation of equipment, medical supplies, and other supplies for hospitals (Unicef, 2020).

In Bolivia, the COVID-19 pandemic has overwhelmed clinics, hospitals, and care centers in cities, causing the death of many infected individuals who required medical care. Moreover, entire families became infected, lost family members, and were left with significant debts. Consequently, local cemeteries and crematoriums in cities could not satisfy demand, and local governments around Bolivia had to build mass graves and buy additional incinerators for low-income individuals (Fundación Connectas, 2020; El Pais, 2020; New York Times, 2020). Moreover, private cemeterics and funeral service firms experienced the loss of their personnel due to COVID-19 infections (El Pais, 2020). In the Bolivian countryside, individuals lacked access to medical care and thus mainly used untested traditional medicine promoted by the government. Hence, death statistics in the Bolivian rural area are practically unknown. Moreover, exacerbating infections, the Bolivian senate authorized the production, distribution, and use of a potentially toxic chemical known as chlorine dioxide (a bleach derivative) as a COVID-19 treatment (Saavedra et al., 2021). Overall, the COVID-19 pandemic highlighted the improvised response of the Bolivian government, which increased poverty by reducing household income, widening the gap in access to information and communication technologies, interrupting school attendance, increasing unemployment and forced inactivity, increasing job insecurity, increasing unpaid work at home, and decreasing access to health care (CEDLA, 2021).

Another negative impact of the pandemic in Bolivia was a lack of staff, infrastructure, and health supplies needed to care for patients in hospitals (Mahmud et al., 2021; Singh & Misra, 2021; Yulita & Hidajat, 2020; Calla et al., 2020). This situation showed that the Bolivian health-care system's painful condition had been historically neglected (WHO, 2021; France 24, 2020). Since the beginning of the COVID-19 pandemic, the actual death toll is still unknown to this day. However, a recently published study by Uzin (2022) (not peer-reviewed) estimates that the death toll in Bolivia due to COVID-19 infections might be three times (64,542 deaths) higher than the official records of the Ministry of Health. Therefore, the available information in Bolivia indicates that the COVID-19 pandemic overwhelmed the health-care system, causing significant uncounted deaths.

2.4   Stakeholder Theory and Hospitals

The stakeholder perspective provides a practical method for understanding how firms might flourish in today's and tomorrow's uncertain environments (Austen, 2012). The stakeholder theory emerged in the private sector due to a desire to comprehend the firms and their surroundings (Mitchell et al., 1997). In addition, stakeholder theory can be considered as the management and ethical theory that focuses on more than just increasing shareholder value (Phillips, 2003). Stakeholder theory is the most comprehensive and widely used theory for CSR practices (Frynas & Yamahaki, 2016). According to Freeman (1994), stakeholder theory relates to the study of stakeholders and how they can be affected by a firm's activities. Specifically, stakeholder theory suggests that firms are interested in how their customers react to their behavior and how this contributes to their financial performance (Berman et al., 1999). Some studies have shown results where practices such as volunteering in the local community or environmental protection have positively influenced consumer behavior (Handelman & Arnold, 1999; Bhattacharya & Sen, 2004; Maignan et al., 1999). Similarly, customers' ethical expectations about a firm's CSR practices influence managerial attitudes (Mandhachitara & Poolthong, 2011). Therefore, the literature suggests that stakeholder theory relates consumers, stakeholders, and their relationship with a firm's governance (Berman et al., 1999).

Regarding hospitals and CSR practices, Ramachandran (2019) suggests that the market mechanism that comprises buyers, sellers, and other stakeholders was developed at the beginning of the modern health-care system. Many beneficial aspects of the free market, such as the power of invention, innovation, and entrepreneurship, are overlooked by a restricted view of markets centered exclusively on costs and profits. Hence, health-care firms such as hospitals must become more flexible, adaptive, and educated to fulfill demands to reduce costs and mistakes while increasing efficiency and quality of the market they offer their health-care services. Notably, for hospitals, stakeholders are the people or firms directly or indirectly affected by their operations (Ramachandran, 2019). Patients (clients), hospital staff, creditors, investors, insurance companies, and the government are among their stockholders. The stockholders may interact with health professionals regularly or irregularly, impacted by changes in health-care systems, policies, and practices. Patients are the ultimate users of a hospital services, and their opinions are essential. Notably, patients care about clinical excellence, but service quality and low prices are important too. On the other hand, high clinical quality (typically articulated in terms of innovative and technologically sophisticated services and facilities) and appropriate support services are the primary demands of hospital staff. Cost minimization, profitability, consistent income, a continuous cash flow stream, and effective utilization of all available resources are the most appealing to hospital management.

When dealing with stakeholders, hospitals are a unique example compared to other commercial firms since they have to deal with more complex issues. They operate in a highly complex environment, as they do not function in a completely free market and must contend with numerous government laws. Hospitals can be considered multidimensional firms with few vocal and influential players pursuing various objectives. Balancing the many stakeholder interests is difficult for hospital governance. Therefore, the available literature suggests the suitability of stakeholder theory to study CSR practices in hospitals in Bolivia. In particular, our study will describe CSR practices related to three stakeholders from Bolivian hospitals: hospital management, hospital staff, and patients.

3   Methodology

To examine whether Bolivian hospitals have implemented CSR practices into their operations during the COVID-19 pandemic, according to Rohini and Mahadevappa (2010) and Hung (2011), we developed three measurement instruments for hospital management, hospital staff, and patients. First, we created an interview guide with five sections to learn about CSR practices in hospitals from hospital managers. The interview guide included the manager's basic information, prior knowledge of CSR practices, stakeholder degree of importance, COVID-19 pandemic management, staff safety measures, patient and suppliers safety measures, and health-care CSR practices. Due to a nationwide lockdown, we conducted telephone and online interviews to collect data from November 2021 to March 2022. Second, we developed an online survey for hospital staff (attending physicians, fellows, nursing staff, residents, and interns). The survey included sections related to sociodemographic questions, perception of CSR practices (socioeconomic responsibility, labor responsibility, and ethical responsibility), health-care quality, and job satisfaction. Next, we collected data from December 2021 to February 2022 from hospital staff in Bolivia. Finally, we developed a new online survey for patients and collected data from November 2021 to February 2022. The survey included sociodemographic questions, knowledge about CSR practices, COVID-19-related questions, perception of CSR practices (socioeconomic responsibility, labor responsibility, and ethical responsibility), health-care quality, and health-care satisfaction.

4   Results

In the following sections, we will present the results of our analyses for hospital management, hospital staff, and patients.

4.1   Hospital Management Results

The hospital management of health-care firms is one of the most critical stakeholders in CSR. Hospital managers are important actors since they cultivate connections with other stakeholders (Elkington, 1998). Moreover, according to Russo (2016), management at hospital facilities focuses on enhancing parts of the service that improve relationships with other stakeholders. Based on the implementation of our interview guide, we found the following results.

We interviewed five hospital managers, of which four belong to the private sector and one from the public sector. Moreover, all were attending physicians (80% men and 20% female), with an average age of 39 years. According to our results, 80% knew or had previously heard about CSR practices. Mainly, CSR practices are related to the commitments, actions, measures, and responsibilities that a hospital assumes toward society for these hospital managers.

Next, hospital managers indicated that patients are the most critical stakeholders of hospitals (100% of managers), followed by society (60% of managers), the environment (60% of managers), the community (60% of managers), suppliers (100% of managers), shareholders (100% of managers), and hospital staff (100% of managers). Moreover, hospital managers specified that during the COVID-19 pandemic, they took the following steps to protect their staff: providing personal protective equipment, regularly diagnosing symptomatology, and establishing new health-care protocols (e.g., office disinfection between consultations, telemedicine, and isolation wards). Next, hospital managers indicated that the practices implemented to protect their patients and suppliers were related to maintaining recommended COVID-19-preventing methods such as measuring patient's temperature upon arrival to the hospital, controlling the use of face masks and alcohol-based hand rubs, using checklists with inquiries about COVID-19 and possible exposure to the virus for triage, and maintaining social distancing. In addition, the public hospital manager commented that they worked to increase the number of patients that they serve every day.

Additionally, hospital managers indicated that the main actions taken to face the COVID-19 pandemic were enforcing the compliance of health-care protocols, implementing the use of nasal swab rapid antigen tests and PCR tests before delicate health-care treatments (e.g., surgeries, significant interventions), and sorting and caring COVID-19-infected patients in isolation wards according to the disease progression and their degree of infection. Next, all hospital managers agreed that these practices helped reduce infections among their staff and improve their patient care. Furthermore, according to 80% of hospital managers, the three CSR practices most commonly used in their hospitals were: (1) receiving donations of medicines, drugs, medical supplies, and medical tests for patients; (2) receiving gifts from private and public firms to improve their services, equipment, and patient care capacity; and (3) promoting volunteer campaigns in rural areas (e.g., primary medical care, COVID-19 vaccines supply). Next, 60% of the managers pointed out that other CSR practices in their hospitals included: (a) donating medicines, drugs, medical supplies, and medical tests for people in rural, remote, and low-income areas; (b) implementing campaigns for collecting money, materials, supplies, medicines, equipment, and other practices to help low-income patients; (c) increasing new spaces for intensive care therapy, intermediate care, and other CSR practices to ensure health care for all patients; (d) using hospital's owned means of transportation for the movement of medicines, supplies, and patients with other branches and other hospitals to support emergencies and critical situations; and (e) implementing counseling, financing, and subsidy programs for low-income patients.

In conclusion, hospital managers in Bolivia associate CSR with the hospital's commitment to the society in which it operates, emphasizing economic and charitable endeavors. This finding is comparable to the study by Ibrahim et al. (2000), who found that board members in commercial firms are more concerned about financial performance and the legal aspect of Corporate Social Responsibility. Likewise, our results suggest that the efforts taken in response to the COVID-19 pandemic are similar to commercial firms and connected to the actions taken for caring physicians, providers, and patients recommended by the WHO protocols. Hence, as Herbas-Torrico et al. (2021) suggested, our analysis indicates that CSR practices in Bolivian hospitals are mainly related to philanthropy and charitable activities from the perspective of the hospital manager.

4.2 Hospital Staff Results

Staff is crucial to a firm's success because they provide services or goods that match client expectations and foster customer loyalty (Rangan et al., 2012). As a result, understanding their perception of CSR practices is crucial for following the firm's desired image to consumers. Next, based on these principles, we show the results we obtained from our survey to hospital staff from different hospitals in Bolivia. We received 66 survey responses from hospital staff (37.9% male, 62.1% female), with an average age of 30 years. Moreover, the majority of respondents were from private hospitals (66.7%), mainly attending physicians (48.5%) and interns (28.8%), and about half of them indicated they knew about CSR practices (48.5%).

Our results showed that hospital staff experienced job satisfaction in their jobs (72.7/100). This result suggests that hospital staff experience a positive impact on situational job factors, such as nature of the job, resources, and hospital environment (Celik, 2011). Moreover, according to Christen et al. (2006), this result also suggests that Bolivian hospital staff believed that, during the COVID-19 pandemic, their job performance and effort were high, and thus they felt satisfied with it. We also found that respondents think their hospitals offered high-quality health-care services (72.6/100). According to hospital staff, this result indicates that the health-care quality of their hospitals is related to the effectiveness of care, patient safety, and patient experience.

Regarding CSR practices, we found that hospital staff believe that current CSR practices in Bolivian hospitals are mostly related to patient privacy (77%), patient rights (74%), health-care improvement (73%), and labor responsibility (52%). However, we also found that hospital staff believe that current hospital CSR practices are not related to socioeconomic (42%) and ethical responsibility (39%). Moreover, we also found that CSR practices best perceived by hospital staff are complying with labor regulations (e.g., work schedules, salaries) (54%), providing facilities for training (58%), availability of in-house training to improve their skills (56%), and recording and resolving complaints from internal and external clients (34%).

Lastly, regarding support practices during the COVID-19 pandemic, hospital staff indicated that the hospital provided them with personal protective equipment (PPE) (67%), patient management and training about the correct use of PPE (62%), and design of adequate procedures for triage (66%). However, hospital staff disagreed that the hospital provided proper infrastructure for patient reception, assessment, and isolation (63%).

As a result, according to hospital staff, these results indicate that current CSR practices in Bolivian hospitals are mostly related to fulfilling existing legislation rather than having a social orientation. Moreover, we found that hospital staff mainly value CSR practices associated with fulfilling labor legislation, staff training, and conflict solutions. These results also indicate that during the COVID-19 pandemic, hospital CSR practices were mainly related to triage training and PPE supply and use. Finally, hospital staff demonstrate that hospitals lack the adequate infrastructure to care for COVID-19-infected patients.

4.3 Hospital Patients Results

Patients are an essential element of CSR practices since they are the ones who give a hospital a purpose to exist. They are also the ones who assess and analyze CSR practices and, as a result, give the hospital its reputation (Rettab et al., 2009; Asatryan & Asamoah, 2014). According to Stanaland et al. (2011), patients benefit from good CSR practices since they build a good hospital reputation that favorably influences trust, perceived quality, and patient happiness. As a result, this critical stakeholder determines the efficacy of management CSR practices and their impact on society. In the following paragraphs, we provide the findings of the last survey of our study.

A total of 358 patients were surveyed, and 230 valid responses were received. Male patients made up 40.9% of these surveys, while female patients made up 59.1%. Notably, 67% of the respondents indicated that they had heard about CSR practices or were familiar with them. In addition, 67% of respondents stated that they had COVID-19, and among the infected, 53.5% did not seek medical help in hospitals, 55.7% went to a private hospital, 22.6% visited primary care, and 33.7% went to a tertiary referral hospital. Mainly to explain these last results, we should consider the Bolivian context, where public health care has a bad reputation, and thus patients from higher income echelons instead seek medical care in the private sector (Hummel et al., 2021).

Our results showed that patients believe Bolivian hospitals have acceptable levels of service quality (67.1/100). Moreover, patients indicated they were somewhat satisfied with the hospital service during the COVID-19 pandemic (65.6/100).

Overall, both results suggest that around two in every three patients were almost happy with the hospital services; however, one in every three patients was not. Hence, this last result indicates that the quality of Bolivian hospitals is not good enough for all Bolivian patients (Brown & Dacin, 1997). Consequently, Bolivian hospitals should consider improving their service quality, understanding that satisfied patients are an added advantage for the community) and enhance the hospital's reputation.

Additionally, our results suggest that CSR practices positively influence patients' perceived value of the hospital service. Servera and Piqueras (2019) found similar results for other contexts, suggesting that CSR practices, policies, and initiatives improve the perceived value of a service, mainly when they are oriented toward the organization's core business. Notably, as Manuel and Herron (2020) suggest, as a result of the pandemic, firms in general, and hospitals in particular, have decided to increase their philanthropic actions, resulting in increased perceived value among patients. Consequently, as predicted by the theory and supported by our findings, hospital managers in Bolivia increased the use of philanthropic and charitable CSR practices, positively affecting patients' perceived value of the hospital services.

Overall, these last results complement our previous findings, suggesting that philanthropic and charitable CSR practices in Bolivia were perceived positively by patients and favorably increased the quality and value of their services. Moreover, most patients believe that the quality of hospital health care is acceptable and are thus somewhat satisfied.

5 Conclusions and Discussion

According to the findings reported in the previous sections, CSR practices in Bolivian hospitals consist primarily of philanthropic and charitable activities. This result is not surprising because similar results were proposed by Herbas-Torrico et al. (2021) for commercial firms, indicating that private and public firms use philanthropy as their primary type of CSR practice in Bolivia. Hence, this result has two implications. First, service industry sectors, such as the health-care industry, similar to other sectors, considered philanthropic and charitable activities as their primary CSR practices during the COVID-19 pandemic. On the other hand, this result also suggests that in Bolivia in general, and the health-care industry in particular, CSR practices are still lagging from developed countries in becoming sustainable practices to improve Bolivian people's lives.

According to the findings of the three stakeholders studied, Bolivia's primary CSR practices focus on financial aid for low-income patients and offering different COVID-19 screening tests at low costs. Moreover, for hospital managers, patients are the most important stakeholders. Moreover, despite hospital staff job satisfaction, patients believe that hospital service quality is acceptable and not excellent, and hence they are somewhat satisfied with it. Hospital staff responses back up this result because they think hospitals lack the adequate infrastructure to care for COVID-19 patients. These results entail putting all of the focus on the patients to help them during difficult times (He & Harris, 2020), such as the COVID-19 pandemic. Following Takahashi et al. (2013), we recommend that Bolivian hospitals implement innovative programs integral to their primary health-care function to reap patient CSR benefits and improve their infrastructure. Similarly, following Kakabadse and Rozuel (2006) recommendations, we suggest that hospitals in Bolivia promote greater hospital staff involvement and interaction to improve stakeholder relationships. For example, Bolivian hospitals can take CSR practices in Bangladesh as an illustration. Specifically, health-care institutions in Bangladesh provide CSR services by delivering medicines, primary nursing care, food distribution, and ambulance services with essential medical support (Werner, 2009). These CSR practices increased patient satisfaction and loyalty through genuine CSR initiatives (Hossain et al., 2019).

As our results shown above, as long as Bolivian patients believe that the CSR practices mentioned above are mostly related to reaping economic benefits instead of improving health-care quality, health-care capacity, and fulfilling the hospital's social orientation, they will not be sustainable in the long run, nor will they improve hospital's reputation in Bolivian society. Unlike Bolivian hospitals, other Latin American hospitals have not embraced CSR practices from a philanthropic or charitable perspective. Instead, they used CSR practices to fulfill their social duty and improve the lives of communities where they offer their health-care services. For example, in Colombia, the Hospital Universitario San José, among its CSR practices, set up houses to accommodate patients arriving from distant communities while they complete their treatments and consultations (Hospital San Jorge, 2021). Another example is the British Hospital of Buenos Aires (Argentina), with the *AMTENA* program, which sends a group of doctors to neighboring areas to provide free clinical and surgical health care (Hospital Británico de Buenos Aires, 2022). Hence, Bolivian hospitals can learn from these examples and create new sustainable CSR practices evolving beyond their current philanthropic and charitable practices.

In conclusion, hospitals in Bolivia should evolve their CSR practices from philanthropy and charitable practices to sustainable CSR practices to fulfill their social purpose. They should also implement new CSR practices that increase the quality and value of their health-care services. Likewise, Bolivian hospitals should keep improving the working conditions of their staff without disregarding legislation and implementing voluntary and sustainable CSR practices. Finally, Bolivian hospitals should seek more significant interaction among hospital managers, staff, and patients to adequately coordinate CSR practices to face future pandemics. As the COVID-19 pandemic changes Bolivian society, hospitals must reaffirm their commitment to CSR practices: *"We can only accomplish so much when working together."*

References

Agrawal, S., & Sharma, E. (2022). Identifying dimension of CSR in developed and developing economies: A case study of India and USA. *Asia Pacific Journal of Multidisciplinary Research, 8*(1), 117–125.

Alsharif, A. (2020). A framework for e-health strategy for managing pandemics in Saudi Arabia: In the context of COVID-19. *JMIR Medical Informatics*.

Argote, P., et al. (2021). Messages that increase COVID-19 vaccine willingness: Evidence from online experiments in six Latin American countries. *SSRN*.

Asatryan, R., & Asamoah, E. (2014). *Perceived corporate social responsibility activities and the antecedents of customer loyalty in the airline industry*. Scientific Papers of the University of Pardubice.

Austen, A. (2012). Stakeholders management in public hospitals in the context of resources. *Management, 16*(2), 217–230. https://doi.org/10.2478/v10286-012-0067-8

Baker, A., Berens, S., Feierherd, G., & Gonzales, I. M. (2020). Informalidad laboral y sus consecuencias políticas en América Latina. *Vanderbilt University Report, 144*, 1–16.

Balog-Way, D., & McComas, K. (2020). COVID-19: Reflections on trust, tradeoffs, and preparedness. *Journal of Risk Research, 23*(7–8), 838–848.

Bargain, O., & Aminjonov, U. (2020). Trust and compliance to public health policies in times of COVID-19. *Journal of Public Ethics, 192*, 104316.

BBC. (2020a, March). Coronavirus: Vodafone offers 30 days free mobile data. *BBC*. https://www.bbc.com/news/technology-52066048

BBC. (2020b, May 20). Coronavirus en Bolivia. *BBC*. https://www.bbc.com/mundo/noticias-5274 7870

Berman, S. L., Wicks, A. C., Kotha, S., & Jones, T. M. (1999). Does stakeholder orientation matter? The relationship between stakeholder management models and firm financial performance. *Academy of Management Journal, 42*(5), 488–506.

Bhattacharya, C. B., & Sen, S. (2004). Doing better at doing good: When, why, and how consumers respond to corporate social initiatives. *California Management Review, 47*(1), 9–24. https://doi.org/10.2307/41166284

Brown, P. (2000). *Ethics, economics and international relations. Transparent sovereignty in the commonwealth of life*. Edinburgh University Press.

Brown, T., & Dacin, P. (1997). The company and the product: Corporate associations and consumer product responses. *Journal of Marketing, 61*(1), 68–84.

Calla, H., Velasco, X., Nelson-Nuñez, J., & Boulding, C. (2020). Bolivia: Lecciones sobre los primeros seis meses de la pandemia de SARS-CoV-2. *Temas Sociales, 47*, 98–129.

Callejas, D., et al. (2020). The SARS-CoV-2 pandemic in Latin America: The need for multidisciplinary approaches. *Current Tropical Medicines Reports, 7*(4), 120–125.

Cameron, R. (2011). Community and government effects on mining CSR in Bolivia: The case of Apex and Empresa Huanuni. In *Governance ecosystems* (pp. 170–186). Palgrave Macmillan UK. https://doi.org/10.1057/9780230353282_11

CEDLA. (2021). *Desigualdades y pobreza multidimensional. Pobreza multidimensional y efectos de la crisis del COVID-19 en Bolivia 2021*. La Paz, Bolivia. Retrieved from https://cedla.org/publicaciones/obess/desigualdades-y-pobreza-multidimensional/serie-desigualdades-y-pobreza-multidimensional-pobreza-multidimensional-y-efectos-de-la-crisis-del-covid-19-en-bolivia-2021-2/

Celik, M. (2011). A theoretical approach to the job satisfaction. *Polish Journal of Management Studies, 4*, 7–14.

Cerveceria Boliviana Nacional. (2021, October 12). *Cerveceria Boliviana Nacional*. https://www.cbn.bo/noticias/cbn-dona-una-planta-generadora-de-oxigeno-para-cobija-y-destaca-el-trabajo-coordinado-con-las-autoridades/#:~:text=A%20partir%20del%20a%C3%B1o%202021,campa%C3%B1a%20de%20inmunizaci%C3%B3n%20con%20la

Chakraborty, I., & Maity, P. (2020). COVID-19 outbreak: Migration, effects on society, global environment and prevention. *Science of the Total Environment, 728*, 138882. https://doi.org/10.1016/j.scitotenv.2020.138882

Christen, M., Iyer, G., & Soberman, D. (2006). Job satisfaction, job performance, and effort: A reexamination using agency theory. *Journal of Marketing, 70*(1), 137–150. https://doi.org/10.1509/jmkg.2006.70.1.137

Coffie, I., & Hinson, R. (2021). Types of corporate social responsibility initiatives as response to COVID-19 pandemic in emerging economies. In T. From Aning-Dorson et al. (Eds.), *Marketing communications in emerging economies*. Palgrave Macmillan.

Creixans-Tenas, J., Gallardo-Vasquez, D., & Arimany-Serrat, N. (2020). Social responsibility, communication and financial data of hospitals: A structural modelling approach in a sustainability scope. *Sustainability, 12*(12), 4857.

De los Salmones, A., & Del Bosque, I. (2005). Influence of corporate social responsibility on loyalty and valuation of services. *Journal of Business Ethics, 61*(4), 369–385.

Diaz, R., & Sánchez, P. (2021). Corporate social responsibility response during the COVID-19 crisis in Mexico. In *Corporate responsibility and sustainability During the Coronavirus crisis*. Palgrave Macmillan.

Duan, J., et al. (2020). A content-analysis based literature review in blockchain adoption within food supply chain. *International Journal of Environmental Research and Public Health, 17*(5), 1784.

El País. (2020, Jun 18). *Diario El País*. https://elpais.com/internacional/2020-06-18/la-pandemia-y-la-precariedad-del-sistema-sanitario-dividen-a-bolivia-sobre-la-fecha-de-las-elecciones.html

Elkington, J. (1998). Partnerships from cannibals with forks: The triple bottom line of 21st-century business. *Environmental Quality Management, 8*(1), 37–51. https://doi.org/10.1002/tqem.3310080106

Estabrooks, C., et al. (2020). Restoring trust: COVID-19 and the future of long-term care in Canada. *FACETS, 5*(1), 651–691.

France 24. (2020, August 8). La crisis sanitaria y la pugna política dibujan una Bolivia camino "al desastre". *France 24*. https://www.france24.com/es/20200801-bolivia-covid-crisis-sistema-salud-politica

France 24. (2021, May 28). Entre records de contagios y decesos, se complica la crisis sanitaria en Bolivia. *France 24*. https://www.france24.com/es/am%C3%A9rica-latina/20210528-bolivia-record-contagios-muertos-covid19-cochabamba

Freeman, R. E. (1994). The politics of stakeholder theory: Some future directions. *Business Ethics Quarterly, 4*(4), 409–421.

Frynas, J., & Yamahaki, C. (2016). Corporate social responsibility: Review and roadmap of theoretical perspectives. *Business Ethics: A European Review, 25*(3), 258–285.

Fundación Connectas. (2020, June 18). *Los endeudados de la pandemia*. Connectas. https://www.connectas.org/pandemia-en-bolivia-victimas-hospitales/

Fundempresa. (2021, January). *Estadísticas del registro de comercio en Bolivia*. Fundempresa. https://www.fundempresa.org.bo/docs/contents/es/269_enero-2021.pdf

Garrett, J., & McNolty, L. (2020). More than warm fuzzy feelings: The imperative of institutional morale in hospital pandemic responses. *American Journal of Bioethics, 20*(7), 92–94.

Gharaee, H., et al. (2013). The relationship of organizational perceived justice and social responsibility in Yazd hospitals. Iran. *Journal of Management and Medical Informatics School, 1*(1), 26–37.

Handelman, J., & Arnold, S. (1999). The role of marketing actions with a social dimension: Appeals to the institutional environment. *Journal of Marketing, 63*(3), 33–48.

He, H., & Harris, L. (2020). The impact of COVID-19 pandemic on corporate social responsibility and marketing philosophy. *Journal of Business Research, 116*, 176–182.

Herbas-Torrico, B., Bjorn, F., & Arandia-Tavera, C. (2018). Corporate social responsibility in Bolivia: Meanings and consequences. *International Journal of Corporate Social Responsibility, 3*(1), 1–13.

Herbas-Torrico, B., Frank, B., & Arandia-Tavera, C. (2021). Corporate social responsibility in Bolivia. In S. Idowu (Ed.), *Current global practices of corporate social responsibility in the era of sustainable development goals*. Springer.

Hospital Británico de Buenos Aires. (2022). *Hospital Británico*. https://www.hospitalbritanico.org.ar/Page/PageContent/sustentabilidad

Hospital San Jorge. (2021,de 03). *Hospital San Jorge*. https://www.hospitalsanjose.gov.co/publicaciones/708/programa-de-responsabilidad-social-empresarial/

Hossain, S., Yahya, S., & Khan, M. (2019). The effect of corporate social responsibility healthcare services on patients' satisfaction and loyalty—A case of Bangladesh. *Social Responsibility Journal, 16*(2), 145–158.

Hummel, C., Knaul, F. M., Touchton, M., Guachalla, V. X. V., Nelson-Nuñez, J., & Boulding, C. (2021). Poverty, precarious work, and the COVID-19 pandemic: Lessons from Bolivia. *The Lancet Global Health, 9*(5), e579–e581. https://doi.org/10.1016/S2214-109X(21)00001-2

Hung, H. (2011). Directors' roles in corporate social responsibility: A stakeholder perspective. *Journal of Business Ethics, 103*, 385–402.

Ibrahim, N., Angelidis, J., & Howard, D. (2000). The corporate social responsiveness orientation of hospital directors: Does occupational background. *Healthcare Management Review, 25*(2), 85–89.

Instituto Boliviano de Comercio Exterior—IBCE. (2019). *Responsabilidad social empresarial: Casos de éxito en Bolivia*. IBCE.

Kakabadse, N. K., & Rozuel, C. (2006). Meaning of corporate social responsibility in a local French hospital: A case study. *Society and Business Review, 1*(1), 77–96. https://doi.org/10.1108/17465680610643364

La Republica. (2019, November 29). Bolivia y Colombia son los paises latinoamericanos con la mayor tasa de pobreza según la CEPAL. *La Republica*. https://www.larepublica.co/globoeconomia/bolivia-y-colombia-son-los-paises-latinos-con-la-mayor-tasa-de-pobreza-segun-la-cepal-2938772

Loureiro, S., Dias, S., & Reijnders, L. (2012). The effect of corporate social responsibility on consumer satisfaction and perceived value: The case of the automobile industry sector in Portugal. *Journal of Cleaner Production, 37*, 172–178.

Mahmud, A., et al. (2021). *Corporate social responsibility: Business responses to coronavirus (COVID-19) pandemic*. Sage.

Maignan, I., Ferrell, O. C., & Hult, G. T. M. (1999). Corporate citizenship: Cultural antecedents and business benefits. *Journal of the Academy of Marketing Science, 27*(4), 455–469.

Mandhachitara, R., & Poolthong, Y. (2011). A model of customer loyalty and corporate social responsibility. *Journal of Services Marketing, 25*(2), 122–133. https://doi.org/10.1108/08876041111119840

Manuel, T., & Herron, T. (2020). An ethical perspective of business CSR and the COVID-19 pandemic. *Society and Business Review, 15*(3), 235–253.

Martinez, A. (2021). Public health matters: Why is Latin America struggling n addressing the pandemic? *Journal of Public Health Policy, 42*(1), 27–40.

Ministerio de Desarrollo Productivo y Economía Plural. (2021, April 8). *Guía de responsabilidad social empresarial*. Ministerio de Desarrollo Productivo y Economía Plural. https://siip.produccion.gob.bo/repSIIP2/documento.php?n=2758

Mitchell, R. K., Agle, B. R., & Wood, D. J. (1997). Toward a theory of stakeholder identification and salience: Defining the principle of who and what really counts. *Academy of Management Review, 22*(4), 853–886. https://doi.org/10.5465/AMR.1997.9711022105

New York Times. (2020, August 22). Bolivia y el coronavirus: La tasa de mortalidad al alza durante la crisis política. *New York Times*. https://www.nytimes.com/es/2020/08/22/espanol/america-latina/bolivia-coronavirus.html

Pagina Siete. (2021a, March 22). Pobreza subió 6,4 puntos y afecta al 37,5% de la población. *Pagina Siete*. https://www.paginasiete.bo/economia/2021/3/22/cepal-pobreza-subio-64-puntos-afecta-al-375-de-la-poblacion-288220.html#!

Pagina Siete. (2021b, March 12). En 2020 la tasa de desempleo cerró en 8,4% por la pandemia. *Pagina Siete.* https://www.paginasiete.bo/economia/2021/3/12/en-2020-la-tasa-de-desempleo-cerro-en-84-por-la-pandemia-287144.html

Puaschunder, J. M. (2018). Intergenerational leadership: An extension of contemporary corporate social responsibility (CSR) models. *SSRN Electronic Journal.* https://doi.org/10.2139/ssrn. 3175656

Phillips, R. (2003). Stakeholder Legitimacy. *Business Ethics Quarterly, 13*(1), 25–41. https://doi.org/10.5840/beq20031312

Ramachandran, R. (2019). Stakeholder management in health sector. *SSRN Electronic Journal.* https://doi.org/10.2139/ssrn.3511454

Rangan, K., Chase, L., & Karim, S. (2012). Why every company needs a CSR strategy and how to build it (No. 12–088). https://www.hbs.edu/ris/Publication Files/12-088.pdf

Rettab, B., Brik, A., & Mellahi, K. (2009). A study of management perceptions of the impact of corporate social responsibility on organizational performance in emerging economies: The case of Dubai. *Journal of Business Ethics, 89*(3), 371–390.

Rohini, R., & Mahadevappa, B. (2010). Social responsibility of hospitals: An Indian context. *Social Responsibility Journal, 6,* 266–285.

Rosenbaum, L. (2020). Harnessing our humanity—How Washington's health care workers have risen to the pandemic challenge. *The New England Journal of Medicine, 382*(22), 2069–2071.

Russo, F. (2016). What is the CSR's focus in healthcare? *Journal of Business Ethics, 134*(2), 323–334.

Saavedra, V. V., López, M. A., & Dauby, N. (2021). The heavy toll of COVID-19 in Bolivia: A tale of distrust, despair, and health inequalities. *American Journal of Tropical Medicine and Hygiene, 104*(5), 1607–1608. https://doi.org/10.4269/ajtmh.21-0097

Servera, D., & Piqueras, L. (2019). The effects of corporate social responsibility on consumer loyalty through consumer perceived value. *Economic Research-Ekonomska Istraživanja, 32*(1), 66–84. https://doi.org/10.1080/1331677X.2018.1547202

Sharma, E., & Swati. (2020). Identifying dimension of CSR in developed and developing economies: A case study of India and USA. *Asia Pacific Journal of Multidisciplinary Research, 8*(1), 117–125.

Sharma, K., & Yogi, D. (2020). A study of corporate social responsibility in India: A special context of COVID-19. *UGC Care Journal, 43*(4).

Singh, K., & Misra, M. (2021). *Corporate social responsibility as a tool for healthtech startups: Modelling enablers of healthcare and social support system to fight coronavirus pandemic.* IEEE.

Stanaland, A. J. S., Lwin, M. O., & Murphy, P. E. (2011). Consumer perceptions of the antecedents and consequences of corporate social responsibility. *Journal of Business Ethics, 102*(1), 47–55. https://doi.org/10.1007/s10551-011-0904-z

Statista. (2021, August 2). *Evolución annual de la tasa de inflación en Bolivia desde el 2015 hasta el 2026.* Statista. https://es.statista.com/estadisticas/1189942/tasa-de-inflacion-bolivia/

Statista. (2022, April 22). *Number of cumulative cases of coronavirus (COVID-19) worldwide by day.* Statista. https://www.statista.com/statistics/1103040/cumulative-coronavirus-covid19-cases-number-worldwide-by-day/

Su, Z., et al. (2021). Mental health consequences of COVID-19 media coverage: The need for effective crisis communication practices. *Globalization and Health, 17*(1), 1–8.

Takahashi, T., et al. (2013). Corporate social responsibility and hospitals: U.S. theory, Japanese experiences, and lessons for other countries. *Healthcare Management Forum, 26*(4), 176–183.

UNICEF. (2020, September). *UNICEF dona materiales, equipos e insumos médicos para la respuesta a la pandemia del COVID-19, para beneficiar a 3,172,910 de personas.* UNICEF. https://www.unicef.org/nicaragua/comunicados-prensa/unicef-dona-materiales-equipos-e-insumos-m%C3%A9dicos-para-la-respuesta-la-pandemia

Uzin, A. (2022). *Seguimiento a las muertes en exceso y muertes por COVID-19*. Retrieved April 20, 2022, from https://www.upb.edu/es/contenido/seguimiento-las-muertes-en-exceso-y-muertes-por-covid-19

Virani, A. K., Puls, H. T., Mitsos, R., Longstaff, H., Goldman, R. D., & Lantos, J. D. (2020). Benefits and risks of visitor restrictions for hospitalized children during the COVID pandemic. *Pediatrics, 146*(2). https://doi.org/10.1542/peds.2020-000786

Werner, W. (2009). Corporate social responsibility initiatives addressing social exclusion in Bangladesh. *Journal of Health, Population and Nutrition, 27*(4), 545–562.

WHO. (2021, March 29). *Coronavirus. Health topics*. World Health Organization. https://www.who.int/health-topics/coronavirus

Xie, J., et al. (2021). Psychological health issues of medical staff during the COVID-19 outbreak. *Frontiers in Psychiatry, 12*, 611223.

Yulita, H., & Hidajat, K. (2020). Implementation of corporate social responsibility (CSR) Menteng Mitra Afia Hospital in educating the public. *Ilomata International Journal of Social Science, 1*(4), 185–195.

Zeler, I., & Capriotti, P. (2019). Communicating corporate social responsibility issues on Facebook's corporate fanpages of Latin American companies. *El profesional de la información*. https://doi.org/10.3145/epi.2019.sep.07

Boris Christian Herbas-Torrico is Professor at the Bolivian Catholic University in Cochabamba, Bolivia. He is a D. Eng. (Ph.D.) and M. Eng. in Industrial Engineering and Management from the Tokyo Institute of Technology (Japan). Most of his research interests are marketing, production, applied statistics, and corporate social responsibility. He currently serves as a researcher at the Exact Sciences and Engineering Research Center (CICEI) from the Bolivian Catholic University.

Carlos Alejandro Arandia-Tavera is an industrial engineer that is currently a production controller in Alicorp Bolivia. He also is a part-time research assistant at the Exact Sciences and Engineering Research Center (CICEI) in Cochabamba, Bolivia. He holds a bachelor's degree in industrial engineering from the Bolivian Catholic University.

Alessandra Villarroel-Vargas is a part-time research assistant at the Exact Sciences and Engineering Research Center (CICEI). She earned his B.S. in Industrial Engineering from the Bolivian Catholic University in Cochabamba, Bolivia. Currently, she works as a route supervisor at the Cerveceria Boliviana Nacional (CBN) in Bolivia, and she is a member of the AB-InBev Company.

Index

A
Accountability, 14, 99, 298
Activism, 211
Adversity, 276
Alkaline nutrition, 19
Alma-Ata, 221
Altruism, 153–160
Amazon, 226
Anchor institution, 5, 194, 196–200, 208–210, 212–214
Angola, 154, 302
Anticipation, 244, 245, 251, 252
Antigen testing, 323, 328
Asset turnover, 178–180
AstraZeneca, 154, 156
Asylum seekers, 152
Australia, 15, 18
Austria, 2, 13, 19, 21–33
Autocracy, 153

B
Barthes, R., 157
Belgrade, 148–151, 153, 154
Belgrade City Fair, 146
Benevolent coercion, 160
Biopharmaceuticals, 252
Biotechnology, 18, 95, 251, 252, 254
Black Pound Day, 228
Board of directors, 4, 175, 177–180, 183, 187
Body/discipline, 156
Bolivia, 8, 9, 355–358, 360, 362, 364, 366, 367
Bolivian hospitals, 9, 355, 356, 362, 364–367

Border control, 152
Bosnia, 154
Business culture, 281
Business ethics, 103, 117

C
Camus, A., 150
Capital, 27, 102, 105, 148, 195, 226, 275, 320, 346
Care, 7, 21, 26, 27, 95, 97–99, 105, 107, 113, 117, 119–121, 131, 136, 141, 173, 187, 197–200, 203–205, 213, 214, 219–222, 225, 227, 230, 232, 244, 248, 269, 270, 282, 286, 304, 305, 319, 323–325, 330, 331, 344, 345, 348, 350, 355, 359–361, 363–367
Carroll, 6, 113, 193–197, 224, 226
Case Study, 5, 13, 129, 194, 196, 197, 200–208, 212, 214, 254, 320
Celebritie(s), 152, 159
Charitable, 208, 224, 233, 299, 357, 364, 366, 367
Charitable activities, 358, 364, 366
Charted Institute of Ergonomics and Human Factors (CIEHF), 230, 233
Charter of Fundamental Rights of the European Union, 33
Chinese vaccine, 154, 156
Chinese vaccine diplomacy, 154, 156
Civil servants, 174
Civil service, 149
Clean water, 275
Climate, 141, 195, 201, 204, 206, 210, 268

© The Author(s), under exclusive license to Springer Nature Switzerland AG 2023
S. O. Idowu et al. (eds.), *Corporate Social Responsibility in the Health Sector*, CSR,
Sustainability, Ethics & Governance, https://doi.org/10.1007/978-3-031-23261-9

Cluster analysis, 4, 179, 184, 186, 188
Coalition against COVID-19 (CACOVID), 284, 289
Collaboration, 28, 208, 231, 251, 252, 254, 270, 320, 348
Communities, 5–8, 15, 20, 28, 32, 96, 98, 100, 113, 114, 117, 127–130, 132, 134, 138–141, 174, 175, 178–180, 184, 185, 188, 194–214, 221, 222, 224–227, 230, 231, 233, 234, 248, 249, 253, 264, 270–273, 275, 276, 279, 280, 283, 286–291, 299–301, 304, 306, 307, 318, 319, 330, 344, 346, 348, 358, 360, 361, 363, 366, 367
Community based testing, 323
Community Gateway Association (CGA), 5, 206–208, 213
Community-oriented CSR, 129, 130, 139, 140
Community Partnership, 252
Community relationship, 290
Community Toilet Sanitisation, 330
Community wealth building, 5, 194, 195, 197, 209–213
Companies, 4, 6–8, 94–97, 100, 101, 103, 107, 111–113, 115, 117, 127–132, 139–141, 173–180, 183, 187, 188, 196, 208, 224, 226–231, 233, 235, 244, 250–254, 272–275, 279–282, 285–292, 298–301, 306, 307, 309, 310, 317–325, 327–335, 348–350, 358, 361
Company board, 175, 177
Comprehensive strategy, 157
Contingency planning, 248
Cooperation, 5, 23, 99, 101, 102, 107, 116, 194, 234, 275, 307, 341
Co-operative principles, 194, 195
Coordination, 14, 22, 23, 233, 269, 343
Coronavirus, 1–3, 7, 8, 15, 93, 94, 104–107, 223, 225, 227, 245, 250, 267, 268, 306, 308, 310, 319, 320, 325, 342, 346
Corporate citizenship, 95, 96, 299, 304, 309, 348
Corporate Health Disclosure Index (CHDI), 8, 318, 322, 327, 328, 331–333, 335
Corporate image, 279, 281, 309
Corporate philanthropy, 273, 348
Corporate reputation, 279, 281
Corporate social responsibility (CSR), 2–9, 94–107, 111–122, 127–130, 132, 134, 136, 138–141, 156, 174–183, 186–188, 193–195, 207, 212, 213, 219, 221, 224–235, 243, 244, 250, 254, 272–275, 279–291, 298–303, 306, 307, 309–311, 317–320, 325, 348, 349, 356–367
Corruption, 150, 284

COVAX, 21, 156, 267
COVID Indemnity Health Cover, 324
COVID testing laboratory, 324, 328
COVID-19, 2, 4–9, 13, 15–20, 22–34, 93, 94, 99–102, 104–107, 117–121, 127, 128, 145–148, 150, 151, 174, 175, 178, 179, 183–188, 208, 221–223, 225–228, 230–233, 244–254, 263–265, 267–274, 276, 280, 281, 283–291, 297, 298, 300–306, 308–310, 317–320, 322–326, 328, 330–332, 335, 339–350, 355–357, 359, 360, 363, 365, 366
COVID-19 Emergency Corporate Social Responsibility (ECSR), 7, 8, 304, 306, 310, 311
COVID-19 pandemic, 1–6, 8, 9, 13, 14, 18, 23, 25, 26, 29–31, 112, 114, 115, 117–122, 127, 129, 131, 132, 138, 140, 141, 146, 158–160, 175, 179, 186–188, 193, 196, 208, 221–223, 225, 226, 229–235, 243, 244, 246, 248–251, 254, 265, 273, 274, 276, 280, 282, 284–288, 290, 301–310, 317–320, 339, 341–343, 348–350, 355–360, 362–367
COVID-19 pandemic response, 318, 320, 322, 326, 330, 332
Crises, 2, 3, 8, 21–33, 94, 104, 107, 128–130, 139, 140, 148, 156–158, 204, 209–211, 213, 223, 233, 247, 249, 254, 275, 280, 284, 300, 301, 306, 309, 317–319, 339, 342, 347, 348, 350
Critical Corporate Social Responsibility, 94, 95, 97, 100, 102, 107
CSR practices, 357, 358, 361, 363
CSR report, 289
Cultural diplomacy, 158
Culture, 150, 156, 158, 160, 187, 195, 225
Curfew, 4, 25, 146, 150, 159
Cyber security, 128, 134, 136, 140
Czech Republic, 154, 155

D
Democracy, 14, 22, 30–33, 210–211, 213
Depression, 2, 93, 106, 107, 223, 224, 249, 319, 320, 344, 359
Detention, 152
Developing countries, 103, 131, 275, 297, 357, 358
Development of CSR in Poland, 114–117
Digital Health consultations, 323
Digital innovation, 247
Digitalization, 128, 131, 134, 136, 139, 140, 349
Disability, 202, 203, 220, 358

Discipline, 112, 150–153, 156, 228–230, 233, 244, 245
Discourse, 3, 145, 152, 156, 157, 298, 303, 306, 356
Disinfection, 121, 343, 347, 363
Disinformation, 343
Djakovica, 147, 149
Donations, 94, 100–102, 225, 232, 233, 251, 272, 273, 275, 281–284, 289, 300, 303, 306–309, 358, 360, 363
Dranjane, 147
Drug development, 250–251

E

Economic, 2, 3, 5, 6, 14, 19, 21, 25–27, 33, 95, 104, 107, 111, 113, 115, 127–130, 146, 151, 153, 155, 159, 178, 194–196, 198, 201, 208, 210–213, 221, 224–226, 228, 229, 234, 246, 249, 250, 265, 271–275, 279–281, 287, 291, 299–301, 317–320, 341, 342, 345, 346, 349–350, 355, 357–359, 364, 367
  diplomacy, 146
  performance, 4, 174
  recovery, 27, 228, 349
  systems, 275
Economy, 2, 6–8, 22–26, 31–33, 94, 96, 104, 106, 113, 115, 116, 130, 195, 208, 211, 213, 224, 226, 246, 266, 271, 274, 283, 286, 300, 302, 317–319, 339–341, 345–346, 349, 350, 356, 358
Education, 159, 160, 173, 195, 199, 202, 203, 206, 220, 225, 230, 248, 264, 266, 287, 319, 339, 347, 356, 358
Electro-mechanical ventilator, 331
Emotional and wellness support, 324
Employee, 2, 3, 94, 96, 97, 100, 102, 103, 112–114, 116–118, 121, 122, 127–134, 136, 138–141, 173–176, 178–180, 183, 203, 204, 224–228, 253, 273, 283, 318, 319, 323, 325, 328, 330–332, 355, 357, 359
Employee-oriented CSR, 128, 139, 140
Employment, 5, 115, 118, 127, 198, 199, 201–203, 206, 210, 213, 224, 226, 227, 346, 350, 355, 359
Environment, 6, 16, 17, 96, 102, 112, 113, 117, 127, 131, 134, 138–140, 199, 204, 206, 210, 230, 249, 270, 280–284, 287, 289, 290, 300, 301, 310, 319, 346, 357, 358, 361, 363, 364
Environmental activities, 291
Environmental sustainability, 320

Epidemic, 14, 22, 31, 32, 148, 150, 251, 264, 267, 305, 319, 342
Equal rights, 152
Ethical behaviour, 7, 279, 290, 291
Ethical obligations, 282, 300
Ethics, 3, 20, 23, 32, 174, 283, 287, 290
Ethnic minorities, 6, 221–223, 227
European Union (EU), 13, 15, 21, 22, 31, 107, 115, 130, 153, 154, 156, 159, 196
Eurovision [Song Contest], 159

F

Face mask, 33, 229, 231, 246, 266, 273, 363
Factor analysis, 4, 179, 185
Fair operating environment, 283
50th Anniversary Summit of the NAM, 155
Financial performance, 177, 229, 318, 361, 364
Financial problems, 128
Financial services, 326, 328, 330–332
Floyd, G., 227
Former Yugoslavia, 149, 158
Foucauldian, 156
Foucault, M., 145, 157
Freedom, 14, 24, 27, 30–33, 115, 151, 152

G

Gender diverse boards, 178
Gender diversity, 4, 5, 175, 177–179, 183, 184, 186–188
Ghana, 6, 8, 263–276
Gjakove, 147
Gligić, A. Dr., 148
Global health, 14, 15, 20, 221, 234, 254, 300, 344
  crises, 2, 6, 8, 14–22, 244, 348
  leadership, 2
Goldman Sachs, 228
Goodwill, 154–156
Google, 24, 132, 226, 228
Greece, 2, 93–95, 100–102, 104–107, 247, 301
Green behavior, 136
Greenland, 149
Gross domestic product (GDP), 175, 178–185, 188, 224, 341, 345
Growth, 18, 96, 207, 210, 234, 271, 280, 300, 302, 319, 341

H

Hand sanitiser, 289, 306, 307, 309, 348
Hannover, 148
Hazmat [suit], 156, 158

Health
   amenities, 286
   centers, 173
   crises, 127, 130, 139, 140, 223, 254, 305,
      317, 343, 359
   diplomacy, 156, 158
   inequalities, 5, 198, 213, 223, 232, 234
   inequities, 211, 346
   institutions, 3, 282–288, 297, 298
   investment, 5, 188
   outcomes, 106, 194, 201, 205, 209, 224
   promotion activities, 283
   risk assessment, 325, 331, 332
   service provider, 1–3
   structure strengthening, 97, 100
   systems, 6, 8, 20, 94, 95, 97, 99, 100, 138,
      231, 234, 264, 269–270, 276, 280, 301,
      305, 322, 323, 325, 326, 328, 356
   value, 97–100
Healthcare, 3, 5, 8, 9, 16, 22, 94–96, 98–107,
      114, 117–122, 148, 150, 158, 160,
      173–175, 183, 185, 188, 198, 201, 202,
      204, 219–223, 229, 231–233, 235, 244,
      245, 247–250, 254, 269, 270, 274, 280,
      282–284, 286, 288, 290–292, 298, 301,
      302, 305, 308, 310, 323–325, 328,
      330–332, 335, 342–345, 348, 350, 356,
      357, 359–367
   crisis, 247–249
   institution, 3, 7, 95, 99, 100, 223, 235,
      280–284, 287, 288, 290, 291, 330
   schemes, 28, 208, 231
   service, 117, 118, 120, 174, 187, 322, 323,
      326, 328, 332, 359, 361, 364, 367
   system, 6, 7, 26, 104, 105, 117, 118, 174,
      175, 219, 221, 222, 234, 248, 249,
      281–284, 286, 287, 291, 301, 305, 309,
      326, 328, 332, 339, 341–343, 350, 360,
      361
Health Care Organization, 173, 174
Health Foundation, 198, 210, 213
Health Hub, 328
Henderson, D.A. Dr., 148, 149
Hospitals
   CSR, 359, 364, 365
   infrastructure, 359
   job quality, 366
   job satisfaction, 362
   management perceptions, 362
   service quality, 366
   service value, 175, 361, 366
   social purpose, 356
   staff perceptions, 9, 27, 248, 356, 359, 361,
      362, 364

Housing, 197, 199, 200, 207–210, 212, 221,
      227, 321, 329, 334
Housing association, 5, 194, 196, 198–209, 214
How Variola defeated Corona, 149
Human Factors, 228–230, 233–235
Human resource, 96, 97, 99, 127, 128, 130, 198,
      248, 252, 301
Human resources policy, 134, 140, 141
Human rights, 2, 14, 22–24, 31–33, 115, 116,
      221, 227, 283, 287, 290, 291
Human wealth, 282

I
İbrahim, A., 6, 244, 250–252
ILO, 127, 131
Immunological autonomy, 156
Immunological independence, 156
Inclusion, 8, 202–203, 226, 227, 244–246, 248,
      251–254
Inclusiveness, 24, 32
Income, 128, 130, 131, 134, 174, 179, 186, 188,
      269, 271, 272, 275, 318, 341, 346, 350,
      359–361, 365
[In]considerate behaviour, 152
In Corpore Sano, 160
Indebtedness, 178–180, 183
Ineffective government response, 358
Inequalities, 5, 106, 107, 187, 213, 221, 227,
      234, 249, 265, 274, 275, 302, 319, 346,
      348, 350, 355
Informal economy, 266
Innovation, 6, 103, 116, 157, 210, 228,
      243–250, 254, 267, 283, 307, 361
Inoculation, 147, 148, 154, 156, 159, 267
Intensive care unit (ICU), 3, 29, 93, 94,
      100–102, 104–106, 187, 221, 301
Inter-cooperation, 211, 213
Investment, 97, 102, 107, 131, 174, 177, 188,
      247, 251, 252, 254, 269, 272, 307, 310,
      328, 332, 341, 345, 349, 350
Iran, 154, 247
Isolation centres, 325, 328

K
Kon, P. Dr., 157
Kosovo, 147, 148, 155

L
Labour policies, 282, 283
Lancashire, 197, 200, 205–206, 208, 209, 211,
      214

Lancashire Teaching Hospitals NHS
  Foundation Trust, 5, 194, 197, 200–208,
  213
Latin, 147, 160, 355, 357, 359, 360, 367
Latin America, 21, 355, 357
Lebanon, 154
Legatum Institute, 153
Legitimacy theory, 176, 178, 281
Libertarian, 150–153, 156, 159
Libertarian-authoritarian, 150–153, 156, 159
Libertarianism, 151, 152, 159
Limited healthcare capacity, 99, 269
Listed companies, 7, 187, 287, 335
Living wage, 199, 209
Local authorities, 201, 205, 206, 220, 221, 225,
  234
Lockdown, 4, 13, 25–27, 29–32, 93, 94, 106,
  107, 146, 150, 151, 208, 209, 223–228,
  246, 266, 302, 319, 320, 339, 345–347,
  349, 362
London School of Hygiene and Tropical
  Medicine (LSHTM), 226
Low-income patients, 356, 358, 363, 364, 366

M
Malaysia, 8, 18, 339–350
Management, 7, 21–33, 98, 99, 103, 104, 111,
  113, 114, 119, 122, 132, 133, 157, 174,
  176, 177, 186, 194, 196, 203, 205, 222,
  227, 230, 231, 268, 272, 283, 289–291,
  298, 324, 325, 331, 348, 356, 358,
  361–365
Mandate, 246, 303, 355
Mandating strategies, 159
Mandatory, 20, 27, 106, 120, 151, 157, 177,
  203, 227, 246, 266, 340
Manley, J., 5
Market place activities, 15, 287, 291
Mass-vaccination, 4, 17, 31, 32, 149
Measles, 151
Media text, 151
Medicines, 3, 5, 7, 13, 100–102, 106, 160, 202,
  204, 226, 230, 252, 253, 267, 269, 302,
  304, 306, 309, 310, 328, 330, 345, 360,
  363, 364, 367
Meditation, 330
Mental health, 130, 134, 160, 199, 222–224,
  228, 231, 232, 270, 330
Mental health talks, 324
Mentivity, 228
Miloš Obrenović, 149
Misconceptions, 6, 264, 268–269
Misinformation, 226, 268, 343
Mobile health vans, 324

Mobile Medical Units, 323
Mondragon, 195, 209
Montenegro, 148, 154, 155
Moody's Analytics, 153
Motivation, 106, 128, 140, 150, 157, 195, 197,
  251, 252
Movement Control Order (MCO), 340, 345,
  346
mRNA vaccines, 17, 18
Multiple linear regression, 4
Myths, 6, 264, 268–269

N
Namibia, 154, 302
Narrative, 145, 147, 152, 156–158
National Autistic Society, 228
National Health Service (NHS), 5, 219, 220,
  235
National Health System (NHS), 2, 93, 173, 174
National Institute for Health Research (NIHR),
  197, 214
National Public Health Organization, 105
NatWest, 227
Neurodiversity, 228
Nigeria, 7, 263, 280–293
Non-Aligned Movement (NAM), 154, 155
Non-sustainable CSR practices, 358, 367
North Macedonia, 154, 155
Nurse Dušica Spasić, 150
Nutrition, 16, 29, 33

O
Online counselling, 324, 330
Ontology, 156
Operational ethical dilemma, 102
Organic food, 18, 21
Over 65, 4
Oxygen concentrators, 7, 270, 309, 310, 325,
  328
Ozone therapy, 13, 16, 33–34

P
Pandemic, 2–8, 14–32, 93, 94, 97, 99–107, 119,
  121, 122, 127–132, 134–136, 138–141,
  146, 151, 158–160, 174, 175, 186–188,
  193, 194, 198, 207, 209, 213, 222–224,
  226, 229–231, 233–235, 244–254,
  263–267, 269–276, 284, 286, 297, 298,
  300, 302–306, 310, 317–320, 328, 332,
  335, 339–350, 355, 356, 358–360, 366,
  367
  hospitals, 248

Pandemic (*cont.*)
impact, 128, 130–132, 134, 140
management, 2, 13, 14, 22, 23, 25, 27–29, 31, 249, 362
prevention and preparedness, 14, 22
Panopticism, 156
Partial least squares structural equation modeling (PLS-SEM), 3
Partnership, 117, 134, 141, 201, 205, 207, 225, 252, 254, 268, 271, 274, 276, 310
Patients' perception, 356
Patient Zero, 147
Pentacell, 226
People, 1, 3, 5–8, 15, 17–19, 24–32, 93, 95, 100, 106, 107, 111, 112, 114, 117, 119–121, 130, 141, 148, 151, 153, 154, 173, 187, 188, 193–196, 198–203, 206–208, 213, 220–223, 225, 226, 228–234, 246, 247, 249, 263–268, 271, 273, 275, 276, 279, 280, 282, 283, 289, 297, 301, 305, 306, 317–320, 335, 339, 341, 343, 346–350, 356, 358, 359, 361, 363, 366
Personal hygiene, 266
Personal protective equipment (PPE), 7, 101, 102, 120, 204, 222, 227, 229, 231, 267, 289, 304, 308, 325, 328, 344, 347, 348, 355, 363, 365
Petrović, R. Prof. Dr., 148
Pfizer/BioNtech, 154
Pharmaceutical industry, 17, 102, 250, 252
Philanthropic activities, 367
Physical distance, 32
Physical health, 134, 140
Place, 5, 19, 20, 23, 30, 102, 105, 106, 117, 120, 121, 128, 149, 195, 201, 206, 209, 211, 213, 223, 226, 228, 229, 231–233, 263, 266, 272, 275, 283, 302, 310, 341, 343, 360
Planet, 1, 3, 175, 201, 203–205, 210, 279
PM-CARES Fund, 8, 324, 325, 330, 332, 335
Policy development, 249
Polish healthcare institutions, 3
Population health, 29, 103, 205, 226
Post-pandemic world, 246, 247
Poverty, 198, 211, 223, 232, 249, 274, 302, 348, 356, 359, 360
Precautionary principle, 14, 16, 20–22, 32, 33
Preston, 5, 194–200, 202, 206–213
Preston City Council (PCC), 196, 197, 201, 208–210
Preston Model, 5, 194–199, 208–212
Preston Vocational Centre, 199, 208
Primary health care, 117, 118, 221, 367

Private, 3, 5–7, 28, 94–96, 98–102, 105–107, 115, 118, 132, 133, 173–188, 196, 207, 220, 221, 225, 231, 234, 246, 264, 271–275, 280, 287, 297–311, 345, 356–358, 360, 361, 363, 365, 366
Private hospitals, 4, 5, 95, 100, 102, 174, 175, 178, 186–188, 287, 290, 293, 305, 364, 365
Procurement, 195–198, 201, 205–206, 210, 229, 324, 328
Profitability, 4, 5, 94, 104, 175–178, 180, 183, 186–188, 196, 226, 361
Proportionality, 13, 14, 24, 30–33
Proportionality of political measures, 17
Protective gears, 323, 325, 328, 332
Protest, 150, 151, 159
Public
expenditure, 179, 185
health, 5, 8, 20–24, 26, 28–30, 32, 33, 97, 102, 106, 107, 114, 117, 152, 174, 175, 178–180, 183–186, 188, 194, 197, 210, 212, 214, 221, 223, 225, 226, 232, 234, 248, 250, 251, 269, 271, 280, 298, 300, 301, 303, 305, 317, 340–343, 365
hospitals, 7, 8, 101, 102, 106, 173, 297, 298, 301–310, 363
services, 196, 200, 229–231, 234
speaking, 207, 226
Public Health England, 18, 221, 222, 233
Public Sector Social Responsibility (PSSR), 98
Public Services (Social Value) Act, 196

Q
Quarantine, 146, 148, 152, 246, 269, 305, 306, 319, 357

R
Racial equality, 227
Radovanović, Z. Prof. Dr., 148, 151, 157
Rashford, M., 232
Recite Me, 228
Reflexivity, 244, 245, 251, 253
Refugees, 152, 346
Regression analysis, 4, 183
Remote education, 248
Remote working, 94, 345
Resilience, 3, 127–130, 132, 140, 210, 249, 276, 345, 349
Responsibility, 4, 6, 23, 26, 95, 98, 103, 111–115, 120, 122, 157, 160, 174, 175, 187, 188, 193, 194, 196, 197, 200, 212, 219–221, 224–226, 228–230, 233, 235, 243, 247, 272, 273, 279, 280, 282, 284,

287, 290, 298, 299, 318, 342, 348, 356, 359, 362–364
Responsible innovation, 6, 244–248
Responsiveness, 3, 244, 246, 247, 251, 254, 327, 328
Restrictions, 4, 26, 32, 129, 130, 132, 136, 141, 146, 222, 225, 245, 246, 266, 267, 319, 340, 345, 350
Return on assets (ROA), 178–181, 183, 184
Revenue, 135, 178, 247, 274, 302, 345
Role model(s), 152
Romania, 3, 127, 132, 141
Russian jab, 153

S

Satisfaction, 98, 103, 104, 129, 140, 299, 362, 364, 366, 367
Screening, 301, 305, 330, 341, 343, 350, 356, 366
Senior citizens, 146, 348
Serbia, 3, 4, 18, 145–160
Severe Acute Respiratory Syndrome Coronavirus 2 (SARS-COV-2), 14–20, 22, 29, 32, 34, 93, 118, 119, 356
Shareholders, 96, 176, 177, 196, 281, 361, 363
Sinopharm, 154
Skills, 106, 135, 177, 178, 199, 202, 207–210, 225, 226, 230, 365
Slobodarski, 151
Small and medium enterprises, 140, 227, 274, 301
Smallpox, 145–150, 156, 158
Social
    capital, 97, 208, 211, 213
    distance, 16, 120
    distancing, 93, 94, 106, 159, 223, 245, 266, 267, 302, 320, 349, 363
    enterprise, 206, 224–226
    innovation, 210, 253, 254
    involvement, 114, 116, 279
    life, 19
    obligations, 279
    prescribing, 204
    responsiveness, 95, 195, 281
    value, 5, 194–201, 206, 208–214, 221
Socially responsible, 4, 6, 7, 112, 115, 116, 118–122, 186, 188, 219, 229, 245–247, 252, 254, 273, 279, 281, 282, 287, 290, 299, 301, 357
Social Value Portal, 210
Sociotechnical, 228, 229
Soft power, 154, 155, 157, 158

Southeast Europe, 154
Sovereignty, 152
Spain, 4, 156, 173–175, 177, 180, 184, 187
Spanish private healthcare, 4
Sputnik V, 154, 156
Stakeholders, 4–6, 23, 94, 96, 98–100, 103, 104, 107, 111–113, 118, 122, 128, 176–178, 196, 204, 206, 228, 230, 232, 234, 244, 245, 273, 280–282, 284, 298, 301, 319, 342, 356–359, 361–363, 365–367
Stakeholder theory, 9, 112, 176, 178, 361–362
Standard Operating Procedures (SOP), 341
Stevie Award, 252
Strategy, 3, 4, 13, 14, 20–22, 27, 28, 30, 32, 33, 96, 97, 102, 103, 116, 128, 130, 131, 140, 141, 151, 153–160, 194, 196, 197, 201, 202, 204, 210, 227, 228, 232–234, 244, 247, 252, 264, 266, 267, 269, 275, 284, 341, 342, 344, 350
Supply chain, 4, 94, 208, 222, 319, 328, 332
Sustainability, 2, 13–16, 20, 21, 26, 30, 34, 95–98, 102, 128, 140, 156, 176, 178, 188, 195–197, 204, 229, 280, 282, 286, 291, 299, 320, 335, 358
Sustainable growth, 247
Sustainable services, 203, 280
Sweden, 24, 26, 28–30, 32
Switzerland, 2, 18, 19, 24, 26, 27, 29, 32
Symptom checker, 323, 328

T

Telehealth consultation, 331
Tele-medicine centres, 324, 328
Teleworking, 131, 134
The 1972 smallpox epidemic, 149
The Heritage Foundation, 151, 153
The New Normal, 157
The Non-Aligned Movement (NAM), 153, 155
The Plague, 145, 150, 156
The Spanish Flu, 2, 3, 145, 149
The Virus, 1, 3, 15, 17, 22, 32, 145–148, 222, 246–248, 250, 266, 268, 272, 297, 302, 330, 341, 346, 348, 363
The World Bank, 116, 153, 270, 271
Third sector, 5, 6, 207, 222, 233, 234
1000 Roses, 148
Toastmasters International, 226
Torlak [virology institute], 149, 156
Tourism, 94, 129, 131, 247, 320, 341, 344–346, 349
Track/trace, 157

Trade bloc, 146, 266
Travel restrictions, 146
Turkey, 246
2020, 3, 14, 93, 117, 127, 146, 175, 207, 222,
    247, 263, 284, 297, 317, 339, 355

U
UNICEF, 267, 268, 330
United Nation (UN), 18, 20, 32, 97, 303, 304
Universal Health Coverage (UHC), 97, 98, 221
University of Central Lancashire, 5, 206
Utilitarian theory, 281, 282

V
Vaccination, 8, 13, 14, 17, 19, 20, 27, 31, 32,
    106, 107, 118, 120, 121, 147–149, 151,
    153–160, 226, 246, 247, 268, 273, 323,
    325, 328, 332, 335, 343
Vaccination program, 13, 17, 107
Vaccine Confidence, 226, 343
Vaccine diplomacy, 153, 155, 156
Vaccine production, 250–251, 254
Vaccines, 3, 13, 17, 19–21, 24, 27–29, 31, 106,
    121, 130, 148, 151, 153–156, 159, 226,
    234, 246, 250, 251, 254, 267, 268, 273,
    286, 343, 363
Variola Vera, 145, 147–150, 156, 160
Virus, 14–16, 18, 22, 29, 31, 32, 118, 148, 149,
    158, 248, 297, 319, 342, 348, 356, 359
Voluntary, 96, 112, 116, 177, 204, 206, 207,
    219, 220, 225, 230, 235, 300, 301, 309,
    367
Vulnerable, 6, 19, 24, 129, 207, 219, 220, 225,
    227, 230, 232, 233, 264, 269, 272, 282,
    289, 318, 323, 331, 335, 348, 357

Vulnerable groups, 99, 100, 225, 246, 267, 275,
    339, 345–347, 350

W
Ward's method, 179
Wastage, 7, 98, 198, 204, 205, 222, 230, 283
Wellbeing, 15, 20, 96, 97, 129, 130, 195, 197,
    202, 204, 211, 213, 221–223, 225, 226,
    229, 230, 232, 249, 280, 282, 330, 345,
    347
Wellness and Health programs (WHP), 322,
    324, 326–328, 332
Western Balkans, 154, 155
Western jabs, 153
WHO guidance, 13, 22
[WHO] smallpox eradication [programme], 148
Women directors, 131, 177, 183, 187
Workforce, 5, 94, 95, 202, 222, 226, 274, 279,
    280, 299, 330, 342, 349
Workplace, 7, 121, 122, 203, 213, 228, 230,
    283, 287, 290, 291, 311
Work precariousness, 356
World Bank, 117, 153, 249, 267, 269–271
World Health Organisation (WHO), 2, 8,
    13–23, 29, 32–34, 93, 97, 147–150, 222,
    246, 248, 267, 270, 271, 297, 303–305,
    317, 318, 342, 344, 355, 359, 364

Y
YouTube, 160, 226

Z
Zambia, 7, 8, 297, 298, 301–311